Chemical Exposures

Chemical Exposures

Low Levels and High Stakes

Second Edition

Nicholas A. Ashford
Claudia S. Miller

JOHN WILEY & SONS, INC.
New York • Chichester • Weinheim • Brisbane • Singapore • Toronto

Printed in the United States of America

Library of Congress Cataloging-in-Publication Data

Ashford, Nicholas Askounes.
    Chemical exposures : low levels and high stakes / Nicholas A.
  Ashford & Claudia S. Miller — 2nd ed.
        p.   cm.
    Includes bibliographical references and indexes.
    ISBN 0-471-29240-0
    1. Multiple chemical sensitivity. I. Miller, Claudia.
  II. Title.
  RB152.6.A84   1997
  615.9'02—dc21   97-19690
                                                       CIP

10 9 8 7 6 5 4 3 2

To my mother, Venette Askounes Ashford, from whom I learned to have compassion for those less fortunate than I, and who inspired me to help.

——N.A.A.

To my parents, Constance Lawrence Schultz and Ernst William Schultz, who patiently read stories to me as a child, encouraged my interest in science, and taught me to keep the two separated.

——C.S.M.

Contents

Part II.
Mechanisms, Diagnosis, and Treatment / 85

Part III. Responding to the Problem / 145

Part IV. Update since the First Edition / 169

Preface to the Second Edition

In 1990, the report for the New Jersey State Department of Health, which formed the basis for the first edition of *Chemical Exposures*, won the prestigious Macedo Award of the American Association for World Health (representing the World Health Organization) for the most outstanding contribution to public health funded by a state health department. Whereas the first edition discussed in detail the clinical observations and writings of controversial clinical ecologists—physicians who first brought attention to this condition in the 1950s and who employ nontraditional therapies to treat what they see as chemically induced illness—the second edition draws from the considerable literature published since 1991, almost all from mainstream, peer-reviewed scientific publications often written by environmental scientists, toxicologists, and occupational medicine physicians. Problems related to chemical sensitivity have now been reported in nearly a dozen Western European countries, and this volume captures some of the observations made by European scientists and physicians. Notably, whether such patients have been seen in North America or in Europe, by university physicians or by clinical ecologists, their patterns of symptoms and chemical intolerances have been remarkably similar.

As with the first edition, our emphasis is on origins and possible mechanisms for the condition. A rationale for intrusive therapies, whether physiological or psychological, is lacking. Physicians and their patients await a better understanding of chemical sensitivity. Continuing the tone set in the first edition, we show that it is not only possible but is necessary to delve into the scientific exploration of origins and mechanisms without becoming embroiled in questions concerning therapy. Unfortunately, many of those involved in the debate that surrounds chemical sensitivity continue to confound these issues. It is important to separate concerns about unproven diagnostic and treatment practices from the fundamental question of whether chemical sensitivity is toxicogenic or psychogenic. The large financial stakes and liability associated with sensitivity to chemicals in no small way drive contentious debate and action. However, it is clear that interest in chemical sensitivity has now expanded significantly beyond a few hundred controversial practitioners.

We deliberately chose to add updated material to, rather than rewrite, the original edition of *Chemical Exposures*, both because we believe that our earlier observations remain accurate and because the earlier material provides a useful "time capsule" against which to evaluate progress in understanding the condition(s) known as chemical sensitivity. Since the first edition, two new patient exposure groups have emerged with features strikingly similar to those of chemically sensitive persons: sick veterans from the Persian Gulf War and women who trace a myriad of adverse health symptoms to silicone breast implants. Both fit what we and others believe to be a two-step model of chemical sensitivity: an identifiable initiating chemical exposure event, followed after a short period by a myriad of symptoms triggered by low-level exposure to a large number of chemically unrelated substances.

We are increasingly convinced that low-level chemical sensitivity offers scientifically testable hypotheses and may signal an emerging new theory of disease: toxicant-induced loss of tolerance (TILT). This theory posits that a single high-level exposure, as in a chemical spill, or repeated lower level chemical exposures, as in a "sick" building, may cause certain susceptible persons to lose their prior natural tolerance for various chemicals, foods, and drugs. Subsequently, very low levels of these and chemically unrelated substances trigger symptoms, thus perpetuating illness. The converging lines of evidence that support this theory are: (1) similar reports by different investigators in both North America and Europe of multisystem symptoms and new-onset intolerances in different demographic groups following exposure to many different types of chemicals; (2) the internal consistency of patients' complaints of intolerances for not only tiny doses of inhaled chemicals but also for foods, caffeine, alcohol, and medications; (3) the degree to which the illness mimics addiction; (4) the identification of an anatomical substrate (involving the nervous system) whose malfunction might explain these problems; and (5) recent animal models that replicate key features of the condition. Toxicant-induced loss of tolerance might provide an explanation not only for many cases of multiple chemical sensitivities but also for other chronic medical conditions whose prevalences have increased over the past several decades, including some (though certainly not all) cases of chronic fatigue syndrome, asthma, depression, and migraine headache.

What has emerged as a result of the first edition of *Chemical Exposures* and subsequent government-sponsored meetings and workshops on chemical sensitivity in the United States is a consensus among involved scientists and clinicians of the need for double-blind, placebo-controlled challenge studies of chemically sensitive patients who are in an appropriate baseline state, these studies to be performed preferably in a hospital-based environmentally controlled unit (an environmental medical unit). Patients sharing the same initiating event, such as a pesticide or remodeling exposure, would be prime candidates for these studies. Such studies should receive priority for funding for a variety of reasons, but, most important, to resolve the debate over whether there is a subset of the population that is sensitive to extremely low levels of many chemicals. Given scarce funds for research in this area, such challenge studies should take precedence over hit-or-miss attempts to find biomarkers for the condition(s). Biomarkers may prove elusive, and the incentive to fund basic research on mechanisms and biomarkers would greatly increase if appropriately blinded and controlled challenge studies were to confirm the patients' sensitivities. At the same time, it is recognized that research enabling us to understand the nature of the initiating events causing persons to become sensitized is urgently needed as well, as this may shed light on public health preventive measures that may need to be taken. It is hoped that this edition of *Chemical Exposures* offers the reader a fresh perspective on chemical sensitivity and its potentially major role in many chronic illnesses, and helps clarify the direction of research needed to further our understanding of this difficult condition.

This edition selectively updates the first one and doubles the size of the original volume by providing a new Part IV consisting of four chapters. Chapter 7 (Recent Developments) discusses the evolving terminology associated with chemical sensitivity, major workshops, government interest and activity, legal developments, and Canadian and European developments since the publication of the first edition. Chapter 8 (Key Research Findings since the First Edition) describes important new clinical observations and research on the origins and mechanisms of chemical sensitivity, including a discussion of overlap conditions, biomarkers, and animal models. Chapter 9 (Reviews, Commentaries, and Polemics) offers a critical analysis of a number of interpretative writings that have attempted to shed light on the nature, origins, and mechanisms of chemical sensitivity. Chapter 10 (Research and Medical Needs) summarizes the authors' current thinking on chemical sensitivity, offers directions for further research, and discusses medical and patient needs. Finally, two new appendixes are added. Appendix B is a compilation of the results of various laboratory and clinical tests used in studies of chemically sensitive patients, and Appendix C consists of a questionnaire used by one of the authors to collect histories from patients concerning their self-reported sensitivities to environmental exposures.

Preface to the First Edition

Given the current controversies concerning the nature of chemical sensitivity and the fact that many physicians and scientists doubt that it is physical in origin, some words of explanation are in order for the readers of this book. At the beginning of this book, we describe patients who claim to be "chemically sensitive," that is, suffer acute adverse reactions to low levels of chemicals commonly found in homes, schools, places of employment, and other environments. In the ensuing pages, we drop the quotation marks and avoid terms such as "allegedly affected individuals" because their continual use would be awkward. Sufficient "proof" is not available to satisfy the most skeptical critic that chemical sensitivity exists as a physical entity, nor is there convincing proof that it does not. However, we are persuaded that the collective evidence, in part anecdotal and in part based on good scientific studies, does present a sufficiently compelling case to warrant further study. We cannot assert that millions of people are affected, although chemicals are ubiquitous and exposures are expected to continue. The size of the public health problem is unknown, but the scale of potential exposure suggests that the problem could be significant. We ask that readers approach this book with open minds and withhold judgment on these issues until they have read the entire book. A more focused second reading might also be needed.

Our purposes in undertaking the research underlying this book were: (1) to clarify the nature of chemical sensitivity and (2) to identify ways federal and state government can assist those who are affected. In undertaking this task, we reviewed much of the available scientific and medical literature relating to low-level chemical exposure and resulting disease. We also interviewed key individuals in various medical disciplines including allergy, clinical ecology, and occupational medicine. We found scientific and clinical evidence to support plausible hypotheses concerning this disorder. The evidence also offers fruitful areas for further research. In addition, we found areas of significant interprofessional conflict as well as areas of agreement. We noted an increasing desire by all parties to find a common ground from which the issues can be objectively and cooperatively addressed.

Much, but by no means all, anecdotal evidence for chemical sensitivities has been reported by clinical ecologists—physician practitioners whose clinical practices have come under intense criticism. However, chemical sensitivity is by no means the exclusive property of clinical ecology. The fields of occupational and environmental medicine contain sufficient examples to suggest a real medical problem. Our focus was on the problem of chemical sensitivity, not on the history of interprofessional conflict surrounding clinical ecology.

Some readers may be concerned that the lack of sufficient data in this area may render our conclusions speculative and hence biased. Certainly this book contains speculation. However, there is a difference between constructing rational hypotheses concerning the existence of chemical sensitivity based on all the evidence and engaging in unfounded conjecture. Finally, we hope that our efforts will stimulate others to undertake serious scientific inquiry into this fascinating and rapidly evolving area.

This book is divided into three parts. Part One defines the problem of chemical sensitivity. It discusses sensitive populations, low-level exposures to chemicals, the history of clinical ecology and its relationship to other disciplines, and the magnitude and nature of the problem. Part Two describes possible mechanisms, diagnostic approaches and therapies, and the areas of agreement and disagreement between allergists and clinical ecologists.

Part Three addresses research needs, patient and community concerns, health care, insurance, and compensation needs, the role of medical practitioners and their societies and recommendations for both research and action, given our present state of knowledge.

Acknowledgments

We are indebted to a number of people who gave generously of their time and energy to educate us on various issues and critique prior versions of the text for the first edition of this book. They included Iris Bell, Eddy Bresnitz, Mark Cullen, Linda Lee Davidoff, Nancy Fiedler, Ted Kniker, William Meggs, David Ozonoff, Theron Randolph, William Rea, Abba Terr, Lance Wallace, and Grace Ziem.

We thank Donald Beavers for research assistance on provocation-neutralization therapies and Susan Kaplan for research assistance on legal issues in the second edition.

In addition, we wish to thank the many people we contacted or interviewed in the course of our research for both editions. They include Yves Alarie, Rosalind Anderson, Emil Bardana, Iris Bell, Nathan Brautbar, Eddy Bresnitz, Dan Costa, Mark Cullen, Linda Lee Davidoff, Earon Davis, Albert Donnay, Roy Fox, Kendall Gerdes, Bill Hirzy, Donald Jewett, Alfred Johnson, Howard Kehrl, William King, William T. Kniker, Hillel Koren, Richard Kreutzer, Mary Lamielle, Alan Levin, William Meggs, Dean Metcalfe, Joseph Miller, Susan Molloy, David Ozonoff, Theron Randolph, Doris Rapp, William Rea, Gerald Ross, John Salvaggio, Val Schaeffer, John Selner, Ray Slavin, John Spengler, Morton Teich, Abba Terr, Lance Wallace, Laura Welch, Grace Ziem, and many patients. We are especially grateful to Kathleen Rest and Bob Miller, who both supported our effort and were our toughest critics.

Finally, we thank Tom Burke and the New Jersey State Department of Health, for providing us the opportunity to investigate chemical sensitivity.

About the Authors

Nicholas A. Ashford, Ph.D., J.D., is Professor of Technology and Policy at the Massachusetts Institute of Technology, where he teaches courses in Environmental and Occupational Health Law and Policy. He has served as Chairman of the National Advisory Committee on Occupational Safety and Health, Chairman of the Committee on Technology, Innovation, and Economics of the EPA National Advisory Council for Environmental Policy and Technology, and as a member of the EPA Science Advisory Board. He is a Fellow of the American Association for the Advancement of Science and is an advisor to the United Nations Environmental Program. He is also the author of *Crisis in the Workplace: Occupational Disease and Injury* (1976, MIT Press) and coauthor of *Monitoring the Worker for Exposure and Disease: Scientific, Legal and Ethical Considerations in the Use of Biomarkers* (1990, Johns Hopkins University Press), and *Technology, Law and the Working Environment* (second edition, 1996, Island Press). He holds both a doctorate in chemistry and a law degree from the University of Chicago, where he received graduate training in economics.

Claudia S. Miller, M.D., M.S., is Assistant Professor in Environmental and Occupational Medicine in the Department of Family Practice at the University of Texas Health Science Center at San Antonio. She is board-certified in Internal Medicine and in Allergy and Immunology and holds a master's degree in environmental health. Prior to entering medical school, she worked as an industrial hygienist and served on the National Advisory Committee on Occupational Safety and Health. Since 1993, she has been an environmental consultant to the Houston VA Regional Referral Center for Gulf War veterans. She is a member of the Department of Veterans Affairs' Persian Gulf Expert Scientific Committee and served on the National Toxicology Board of Scientific Counselors. Her research focuses on the health effects of low-level environmental chemical exposures including indoor air pollutants, pesticides, and chemicals used during the Persian Gulf War.

Introduction

Chemical exposures are endemic to our modern industrial society. Patients who believe they are chemically sensitive are caught up in an acrimonious cross fire among several different groups of physicians—traditional allergists; clinical ecologists; and in some cases, ear, nose and throat specialists; occupational physicians; and others. This acrimony is fueled by different medical paradigms of the definition, diagnosis, and treatment of disease or symptoms associated with exposure to low levels of chemicals in food and water, the outdoor environment, the work environment, indoor air, and consumer products. Legal conflicts further complicate the associated scientific and medical differences as attempts by "chemically sensitive" persons to obtain workers' compensation, disability payments, and damage awards from employers and from the producers and users of chemical products result in an adversary system that draws medical practitioners unwillingly into the center of the conflict. Further exacerbating the situation are the insurance industry and employers, who seek to reduce costs for medical care; their involvement continues the volatile history of economic tugs-of-war characteristic of health care in general. "Chemically sensitive" patients seek medical care and consideration from traditional medical practitioners, many of whom are ill-equipped or reluctant to provide the painstaking and time-consuming attention that is required for this condition.

The research underlying the first edition this book was commissioned by the New Jersey State Department of Health in order to clarify the nature of chemical sensitivity and identify ways in which a state department of health can assist the chemically sensitive person and disengage the patient from the medical cross fire and its attendant conflicts. In this book we argue that both federal and state initiatives are needed. In undertaking this task, we reviewed much of the available scientific and medical literature relating to low-level chemical exposure and resulting disease. We interviewed key individuals in various medical disciplines including allergy, clinical ecology, and occupational medicine. This effort was facilitated by the fortuitous scheduling of national conferences by the allergists and by the clinical ecologists in the same 7-day period in Texas in February 1989. Physicians involved with the chemically sensitive patient are concerned about being drawn into a legal and political struggle that ultimately may not help the patient. Through our interviews we were able to identify not only areas of conflict between the allergists and clinical ecologists but also unexpected areas of common ground.

This book comes at a critical time. Since the government of Ontario completed a report on "environmental hypersensitivity disorders" (Thomson 1985) in 1985, sensitivity to chemicals has received unprecedented attention from many quarters in the United States. A "Workshop on Health Risks from Exposure to Common Indoor Household Products in Allergic or Chemically Diseased Persons" held by the National Academy of Sciences (NAS) on July 1, 1987, recommended an 18-month study to address the "15 percent of the U.S. population [who] have an increased allergic sensitivity to chemicals commonly found in household products, such as detergents, solvents, pesticides, metals and rubber, thus placing them at increased risk [of] disease" (National Research Council 1987). Although that study has not yet been funded, in 1989 the NAS convened a panel to examine the interrelationships of toxic exposures and immune response. Later the same year, the U.S. Office of Technology Assessment (OTA) began a study of noncancer risks of chemicals, including immunotoxicity. OTA completed a neurotoxicity study in 1990 [OTA 1990]. A Canadian national advisory committee held a workshop on "environmental sensitivities" in May 1990 (Canada 1991). The NAS, in response to a request from the EPA's Office of Indoor Air, conducted a multiple chemical sensitivity workshop in early 1991 to develop research protocols for the syndrome (Hileman 1991; National Research Council 1991).

The U.S. Congressional Research Service has issued a report on indoor air pollution in which chemical sensitivity is explicitly recognized (Courpas 1988, p. CRS-9). The Environmental Protection Agency (EPA) acknowledges that health problems exist with low-level exposures well below those allowed by existing regulations (Claussen 1988); in its *Report to Congress on Indoor Air Quality,* EPA identifies multiple chemical sensitivities as a health concern (EPA 1989, p. 16). The Superfund Amendments, SARA, Title IV mandate a vigorous investigation of the problems of indoor air pollution by EPA. John D. Spengler of Harvard's School of Public Health, a leading authority on indoor air pollution, has testified (1988):

> There is growing evidence that there are chemically sensitive individuals in our society. Many, it is believed, may have acquired the sensitivity due to chronic exposures. But even without frank illness, the syndrome of irritation, fatigue, shortness of breath and nausea associated with building-related problems results in lost productivity and wasteful investigations and litigation.

Legislation introduced in Congress (S.1629, H.R. 5373) explicitly recognized multiple chemical sensitivities resulting from indoor air pollutants as a serious threat to public health. Maryland has completed a study of "chemical hypersensitivity syndrome" (Bascom 1989). Legislation establishing a demonstration program to provide services and assistance to chemically hypersensitive persons (S.696) was considered in New Jersey. These activities underscore a ground swell of activity that requires in-depth and thoughtful attention to chemical sensitivity.

We are at a critical crossroads. We have at this time a small window of opportunity that may close if we do not take action to address the problems of the chemically sensitive individual in a caring and equitable way. The recommendations in this book result from our interviews, literature review, and examination of the issues, and we suggest that their adoption is necessary for making substantial progress in this area. As the second printing of this book goes to press, both a Canadian national advisory committee (Canada 1991) and The U.S. National Research Council (National Research Council 1991) have formulated specific recommendations that have the potential to take the issue of chemical sensitivity into mainstream medicine and public health.

The second edition of this book provides a significant update of the original material of the first edition, presented in Part IV. We have chosen to leave the original chapters from the first edition essentially unaltered because the material therein remains accurate. In this way the reader is provided a "time capsule" against which new research findings and understanding may be compared.

PART · I

Defining Chemical Sensitivity

CHAPTER 1

Chemical Exposures and Sensitive Populations

Groups Sensitive to Low-level Chemical Exposure

A review of the literature on exposure to low levels of chemicals reveals four groups or clusters of people with heightened reactivity:

1. Industrial workers
2. Occupants of "tight buildings," including office workers and school-children
3. Residents of communities whose air or water is contaminated by chemicals
4. Individuals who have had personal and unique exposures to various chemicals in domestic indoor air, pesticides, drugs, and consumer products

These four groups are listed for comparison in Table 1-1. Note that they differ in professional and educational attainment, age and sex, and the mix and levels of chemicals to which they are exposed, but that all have multiple symptoms involving multiple organ systems with marked variability in the type and degree of those symptoms. Symptoms are often "subjective." For example, central nervous system (CNS) symptoms such

3

TABLE 1-1. Chemically Sensitive Groups

Group	Nature of Exposure	Demographics
Industrial workers	Acute and chronic exposure to industrial chemicals	Primarily males; blue collar; 20 to 65 years old
Tight-building occupants	Off-gassing from construction materials, office equipment or supplies; tobacco smoke; inadequate ventilation	Females more than males; white-collar office workers and professionals; 20 to 65 years old; schoolchildren
Contaminated communities	Toxic waste sites, aerial pesticide spraying, ground water contamination, air contamination by nearby industry and other community exposures	All ages, male and female; children or infants may be affected first or most; pregnant women with possible effects on fetuses; middle to lower class
Individuals	Heterogeneous; indoor air (domestic), consumer products, drugs, and pesticides	70–80% females; 50% 30 to 50 years old (Johnson and Rea 1989); white, middle to upper middle class and professionals

as difficulty concentrating or irritability are common, and physical examinations are frequently unremarkable for individuals in each category. Careful analysis of these groups may reveal differences that can illuminate the etiologies and suggest effective therapeutic options for the myriad problems comprising chemical sensitivity. These differences also may create a referral or selection bias such that members of the four groups present themselves preferentially to different medical practitioners; some may consult occupational health physicians, others primary care physicians, and still others clinical ecologists or allergists (see Chapter 6).

Problems experienced by people in tight buildings, by industrial workers in a particular workplace, or by the residents of a contaminated community often occur within a relatively short time period—perhaps weeks or a few months. These problems may occur after a recognized event such as the installation of new carpeting, relocation to a new workplace, or changes in workplace or community exposures. The temporal cohesiveness of exposures and problems can contribute to the recognition of the problem as real. Acceptance of these problems as bona fide physical disease may also be facilitated by the recognition that these problems are widespread in nature and simply are not limited to what some observers would describe as malingering employees, hysterical housewives, and workers experiencing mass psychogenic illness. We are struck by the fact that individuals in such demographically divergent groups as those in Table 1-1, including industrial workers, office workers, housewives, and children, report similar polysymptomatic complaints triggered by chemical exposures. Perhaps some common thread unites these individuals. The similarities of both their medical complaints and their exposure histories may be more than coincidental.

In a survey of some 6,800 persons claiming to be chemically sensitive, 80 percent asserted they knew "when, where, with what, and how they were made ill" (National Foundation for the Chemically Hypersensitive 1989). Of the 80 percent, 60 percent (that is, almost half of those who replied) blamed pesticides. The respondents to the survey were self selected, and the results must be interpreted with caution. Nevertheless, the results indicate that future surveys of persons with different exposure histories and symptoms might contribute to an understanding of underlying mechanisms and causes.

In some chemically sensitive patients, no single, identifiable, "high-level" exposure seems to have been associated with the onset of their difficulties. Exposures may have occurred but were not recognized or remembered. Some observers suggest that repetitive or cumulative lower-level exposure events may lead to the development of sensitivities. Still others implicate genetic predisposition, pregnancy, major surgery with anesthesia, physical trauma, or major psychological stress as contributors to the illness (see Chapter 4).

Types of Sensitivity

The different meanings of the term *sensitivity* are at least partially responsible for the confusion surrounding chemical sensitivity.

Individuals differ in their responses to increasing doses of a toxic substance. The underlying causes of interindividual variability include age, sex, and genetic makeup; lifestyle and behavioral factors, including nutritional and dietary factors; alcohol, tobacco, and drug use; environmental factors; and preexisting disease (Ashford et al. 1984). In the classical, toxicological use of the word *sensitivity*, those individuals who require relatively lower doses to induce a particular response are said to be more sensitive than those who would require relatively higher doses before experiencing the same response (Hattis et al. 1987). A hypothetical distribution of sensitivities, that is, the minimum doses necessary to cause individuals in a population to exhibit a harmful effect, is shown in curve *A* in Figure 1-1. (If we plot the cumulative number of individuals who exhibit a particular response as a function of dose, we generate a population dose-response curve; see curve *A* in Figure 1-2.) This distribution describes the traditional toxicological concept of sensitivity. Curve *A* in Figure 1-1 illustrates that health effects of classical diseases

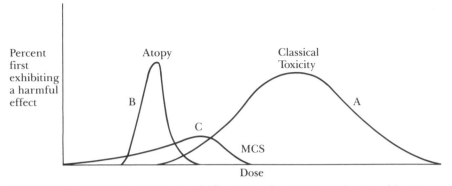

FIGURE 1-1.   *Hypothetical distribution of different types of sensitivities as a function of dose. Curve A is a sensitivity distribution for classical toxicity, e.g., to lead or a solvent. Sensitive individuals are found in the left-hand tail of the distribution. Curve B is a sensitivity distribution of atopic or allergic individuals in the population who are sensitive to an allergen, e.g., ragweed or bee venom. Curve C is a sensitivity distribution for individuals with multiple chemical sensitivities who, because they are already sensitized, subsequently respond to particular incitants, e.g., formaldehyde or phenol.*

are seen in a significant portion of the normal population at a certain dose; the sensitive and resilient populations are found in the tails of the distribution. (Of course, not all toxic substances have large variances or significant tails.) Painstaking scientific research and removing the effects of confounding variables have resulted in the discovery of sensitive individuals at levels heretofore considered safe. Recent work on lead (Bellinger et al. 1987) and benzene (Rinsky et al. 1987) are just two examples. For the sensitive person, avoidance of low-level exposures generally leads to improvement, or at least to the arrest of the development of the disease.

A second meaning of the word *sensitivity* appears in the context of classical IgE-mediated allergy (atopy). IgE is one of five classes of antibodies made by the body, and is, from the perspective of classically allergic individuals, the most important antibody. Atopic individuals have IgE directed against specific environmental incitants, such as ragweed or bee venom. Positive skin tests in these individuals correlate with a rapid onset of symptoms when they are actually exposed to those allergens. The atopic individual exhibits a reaction whereas nonallergic persons do not, even at the highest doses normally found in the environment. A hypothetical sensitivity distribution for an atopic effect is shown in curve *B* of Figure 1-1, and the dose-response curve derived from that distribution is found in curve *B* of Figure 1-2.

Allergists include in the term *allergy* well-characterized immune responses that result from industrial exposure to certain chemicals, such

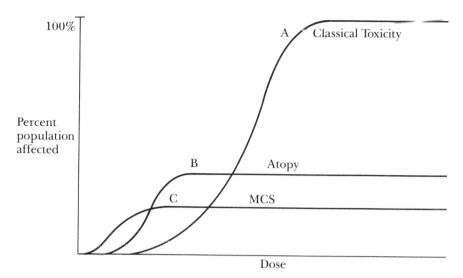

FIGURE 1-2. Hypothetical population dose-response curves for different effects. Curve A is a cumulative dose-response curve for classical toxicity, e.g., to lead or a solvent. Curve B is a cumulative dose-response curve for atopic or allergic individuals in the population who are sensitive to an allergen, e.g., ragweed or bee venom. Curve C is a cumulative dose-response curve for individuals with multiple chemical sensitivities who, because they are already sensitized, subsequently respond to particular incitants, e.g., formaldehyde or phenol.

as nickel or toluene diisocyanate (TDI). Most allergists refer to such responses as *chemical sensitivity,* and reserve this term for responses that have a distinct immunological basis, preferring to use a term such as *chemical intolerance* for nonimmunological responses to chemicals.

Patients suffering from multiple chemical sensitivities may be exhibiting a third and entirely different type of sensitivity. Their health problems often (but not always) appear to originate with some acute or traumatic exposure, after which the triggering of symptoms and observed sensitivities occur at very low levels of chemical exposure. The inducing chemical or substance may or may not be the same as the substances that thereafter provoke or "trigger" responses. (Sometimes the inducing substance is described as "sensitizing" the individual and the affected person as a "sensitized" person.) Reactions may sometimes be observed at incitant levels similar to those to which classically sensitive and atopic patients respond. Unlike classical toxicity, however, here the effects of low-level exposures are not simply those effects observed in normal populations at higher doses. The fact that normal persons—for example, most doctors—do not experience even at *higher* levels of exposure those symptoms that chemically sensitive patients describe at much lower levels of exposure probably helps to explain the reluctance of some physicians to believe that the problems are physical in nature. (Although this also describes atopy, here the sensitivity is not IgE mediated.) To compound the problem of physician acceptance of this illness, multiple organ systems may be affected, and multiple substances may trigger the effects. Over time, sensitivities seem to spread, in terms of both the types of triggering substances and the systems affected (Randolph 1962, pp. 98 and 119). Avoidance of the offending substances is usually effective but much more difficult to achieve for these patients than for classically sensitive patients because symptoms may occur at extremely low doses and the exposures are ubiquitous. Adaptation to chronic low-level exposure with consequent "masking" of symptoms (discussed more fully later) may make it exceedingly difficult to discover these sensitivities and unravel the multifactorial triggering of symptoms. A hypothetical sensitivity distribution for a single symptom for the already chemically sensitive person in response to a single substance trigger is shown in curve *C* of Figure 1-1, and the corresponding dose-response curve is shown in curve *C* of Figure 1-2. It should be emphasized, however, that individuals who become chemically sensitive may have been exposed to an initial *priming* event that was *toxic*, as classically defined.

Conceivably, exposure to certain substances, such as formaldehyde, might elicit all three types of sensitivities.

The fact that sensitivity means something quite different to toxicologists, allergists, and clinical ecologists reflects the different disease paradigms under which each operates. Neither traditional allergists nor toxicologists fully appreciate the two-step process of induction and triggering that seems to characterize multiple chemical sensitivities.

Those clinical ecologists who reference the literature on classical chemical toxicity to buttress their case for chemical sensitivity may be adding to the confusion and contributing to others' reluctance to accept their ideas. Likewise, allergists who dismiss chemical sensitivity on the grounds that it is not consistent with a recognized immunologic mechanism may be overlooking another kind of sensitivity in their patients. Although chemicals may act in some manner (via a toxic mechanism, for instance), to predispose or cause the body to be reactive to subsequent low-level chemical exposures, the *resulting* hyperreactivity to low levels of chemically diverse and unrelated substances is not toxicity as classically defined or understood at this time (see Chapter 4). Some allergists maintain that the term *chemical sensitivity* should not be used in the context we have used here, but should be reserved only for those responses having an immunological basis. We feel that the term *sensitivity* has broader applicability. A parallel might be the word *resistance*, which is widely understood whether one is talking about electricity, psychiatry, or an infectious disease. Similarly, *sensitivity* is easily understood when used in any of the three contexts illustrated in this section; it is not the exclusive property of the atopist.

Cullen (1987a) proposes that individuals with well-defined clinical entities such as asthma should not be given a diagnosis of multiple chemical sensitivities. Yet asthma may be one of the manifestations of this syndrome. It is important not to confuse diagnosis with etiology. The extent to which occupational asthma may overlap with multiple chemical sensitivity needs study and clarification. Classically, asthma has been divided into two categories: *extrinsic* asthma triggered by allergic (IgE) responses to pollens, dust, mold, and so on, and *intrinsic* asthma in which exposures outside the individual are not felt to play a causative role. The etiology of intrinsic asthma is ill-defined. Physicians have long warned their asthmatic patients to avoid irritants such as cigarette smoke, perfume, and strong cleaning agents, suggesting that such exposures might further irritate vulnerable airways, making their asthma worse. Few physicians, however, would view these irritants as a primary cause of their patients' asthma. Yet chronic irritation of any kind can lead to inflammation. Increasingly, the pathogenesis of asthma is being recognized as inflammation of the airways, and the most effective therapies for asthma are considered to be anti-inflammatory drugs such as cromolyn or steroids. Thus, asthma *is* inflammation, and inflammation can be caused by irritants, chemical or otherwise. Hence, it is quite possible that some asthma formerly designated as *intrinsic* may turn out to be external or *extrinsic* in origin, when the pathways leading to inflammation are delineated. Indeed, some feel that recent upward trends in asthma morbidity and mortality parallel increases in atmospheric pollution. Recently, a new clinical entity called reactive airway dysfunction syndrome (RADS) which shares certain features of multiple chemical sensitivity (Brooks 1985), has been described. Like multiple chemical sensitivity, RADS may be triggered by a single massive chemical exposure, for example, a chemical spill or fire. Subsequently, low levels of many common chemicals, (e.g., cigarette smoke, detergents, or perfume) that had never caused problems before may trigger airway constriction.

Physicians who see more or less random individuals who are not members of an identifiable exposure group are less likely to recognize patterns or similarities among these patients who claim to be chemically sensitive. Now that more attention is being focused on problems of industrial workers, occupants of tight buildings, and families in contaminated communities, these "random" patients (the fourth group in Table 1-1) may be diagnosed more readily. Once physicians recognize a constellation of symptoms that occurs repeatedly in individuals who share similar exposure histories, the disease seems to change its label from "idiopathic" or "psychogenic" to a recognized disorder, such as has occurred in the case of sick building syndrome (Kreiss 1989). Cullen's recent book (1987) on multiple chemical sensitivities was stimulated by his observations of a particular pattern of symptoms among workers which was previously unfamiliar to most occupational physicians. In the future, patterns observed in occupational and other cohesive groups of patients should facilitate a better understanding of what seems to many to be a hopeless confusion of reported symptoms.

Changes in Chemical Production and Use, and the Emergence of Chemical Sensitivity

In the conclusion to his *Workers with Multiple Chemical Sensitivities*, Cullen (1987b, p. 804) writes:

> The health problems of workers who react to low levels of environmental pollutants and chemicals, increasingly reported and recognized in recent years, has [sic] posed a serious dilemma for health providers from a wide array of disciplines, including generalists, internists, family practitioners, allergists, psychiatrists, social workers, and frequently occupational physicians and nurses. The inability of these professionals to provide satisfactory care from the patient's perspective has led to the emergence of new and alternative clinical theories and approaches, challenging traditional views. Unfortunately, the success of these alternative approaches has also not been demonstrated, fueling an ever widening and hostile debate in which *the patient is held hostage and virtually all clinicians are rendered impotent because of widely known intraprofessional disagreements.* [emphasis added]

How did these disagreements arise? Why are more and more problems related to low-level exposure to chemicals being reported in recent years? Is the problem merely increasingly recognized, or are the numbers of individuals being affected actually increasing? We shall try to shed some light on these questions by examining the development of this problem and the changes in chemical production, consumer products, and building design that have accompanied its emergence. We also include a brief history of clinical ecology, noting its split from allergy, subsequent growth, and continued conflicts with traditional allergists.

The increased medical interest in exposure to chemicals, especially low-level exposures, accompanied changes in the production of synthetic organic chemicals, building construction, and indoor air quality.

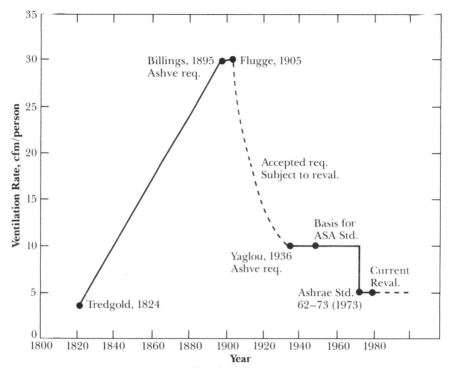

FIGURE 1-3. *Historical development of ventilation standards in the United States. Reprinted with permission from Mage, D. and R. Gammage, "Evaluation of Changes in Outdoor Air Quality Occurring Over the Past Several Decades," in* Indoor Air and Human Health, *R. Gammage and S. Kaye, Eds. (copyright 1985, Lewis Publishers, Inc., Chelsea, MI) p. 13.*

With the concern for energy conservation in the 1970s, homes and office buildings in the United States were constructed more tightly and make-up air (fresh air intake) was cut to a minimum. The historical trend of ventilation standards, used by architects and building designers, can be seen in Figure 1-3. The earliest standard, proposed by Tredgold in 1824 to prevent stuffiness, provided 4 cubic feet per minute (cfm) fresh make-up air per occupant. In 1893 Billings recommended 30 cfm per person, a value subsequently adopted by the American Society of Heating and Ventilation Engineers (ASHVE) and incorporated into the building codes of 22 states by 1925. In 1936 this standard was lowered to 10 cfm per person in response to research by Yaglou on the threshold of detection for human body odors; the American Standards Association adopted this value in 1946. Thus, before the energy crisis of 1973, odor detection was the basis for the ventilation standard. In the mid-1970s, as a result of energy concerns, ASHRAE (American Society of Heating,

Refrigeration and Air Conditioning Engineers) lowered the standard to 5 cfm per occupant, and 45 states adopted this standard into their codes in disregard of studies indicating that more fresh air was needed to dilute human odors and tobacco smoke to comfortable levels (Morey and Shattuck 1989). In 1981 ASHRAE revised the standard to 20 cfm of fresh air per occupant in areas where smoking is allowed. The current standard, issued in 1989, recommends at least 15 cfm per person, regardless of smoking. Fifteen cfm per person in schools, 20 cfm in offices, and 25 cfm in hospital rooms are recommended, with even higher rates if air from the ventilation system does not adequately mix with room air breathed by occupants or if unusual sources of contaminants are present (Morey and Shattuck 1989). However, from the mid-1970s into the 1980s, many commercial buildings were designed in accordance with the 5-cfm-per-occupant ASHRAE standard.

Similarly, beginning in the 1970s homeowners and new home builders caulked and sealed, installed storm windows and extra insulation, and effectively reduced fresh air infiltration. Homes, unlike commercial buildings, do not have ventilation systems that supply fresh make-up air, but rely on infiltration through doors, windows, cracks, and crevices instead. Such repairs were economically advantageous and in part tax deductible. In older homes not given these energy overhauls, the average fresh air infiltration rate is almost twice that of newer homes (0.9 versus 0.5 air changes per hour), but individual homes vary tremendously from 0.1 to more than 3 air changes per hour (Mage and Gammage 1985). See Figure 1-4.

More than 800 different volatile compounds were observed inside four buildings studied by the EPA (Wallace 1985). Wallace summarizes recent studies of indoor air pollutants:

1. Indoor median concentrations of volatile organics are consistently greater, by factors of 2 to 5, than outdoor medians.

2. At higher concentrations, the indoor-outdoor ratio increases, often beyond factors of 10.

3. Concentrations are extremely variable, covering 3 to 4 orders of magnitude, indicating the presence of intense indoor sources.

4. These sources are many, including paints, adhesives, cleansers, cosmetics, and other consumer products and building materials; but also common activities, such as visiting the dry cleaner shop or even taking a hot shower!

EPA conducted TEAM (Total Exposure Assessment Methodology) studies on a variety of volatile organics (1980–1987), carbon monoxide (1982–1983), pesticides (1986–1989), and particulates (1987–present).

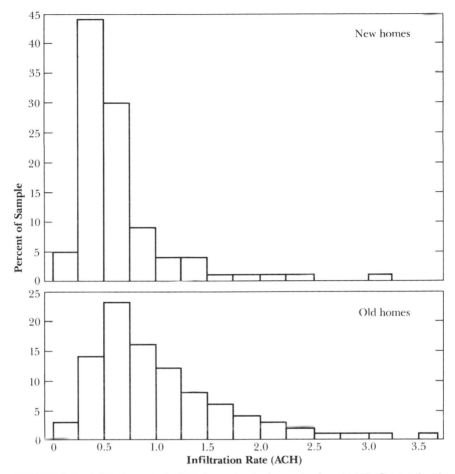

FIGURE 1-4.   Infiltration rates in U.S. houses in air changes per hour (ACH). Reprinted with permission from Mage, D. and R. Gammage, "Evaluation of Changes in Outdoor Air Quality Occurring Over the Past Several Decades, in Indoor Air and Human Health, *R. Gammage and S. Kaye, Eds. (copyright 1985, Lewis Publishers, Inc., Chelsea, MI), p. 16.*

The goals of the studies were to develop methods for measuring individual total exposure and resulting body burden of toxic and carcinogenic organic chemicals and to estimate the exposures and body burdens of urban populations in several U.S. cities. Representative data from a study of 20 volatile organic compounds in the personal (indoor) air, outdoor air, drinking water, and breath of approximately 400 residents of New Jersey, North Carolina, and North Dakota are shown in Figures 1-5 and 1-6 (Wallace 1987).

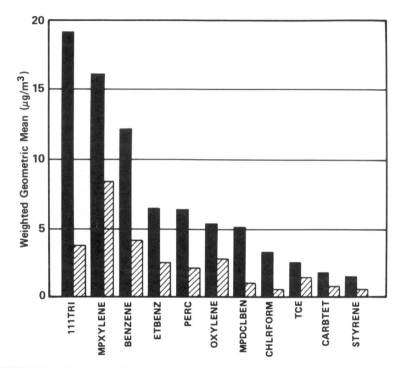

FIGURE 1-5.   *Comparison of indoor and outdoor exposures to toxic compounds. Estimated geometric means of 11 toxic compounds in overnight (6:00 PM to 6:00 AM) air samples for the target population (128,000) of Elizabeth and Bayonne, New Jersey, between September and November 1981. Personal air (i.e., indoor) estimates (solid) are based on 347 samples and outdoor air estimates (hatched) are based on 84 samples. Compound abbreviations are: 1,1,1TRI, 1,1,1-trichloroethane; MPXYLENE, m,p-xylene; ETBENZ, ethylbenzene; PERC, tetrachloroethylene; OXYLENE, o-xylene; MPDCLBEN, m,p-dichlorobenzene; CHLRFORM, chloroform; TCE, trichloroethylene; CARBTET, carbon tetrachloride.*
Source: *Wallace, L. et al., "The TEAM Study: Personal Exposures to Toxic Substances in Air, Drinking Water, and Breath of 400 Residents of New Jersey, North Carolina, and North Dakota,"* Environmental Research *(1987) 43:290–307, Academic Press, San Diego, California, p. 297.*

Ten of the 11 chemicals measured in the breath of New Jersey residents correlated significantly with indoor air exposure levels, which were uniformly higher than outdoor levels (Fig. 1-5). Only for chloroform did breath levels correlate more closely with drinking water concentrations. Breath levels for most chemicals measured were 30–40 percent of indoor air levels, but measured up to 90 percent of indoor air levels in some cases—tetrachloroethylene, for example. A study of non-occupational pesticide exposure (Fig. 1-7) also shows dramatically higher concentrations of pesticides indoors than out of doors [Immerman 1990].

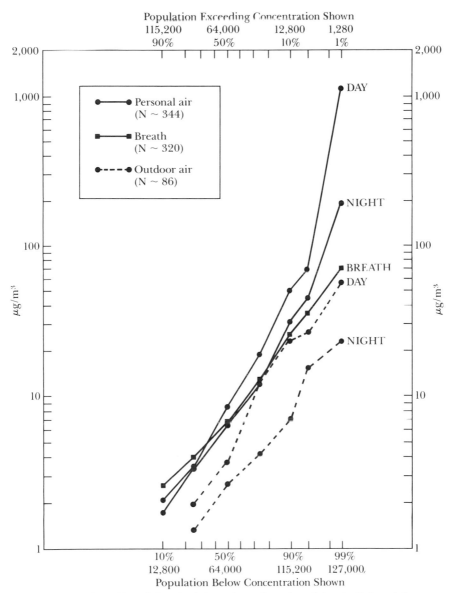

FIGURE 1-6. *Tetrachloroethylene in indoor air, outdoor air and breath. Estimated frequency distributions of tetrachloroethylene in personal air exposures, outdoor air concentrations, and exhaled breath values for the combined Elizabeth-Bayonne target population (128,000). All air values are 12-hr integrated samples. The breath value was taken following the daytime air sample (6:00 AM– 6:00 PM). All outdoor air samples were taken in the vicinity of the participants' homes.*
Source: Wallace, L., et al., "The TEAM Study: Personal Exposures to Toxic Substances in Air, Drinking Water, and Breath of 400 Residents of New Jersey, North Carolina, and North Dakota, Environmental Research *(1987) 43:290–307. Academic Press, San Diego, California, p. 296.*

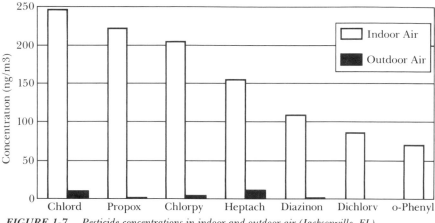

FIGURE 1-7.   Pesticide concentrations in indoor and outdoor air (Jacksonville, FL).
Source: *Immerman 1990.*

Remarkably, these sources present in indoor air are the same ones individuals with multiple chemical sensitivities identify as provoking their vague and seemingly inexplicable symptoms. With their homes and workplaces already filled with synthetic materials that off-gas, gas furnaces, cigarette smoke, and other sources of pollutants, Americans sealed their buildings for energy efficiency. Not surprisingly, indoor air pollution levels rose dramatically, and so did health complaints. In addition, Americans spend many more hours per day indoors at work and at home, in schools, shopping malls, and other buildings than preceding generations (Environmental Protection Agency 1989, Massachusetts 1989).

With indoor air pollution on the rise since World War II and tighter, more energy-efficient construction of schools and workplaces, outbreaks of sick building syndrome appeared in the late 1970s. Chlorine production is felt by some to provide a useful index of the increased quantities of synthetic organics that are found indoors (e.g., polyvinyl chloride). Figure 1-8 shows the dramatic rise in chlorine production in billions of pounds per year that has occurred since World War II, plotted against mean sperm density, a widely recognized and subtle indicator of the toxic effects of a variety of chemicals, for example, lead. Actually, increases in chlorine production underestimate increases in the amount of synthetic organics. Figure 1-9 depicts production changes since 1945. Before World War II, U.S. production of synthetic organic chemicals totaled fewer than a billion pounds per year. By 1976, production had soared to 163 billion pounds annually (Odell 1980, p. 213). Increased sources of indoor air pollution, coupled with decreased fresh make-up

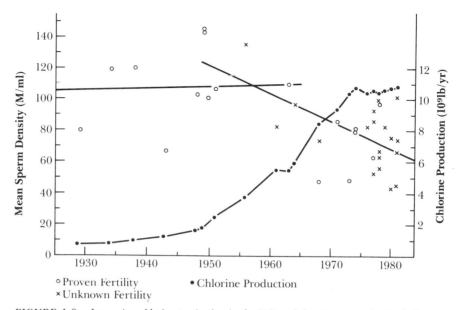

○ Proven Fertility   ● Chlorine Production
× Unknown Fertility

FIGURE 1-8.   Increasing chlorine production in the U.S. and the apparent reduction in human sperm density. Reprinted with permission from Mage, D. and R. Gammage, "Evaluation of changes in Outdoor Air Quality Occurring Over the Past Several Decades," in Indoor Air and Human Health, *R. Gammage and S. Kaye, Eds. (copyright 1985, Lewis Publishers, Inc., Chelsea, MI) p. 10.*

air, have transformed the indoor environment. Community exposures to toxic chemicals, industrial and office exposures, and other episodic exposures of individuals also increased, reflecting the rise in production of coal- and oil-derived chemicals and synthetics.

These changes in chemical production, consumer products, and building design have been accompanied by an increasing number of people who appear to react to low levels of environmental pollutants. Indeed, since World War II certain illnesses, such as asthma (Sly 1988) and depression (Klerman and Weissman 1989), seem to have shown upsurges. Many patients with these conditions and other health problems, frustrated by their lack of success with traditional medicine, sought the care of clinical ecologists, who related their patients' symptoms to environmental exposures.

Theron Randolph, who founded clinical ecology, and a number of other clinical ecologists are board-certified allergists. Randolph received his M.D. degree from the University of Michigan, where he completed his residency in internal medicine. His allergy and immunology fellowship was completed at Massachusetts General Hospital and Harvard

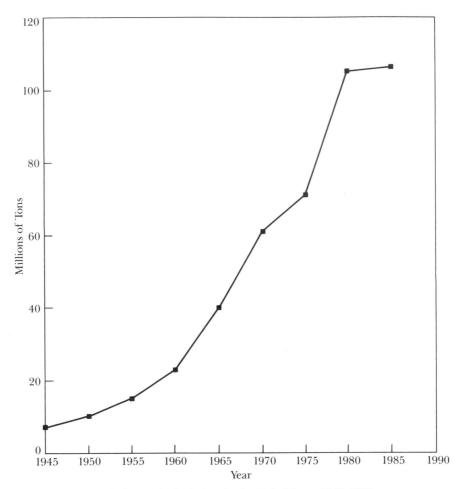

FIGURE 1-9. Synthetic organic chemical production United States, 1945–1985. Source: U.S. Intern. Trade Commission.

Medical School in 1942 to 1944. He entered private practice in Chicago, where he also served as a clinical instructor in allergy at Northwestern University Medical School for several years. Randolph reported that many of his patients reacted adversely to common foods such as corn (a food ubiquitous in the American diet in the form of sugar, starch, and oil, as well as in its unrefined state), wheat, milk, and eggs. Indeed, he later described corn allergy as "the most common food allergy in North America" (Randolph and Moss 1980, p. 109). He used Rinkel's technique (Rinkel 1944; Rinkel et al. 1951) of having patients avoid specific foods for 4 to 6 days (and not much longer) to "unmask" reactions prior to test-feeding them. Herbert Rinkel, another American allergist, had first described the phenomenon of "masking" and its clinical application for food sensitivities.

According to ecologists, most allergists then, as now, do not recognize or employ this avoidance or "unmasking" period prior to testing and thus never see this type of food sensitivity in their patients. Recent guidelines on food testing written for allergists and published in their journal do not address this issue (Bock et al. 1988). Many traditional allergists recognize primarily IgE-mediated, immediate reactions to foods that generally are more readily observable by the investigator and do not require an avoidance period in order to detect them.

In 1951, Randolph realized that not only foods but also chemicals might be responsible for some of his patients' symptoms. A physician's wife who sold cosmetics had been seeing Randolph over a 4-year period for rhinitis, asthma, headache, fatigue, irritability, depression, markedly fluctuating weight, and intermittent episodic loss of consciousness (Randolph 1987, pp. 73–76). At each visit, he recorded almost verbatim on his typewriter the patient's statements about her condition, without editing. By 1951, he had compiled 50 single-spaced typewritten pages concerning this woman. Reading over these pages, he realized there was a common denominator: each event was associated with exposure to gas, oil, coal, or their combustion products. Similar observations in other patients followed. A few years later, Randolph published a series of six abstracts in the *Journal of Laboratory and Clinical Medicine* concerning "allergic type" reactions to industrial solvents and liquid fuels, mosquito abatement fogs and mists, motor exhausts, indoor utility gas and oil fumes, and chemical (coal or petrochemically derived) additives in foods, drugs, and cosmetics (Randolph 1954 a–f).

Randolph and other ecologists often refer to chemical sensitivities as the "petrochemical problem" because the increase in the incidence of this illness seems to parallel the growth of the petrochemical industry and the increased use of synthetic materials such as particleboard, pesticides, synthetic textiles, plastics, and food additives by consumers since World War II. Late in the 1950s, Randolph adopted the term *clinical ecology* in order to describe his practice and its focus on environmental incitants and in order to avoid use of the word *allergy*.

Randolph, who had been hospitalizing patients and testing them for their food sensitivities, found that another critical element in many of his patients' recoveries was avoidance of environmental chemical exposures in their jobs and/or homes while in the hospital. He developed "Comprehensive Environmental Control," a diagnostic approach in which patients avoid exposure to synthetic chemicals in order to facilitate diagnosis of chemical sensitivity. The next chapter describes this method in detail. A description of comprehensive environmental control and its role in diagnosis and therapy first appeared 30 years ago in *Clinical Physiology* (Randolph 1960) and again in the *Annals of Allergy* in 1965 (Randolph 1965); *Human Ecology and Susceptibility to the Chemical Environment* lists the most common chemical exposures Randolph felt provoked symptoms in his patients (Randolph 1962).

Although Randolph reported treating a wide range of illnesses successfully, his and other clinical ecologists' enthusiasm for this approach was not shared by many of their contemporaries. Randolph's early work showing that environmental influences can provoke mental and behavioral disturbances in a demonstrable, cause-and-effect way occurred in the 1950s, at the same time modern psychopharmacology was developing and the use of phenothiazine tranquilizers was expanding. Drugs that could control behavior, so-called chemical restraints, were easy to administer and mass applicable, and drug companies promoted them widely. University research in psychiatry, funded by drug companies, became focused on the development of better drugs. Only later were the long-range complications of many of these drugs realized. The dominance of psychoanalytical, behavioral, and pharmacological approaches to mental illness abrogated any major attempts by psychiatrists to look for food or chemical triggers for their patients' illnesses (Randolph 1987, p. 188). Randolph and other clinical ecologists are critical of these developments in psychiatry: "especially psychoanalysis, despite its wide application, has been devoid of any demonstrable evidence of etiology and has been relatively ineffective therapeutically" (Randolph 1987, pp. 190–191).

As clinical ecologists continued to apply their concepts of environmental illness, many of which did *not* appear dependent upon immune mechanisms, they distanced themselves more and more from traditional allergists. When von Pirquet coined the name *allergy* in 1906, he defined it as "altered reactivity" of whatever origin. Thus the word *allergy* as originally used embraced both immunity and hypersensitivity (Corwin 1985). In 1925 European allergists influenced their American colleagues to redefine *allergy* in the context of antibodies and antigens. Randolph, Coca, and other allergists objected, preferring to call this development the "immunologic theory of allergy," but the new definition prevailed. Thus clinical ecology, which was concerned with heightened reactivity of unknown etiology, did not fit under this new definition. In 1967, when IgE was discovered, it enhanced allergy's credibility as a specialty. (At the time, allergy shots were dimly viewed, even called "witchcraft" or "voodoo medicine" by some medical practitioners.) IgE's discovery provided allergists with a scientific basis for their practice, and some began to look down on areas such as clinical ecology that did not have such a basis. Thus the observations of clinical ecologists, irrespective of their validity or clinical utility, were excluded from allergy, in part because IgE did not appear to be involved. Allergy and clinical ecology have continued to develop and define their separate paradigms. Tables 1-2 and 1-3 present the salient differences in approach and philosophy.

These dueling paradigms have continued to hamper meaningful dialogue between the two groups. Clinical ecologists are not found in allergy departments in medical schools. Their articles seldom appear in premier allergy journals—many times, but not always, because the articles fall short of recognized standards for scientific publications. Peer review in medicine may have both positive and negative consequences (Horrobin 1990). It serves to ensure quality control but it may also deter innovation. In the United States clinical ecologists are absent from academic medicine. In contrast, the Robens Institute, University of Surrey, England, has given ecologist William Rea a chair in environmental medicine. An environmental unit has recently been established at the Beijing Union Medical School in China. In Ontario, the Ministry of Health is funding a $600,000 study of food sensitivities at the University of Toronto as a result of the Ontario study on chemical hypersensitivity.

Feeling shut out of allergy, Randolph and several other allergists founded the Society for Clinical Ecology in 1965 and opened its doors to family physicians, otolaryngologists, and other physicians interested or involved in the area. In 1984 the society changed its name to the American Academy of Environmental Medicine, much to the chagrin of allergists, toxicologists, and other academicians in environmental medicine. Membership in the academy has grown by 225 members in the last 2 years and presently totals 570 (Howard 1989). An examination leading to certification in Environmental Medicine is offered, but the specialty has not been recognized by the American Board of Medical Specialties.

Allergists continue to point to the scientific basis of their practice and their detailed knowledge of immune mechanisms. Clinical ecologists stress the importance of their clinical observations. To some degree, their conflicts are an extension of the traditional tension between academicians and clinicians—a tension that has served neither side well.

Unfortunately, these conflicts may result in adverse economic consequences for patients who are already frustrated by their illness and their attempts to gain help. Allergists successfully persuaded Medicare not to reimburse for provocation-neutralization therapies (the predominant therapy used by ecologists, which is discussed further in Chapter 3) for treating food allergies, and allergists are often asked to provide independent assessments to insurance carriers who are contesting workers' compensation claims for disability associated with chemical exposure. One prominent allergist we interviewed was distressed at finding himself "on

TABLE 1-2. Contrast Between Paradigms of Traditional Medicine and Ecologically Oriented Medicine

Aspect	Traditional Medicine[a]	Clinical Ecology
Focus/approach	Body-centered; diagnosis contingent upon laboratory or clinical findings; symptoms alone generally insufficient for diagnosis	Environmentally oriented; diagnosis based upon temporal relationship between symptoms and chemical/food exposure; testing by avoidance and reexposure, sublingual or cutaneous provocation
Stage at which disease is diagnosed	End-organ damage generally must be present	Diagnosis may be made prior to end-organ damage, i.e., in a subclinical or pre-morbid state
View of patient	Focus on bodily parts and their malfunction. The patient is sick.	Focus on patient's chemical exposures and dietary habits. The patient's environment is sick.
Specialization	Anatomically demarcated[b]	No specialties per se. Concept of specialties considered limiting because environmental exposure may provoke symptoms in several systems simultaneously.
History taking	Review of all body systems by more thorough primary care takers; organ-oriented by specialists. Limited emphasis on dietary factors except in certain diseases, e.g., obesity, diabetes, hypercholesterolemia. Minimal attention to environmental exposures (except smoking) unless issue raised by patient	The more thorough practitioners review symptoms involving each system and search for environmental contributors to patient's illness. Emphasis is placed upon dietary and exposure histories.
Therapies	Drugs, surgery	Avoidance of environmental and food incitants; "neutralization" of symptoms by giving small dose of incitants; nutritional supplements; "detoxification"

[a] Within traditional medicine, allergists are among the practitioners most skilled in exploring environmental factors relating to a patient's illness. They also appreciate multiple organ involvement. Differences between allergy and clinical ecology are summarized in Table 1-3.

[b] Patients become "trained" to limit their complaints to the specialist's organ of interest; thus they may not complain of a headache to their gynecologist or of a rash to their psychiatrist.

22

TABLE 1-3. Contrast Between Traditional Allergy and Clinical Ecology

Aspect	Traditional Allergy[a]	Clinical Ecology[a]
Focus	Search for environmental triggers for symptoms/disease	Same as allergists
Practice priorities (in terms of frequency of diagnosis/relative importance in their patient population)	1. Biological inhalants, e.g., pollen, dust mite, molds, etc. 2. Food sensitivities 3. Chemical incitants	1. Chemical incitants 2. Food sensitivities 3. Biological inhalants
Diagnostic approaches: Biological Inhalants	Skin tests using extracts of pollens, molds, etc. Sometimes in vitro testing.	Skin testing also but using techniques/extract concentrations that differ from allergists
Foods	Skin tests using food extracts or in vitro tests, but usually limited to IgE-mediated diseases, such as eczema. Some practitioners do use elimination diets and food challenges for diagnosis. Double-blind, placebo-controlled challenges preferred.	Elimination diet with removal of suspect foods (or fasting) for 4–7 days followed by feeding challenges; sublingual or cutaneous provocation-neutralization heavily relied upon by majority
Chemicals	Interest in patient's chemical exposures includes occupational asthma, chemical "irritants" that exacerbate asthma and contact dermatitis. Also drug allergies or adverse reactions to drugs. "Chemical testing" limited to: 1. Skin testing for drugs (especially penicillin) 2. Patch testing for contact dermatitis 3. Inhalation challenges in specially constructed exposure chambers by a few practitioners (usually related to disability cases)	Low-level, often subtle chemical exposures (e.g., gas heat, formaldehyde, off-gassing from particleboard) responsible for many diverse symptoms/diseases. Patient avoids incitants. Some practitioners use sublingual or cutaneous doses of certain chemicals to provoke and neutralize symptoms. A few practitioners perform inhalation challenges.

(continued)

Aspect	Traditional Allergy[a]	Clinical Ecology[a]
Therapies used	Avoidance where practical. Antihistamines, topical and systemic steroids, cromolyn, and other drugs. Immunotherapy.	Chemical, food incitants avoided to regain tolerance. Drugs generally are avoided because of potential to sensitize. Some use of pancreatic enzymes, oral cromolyn, transfer factor, nystatin for *Candida* (yeast) sensitivity, etc. "Neutralization" used by most. Rotary/elimination diets often using organically grown food. Detoxification using saunas.
Definition of allergy	Adverse reaction involving antigen-antibody or sensitized lymphocytes	Adverse reaction to a substance
View of multiple chemical sensitivities	Varies greatly but majority of patients felt to have psychiatric disorders or erroneous belief systems	Consequence of exceeding patient's capacity to adapt to total environmental load

[a] Practice styles vary widely and even overlap within these groups; e.g., some allergists treat for *Candida* hypersensitivity, and some ecologists use traditional skin testing methods and immunotherapy for inhalant allergies.

the wrong side" from the patient's perspective as a result of taking referrals from insurance carriers.

While conflicts and antagonisms continue between allergists and clinical ecologists, the fields of toxicology and epidemiology are expanding their recognition that chemicals are harmful at lower and lower levels (Ashford 1987). Both classical toxicology and epidemiology have been invaluable in studies involving a single cause resulting in a single effect. With synergism (multiplicative effects of several toxins) or multiple effects, scientific investigations are more difficult to conduct and interpret. Indeed, the design of epidemiological studies for the discovery of chemical sensitivity requires great care. Because such a variety of inducing substances and, subsequently, triggering substances seems to be involved, several mechanisms could be operating simultaneously, and the "disease" may not be the same in all cases. Performing retrospective epidemiological studies on chemically sensitive persons without carefully defining the group to be investigated may result in a dilution of the prevalence of significant health effects. Then again, stratifying groups too narrowly may not yield statistically significant findings.

Other difficulties may interfere with using epidemiology to detect significant health effects of chemical exposures in certain situations. Spengler and associates (1983a) enumerate some potential problems in using epidemiologic studies to discover health effects of nitrogen dioxide (NO_2) exposure upon residents of homes with gas cooking.

1. Some homes with gas stoves have very low levels of NO_2, as low as those with electric stoves. Such variability could weaken the association between health effects and pollutant categories.

2. Only a subset of the exposed group (those having gas stoves) may be affected by gas stove combustion products.

3. High, short-term, or "peak" concentrations of NO_2 could be the primary cause of health effects, so that only a certain percentage of gas-cooking homes would be at risk.

4. Pollutants other than NO_2 might be more important causes of health effects. If so, variations in these pollutants among gas fuels and over time would tend to weaken any observed effect.

5. Statistical correction for factors that correlate with NO_2, such as socioeconomic status, may interfere with observing health effects. For example, low-income families are more likely to use gas stoves for supplementary heating.

The problems encountered in using epidemiology to identify health effects of NO_2 apply to multiple chemical sensitivity, although one further difficulty may be involved in the case of chemical sensitivity; the *same* exposure may produce different health effects in different individuals. These obstacles may be overcome as more is understood about this problem and the exposures involved.

At present, the allergists do not identify with the clinical ecologists, even though the ecologists are concerned with "altered reactivity." The toxicologists and epidemiologists do not seek to establish communication with the ecologists, even though the ecologists share their concern about exposure to toxic substances. If the model employed by clinical ecologists offers any insight into a cause-and-effect relationship between environmental incitants and illness, its application will be seriously hampered by the present state of affairs. Randolph has called for strengthening the relationship between toxicology and clinical ecology. We believe that some tenets of clinical ecology at its best will contribute to a *dynamic toxicology,* that is, observing the effects of chemical incitants in real time as those effects evolve.

Magnitude and Nature of the Problem

Chemical sensitivity presents a challenging puzzle for the scientist, physician, and public policy decision-maker. The pieces of the puzzle include (1) observations of possible offending or triggering substances and health effects, and (2) plausible mechanisms, diagnostic approaches, and therapies. Although a definitive and accurate picture is yet to come, at this time the pieces—viewed collectively—provide sufficient evidence to conclude that chemical sensitivity does exist as a serious health and environmental problem and that public and private sector action is warranted at both the state and federal levels (see also Massachusetts 1989, p. 1). Just how large a problem exists is not known at this time. The National Academy of Sciences has suggested, without providing documentation, that approximately 15 percent of the population may experience "increased allergic sensitivity" to chemicals (National Research Council 1987). Subsequent clarification by the chairman of the 1987 NAS workshop revealed that this figure is based on occupational studies of hypersensitivity to chemicals where the hypersensitivity was considered immunologic in origin (Lebowitz 1990). Based on the increasing outbreaks of sick building syndrome, increased reporting of symptoms in contaminated communities to state health departments, increased recognition of problems in the industrial workplace, and the increasing numbers of physicians treating chemically related sensitivities, the existing evidence does suggest that chemical sensitivity is on the rise and could become a large problem with significant economic consequences related to the disablement of productive members of society.

For an update on prevalence, see the section "Magnitude of the Problem" in Chapter 8. Also see "Origins of Chemical Sensitivity" in that chapter for a description of two possible new groups of affected persons, sick Gulf War veterans and women with silicone breast implants.

CHAPTER 2

Key Terms and Concepts

Terminology

A wide array of names has been applied to the syndromes suffered by patients with heightened reactivity to chemicals (Table 2-1). Each name has specific implications regarding the underlying cause, mechanism, or manifestations of the disease, and they overlap. A major hindrance in achieving scientific respectability has been the difficulty in agreeing upon a definition for this condition (or conditions). Cullen (1987b) has emphasized the importance of establishing a uniform case definition before meaningful epidemiologic studies can be undertaken, but cautions, "However constructed, the goal of descriptive studies must be refinement of the diagnostic criteria, in particular the very tentative boundaries with other diagnostic entities such as allergic, anxiety, panic and post-traumatic stress disorders, and physiologic sequelae of central nervous system (CNS) intoxication or injury, especially by organic solvents." He acknowledges possible overlap among these entities and offers the following case definition:

> Multiple chemical sensitivities (MCS) is an acquired disorder characterized by recurrent symptoms, referable to multiple organ systems, occurring in response to demonstrable exposure to many chemically unrelated compounds at doses far below those established in the general population to cause harmful effects. No single widely accepted test of physiologic function can be shown to correlate with symptoms. (Cullen 1987a)

TABLE 2-1. Attributes of Names for Heightened Reactivity

Cause	Mechanism	Effect
Environmentally induced illness	Immunologic illness	Multiple chemical sensitivities (MCS)
	Immunotoxicity	
Chemically induced (or acquired) hypersusceptibility	Immune dysfunction	Multiple chemical sensitivity syndrome
	Immune dysregulation	
	Conditioned odor response	Chemical hypersensitivity syndrome
Chemically acquired immune deficiency syndrome (chemical AIDS)	Fear/anxiety	Universal allergy
	Mass psychogenic illness	20th-century illness
	Various psychiatric disorders	Total allergy syndrome
		Environmental allergy or illness
		Cerebral allergy
The petrochemical problem		Environmental maladaptation syndrome
		Food and chemical sensitivity

This case definition, intended for epidemiological use, is intentionally narrow. Cullen excludes persons who react to substances no one else is aware of on the basis that such individuals may be delusional and excludes persons who have bronchospasm, vasospasm, seizures, or "any other reversible lesion" that can be identified and specifically treated. Clinical ecologists, however, would argue that persons with bronchospasm, vasospasm, seizures, and other illnesses excluded by Cullen may well have the chemical sensitivity problem. Each issue of the clinical ecologists' journal, *Clinical Ecology,* contains the following definition:

> Ecologic illness is a chronic multi-system disorder, usually polysymptomatic, caused by adverse reactions to environmental incitants, modified by individual susceptibility and specific adaptation. The incitants are present in air, water, food, drugs and our habitat.

Although the patients the clinical ecologists and Cullen see are demographically divergent, the definitions of their illnesses are remarkably alike. Both *describe* the chemically sensitive patient in similar terms. (See Chapter 1 for a discussion of sensitive populations.) However, what is sorely needed is an objective test that can be applied in each individual case to determine, incontrovertibly, whether a particular person has multiple chemical sensitivities.

Given the multitude of environmental exposures (both chemical and food) that allegedly can result in a seemingly endless array of physical and mental syndromes and the frequent absence of findings on routine physical examination, the practitioner who sees these patients with their divergent and unfamiliar litany of complaints is at great disadvantage in trying to diagnose the condition.

To circumvent this problem, we propose the following *operational* definition of multiple chemical sensitivity, a definition that is based upon environmental testing:

> The patient with multiple chemical sensitivities can be discovered by removal from the suspected offending agents and by rechallenge, after an appropriate interval, under strictly controlled environmental conditions. Causality is inferred by the clearing of symptoms with removal from the offending environment and recurrence of symptoms with specific challenge.

Challenges conducted for research purposes should be performed in a double-blind, placebo-controlled manner. This definition embodies the approach to discovering environmental causation that was developed by Theron Randolph. Randolph originated the idea of an environmental unit employing what he terms "comprehensive environmental control" as both a diagnostic *and* therapeutic tool for dealing with these patients. Briefly, this technique involves placing the patient in a specially constructed environment devoid of materials that off-gas; avoiding the use of drugs, cosmetics, perfume, synthetic fabrics, pesticides, and similar substances; and having the patient fast for a period of days until symptoms resolve. This initial period of avoidance and fasting requires approximately 4 to 7 days on the average. During this time, the patient exhibits withdrawal symptoms such as headache, malaise, irritability, or depression. At the end of this time, the patient's symptoms, if environmentally related, should clear, provided that end-organ damage has not occurred. Clinical ecologists say this clearing does occur in the vast majority of patients. At the end of this avoidance phase, the patient generally has a markedly lower pulse rate and an increased sense of well-being, as well as a resolution of symptoms. Drinking waters from a variety of sources also are tested to find one most compatible with the patient. Next, individual foods are reintroduced, one per meal, over a two- to -three-week period. Following this, the patient is placed on a rotating diet of "safe" foods (i.e., foods that did not provoke symptoms for that particular patient). Finally, the patient is challenged with very low levels (levels routinely encountered in daily living) of common chemicals. Those exposures, both food and chemical, that induce symptoms are to be avoided. (Comprehensive environmental control is discussed in more detail later in this chapter.)

We feel strongly that this operational definition is essential to resolving, once and for all, the debate about whether an individual's symptoms are or are not environmentally induced. An environmental unit is necessary for scientific validation of the concept of chemical sensitivity. Because of the expense and time required by patients and physicians alike, we are not arguing that the unit be used for all patients. Such stringent measures are not necessary for most patients. For severe cases, however, no alternative is available at present, and only from firsthand observation of hospitalized patients can physicians have the opportunity to understand this illness better. In time, as more clinical data on these patients accumulate, physicians may be able to diagnose this disorder on the basis of the patient's history and a few key laboratory tests. For now, reliance must be placed on rigorous study in an environmental unit. Ultimately a *phenomenological* definition may emerge that allows physicians to diagnose, at least tentatively, chemical sensitivity based on a history of a specific sensitizing event (such as a pesticide exposure) followed by evidence of chemical and food sensitivities, multisystem effects, improvement after avoidance of exposure, and similar experiences of persons with like histories.

The environmental unit is the gold standard against which all other diagnostic approaches and screening techniques should be measured. An environmental unit is necessary in more severe cases, such as those who have failed outpatient attempts at management or for patients with seizures, suicidal tendencies, incapacitating migraine headaches, arrhythmias, or other problems requiring continuous vigilance. However, most individuals can remain outpatients while they are guided through an elimination diet, avoidance of possible chemical incitants, and rechallenge with suspected offenders. In a later section, we discuss provocation-neutralization and other office-based techniques that have been adopted by clinical ecologists in order to screen for and treat this illness. An enormous number of diagnostic and therapeutic modalities have been proposed, many of them lacking scientific verification. The gold standard—comprehensive environmental control with the use of an environmental unit—must be separated from the "fool's gold" of some of the more outlandish and untested diagnostic and therapeutic modalities. That is not to say that certain of those approaches are not now efficacious or may not "pan out" in the future, but many await and need critical scientific appraisal.

One aspect of clinical ecology that has repelled many traditional practitioners is the hodgepodge of unscientific, sometimes "new age," and even spiritual approaches patients with this illness have resorted to in a desperate struggle to restore their health. Randolph himself has expressed dismay at this turn of events. A survey of arthritis patients, who (like the chemically sensitive patient) have limited therapeutic options and may lead constricted life-styles, revealed that 94 percent had tried at least one unorthodox therapy (Wasner et al. 1980). From the chemically sensitive patient's viewpoint, searching for alternative therapies is understandable because the available treatments for this problem primarily have been avoidance of exposure and an elimination diet. Restrictive diets and avoidance do not permit full engagement in a modern life-style, and naturally patients seek alternatives. To the outside observer, these patients' practices appear cultist, and members of this supposed cult have been labeled in print as "true believers" and their physicians as "gurus" or "pseudoscientists." We find such terminology unfortunate and counterproductive. It would not appear to reflect the level of intelligence and professional achievements of these patients, many of whom are scientists, physicians, lawyers, teachers, and others from whom one would expect a modicum of common sense. Many are intelligent individuals who are angry at traditional medical practitioners for their unwillingness to study and understand this illness.

As individuals with chemical sensitivities are caught up in the escalating debate among medical practitioners, they find it more and more difficult to obtain unbiased, useful information regarding their condition. This difficulty underscores the importance of the operational definition we have proposed for chemical sensitivity. This definition takes the problem seriously and offers the environmental unit as the objective, scientific means for its study. With regard to other definitions that have been proposed, we agree that Cullen's narrow, descriptive case definition may have utility in some epidemiological investigations, for example, in tight-building syndrome or certain occupational outbreaks. In dealing with such a diversity of agents causing equally diverse effects at extraordinarily low levels with no true unexposed control group, however, such a definition may make engaging in meaningful epidemiologic investigations difficult.

Another term in this controversy that has confused patients and physicians alike is the word *allergy*. In a scholarly review, Alsoph Corwin (1985) of Johns Hopkins discusses the historical consequences that have arisen out of what he calls a faulty definition of *allergy*. He traces the evolution of the term and its consequences for the development of the field.

> Essentially, the fallacy lies in the confusion of hypersensitivity with immunity and the consequent exclusion from consideration of those cases of hypersensitivity which do not exhibit serological abnormalities.

These include many food reactions, drug allergies and reactions to environmental pollutants.

Corwin acknowledges Randolph and other clinical ecologists for not having been hamstrung by the limited definition of allergy as IgE-mediated (atopic) disease and for having attempted to document and elucidate the mechanisms of individual hypersensitivity, a problem he describes as much more prevalent than atopy. According to Corwin, "Estimates of the incidence of hypersensitivity in the general population run from 50–90%, whereas only approximately 6% have atopic allergy." The faulty definition of allergy, by excluding most hypersensitivities, has had devastating consequences, according to Corwin. He points to the work of Randolph in establishing cause-and-effect relationships between environmental factors and disease, saying, "Exclusion of these phenomena [that is, restricting the definition of allergy to IgE-mediated disease] also involves the world in tremendous expenditures for research for the elucidation of disease states when the solutions to the problems lie unused in the great medical libraries of the world." (See Chapter 3 and Appendix A.) Here Corwin is alluding to the writings of Randolph and others.

The field of allergy and immunology today embraces antigen-antibody interactions (including those involving IgE), sensitized lymphocytes, and anaphylactoid drug reactions whose mechanisms remain obscure. Nevertheless, the view of allergy as primarily concerned with IgE-mediated phenomena has prevailed and has had considerable consequences for the practicing traditional allergist. William T. Kniker (the 1985 Bela Schick Lecturer, a professional honor bestowed by the American College of Allergy and Immunology) described the erosion of the allergist's practice by ear, nose, and throat (ENT) physicians, pulmonary specialists, and other groups. Kniker (1985) warned his fellow allergists: "We are not yet comfortable with other hypersensitivity diseases (immunologically triggered or not), adverse reactions to foods and environmental factors (occupational, hobby, home). . . . The narrowness of our specialty makes us extremely vulnerable." He quoted the author of *Megatrends,* who forecast "the triumph of the new paradigm of wellness, preventive medicine, and holistic care over the old model of illness, drugs, surgery, and treating symptoms, rather than the whole person. *The next big shift will be to focus on the environmental influences on health!*" (emphasis by Kniker).

Randolph (1987, p. 292) estimates that nearly 2,000 physicians, including roughly 900 ENT physicians, are applying the techniques of clinical ecology, in contrast to 3,000 to 3,500 conventional allergists. Many communities have a surfeit of traditional allergists, and new allergists find demand for their skills waning (Kniker 1985). In contrast, clinical ecologists are quite busy, Randolph (1987) notes "when there were only a few of us we were treated as gadflies. Now that we are 40% of the total we are perceived as a real threat and dealt with accordingly." Almost all of the traditional allergists we interviewed feel strongly that allergy should embrace patients who have heightened reactivity to chemicals and/or foods, irrespective of the etiology of their problem. Selner (1985b), in particular, has written:

> There is every indication that the problem of chemical intolerance will continue to grow. We view these events as an opportunity for Allergy to appropriately expand its interest and influence into areas to which the public and the medical profession have traditionally turned to allergists for answers. . . . Although this may require fundamental changes in traditional practice priorities as well as allergy training curriculums, we believe the future of allergy practice can be found within this challenge.

Doris Rapp, a board-certified allergist who practiced traditional allergy for 18 years, turned to clinical ecology 14 years ago when she observed a dramatic reaction to food in a friend and became intrigued that clinical ecologists almost never placed asthmatic patients on steroids. She feels it is "ludicrous" to say that what ecologists do is not allergy. In her view (1985) she always was, still is, and will continue to be an allergist:

> I am doing the same things, for example, that I did for the first 18 years, but much better. I use the same extracts to test and treat. What I do, however, requires much more time, and the overhead is discouragingly increased. But the rewards are that patients, not helped by others, or previously not helped by myself, often get well quickly.

Ironically, patients with chemical sensitivity who have seen traditional allergists for what they felt was an "allergy" to tobacco smoke or some other substance have been lectured to on the subject of allergy and what its definition *really* is, that is, IgE-related disease. Some allergists we interviewed told us they attempt to educate these patients by handing them "Clinical Ecology" or "Controversial Techniques," the position papers of the American Academy of Allergy and Immunology (see references). Patients who consult allergists probably do not care whether what they have is, by definition, an allergy or not; what they are interested in is help in treating an adverse reaction to some substance.

For an updated discussion of terminology, see "A Note on Terminology" in Chapter 7.

Adaptation

One of the difficulties the observer encounters in trying to understand multiple chemical sensitivity is the ostensible lack of a central concept or unifying theory. Such a unifying theory does exist and revolves around the concept of adaptation, known in other contexts as *acclimation* or *acclimatization, habituation, developing tolerance* and even *addiction* (which we will explain later). Randolph has used the terms *adaptation* and *addiction* most often. However, reference to one of the other words may make grasping the concept easier. *Acclimatization* is a widely used term in occupational health that refers to workers gradually becoming accustomed to exposures on the job, for example, heat stress. Understanding adaptation is important here for two reasons: (1) adaptation makes difficult the discovery of the effects of a particular exposure on the body and (2) chemical exposures may adversely impact adaptation mechanisms and thus lead to illness.

Relatively little can be discovered about physiological "adaptation" by reading medical textbooks or recent major medical journals, in part because of the absence of Randolph's writings from such publications for a quarter of a century. However, detailed discussions of adaptation appear in all of Randolph's books, in journal articles by him from the 1950s, and in the clinical ecologists' literature. Our impression from interviewing traditional allergists is that many allergists are not aware of this concept and its potential clinical ramifications.

Concerning adaptation (or acclimation), Randolph (1962, p. 5) wrote:

> Human ecology embodies the concept of a person's adaptation to the conditions of his existence. The ecologic effects of chemical incitants are observed most advantageously by first isolating an individual from the *total chemical environment* and then observing his response to *re-exposure* to previously avoided *parts* of it.

That human beings respond to chronic exposure to environmental challenges by adapting, acclimating, acclimatizing, or even becoming addicted is widely recognized for a variety of substances. Most would agree that the use of narcotics, alcohol, nicotine, and even caffeine can be addicting. For example, the first cigarette ever smoked might be associated with eye and throat irritation, but over time, with more cigarettes, most individuals adapt, and primarily the pleasurable effect of nicotine on the brain are experienced. After months or years, more cigarettes (or alcohol or caffeine or drugs) may be required for the same amount of lift. The individuals may exhibit addictive behavior seeking cigarettes more frequently. Subsequently, quitting cigarettes (alcohol, caffeine, or drugs) may lead to withdrawal symptoms including irritability, drowsiness, fatigue, moodiness, and headache. After individuals have quit smoking, they may find themselves supersensitive to others' smoke (the forgotten eye and throat irritation reappear after a period of avoidance). This example parallels the food and chemical adaptation and addiction that ecologists like Randolph have described in their patients: frequent exposure to a substance results in adaptation (irritation and other warning signals may disappear); continued exposure may lead to addiction; reduction or cessation of exposure generally results in withdrawal symptoms.

The difference between chemical exposures and cigarettes, alcohol, or caffeine is that in the former case addiction is an unwitting process. The individual may have no idea it is occurring. But if the offending chemical is removed, withdrawal symptoms may ensue (Table 2-2). With reexposure to the substance following a period of avoidance, symptoms return, often quickly and much more obviously related to the exposure. What confuses many patients and practitioners is that the symptoms for which the individual is most likely to seek a physician's help are those that occur during withdrawal when the person is no longer exposed (or less exposed) to the offending agent! Thus headaches may occur when the individual smokes *fewer* cigarettes than usual or drinks *less* caffeine. Indeed, these unpleasant withdrawal symptoms may be forestalled by smoking another cigarette or taking another drink of coffee or alcohol and thus perpetuating addiction. Patients may report that smoking a cigarette or drinking a cup of coffee in the morning (after 8 or so hours without) *relieves* their headache (a withdrawal symptom) and they feel better, not suspecting that the cigarette or coffee might also be the cause of their headache.

Occupational health presents many examples in which acclimatization, inurement, or tolerance to a substance is known to develop, for example, exposure to ozone, nitroglycerin, cotton dust, welding fumes (containing zinc), and solvents. Note that the incitants mentioned thus far are all quite different from one another: some are ingestants, others inhalants; some are solid, others liquid or gaseous in form; some are organic, others inorganic; some (ozone) are simple inorganic molecular gases, whereas others (welding fumes) are complex mixtures of organic and inorganic substances in solid, liquid, and gaseous phases. The point is that the human organism has the capacity to adapt to an endless array of substances. In the extreme, as described for cigarettes and caffeine, individuals unknowingly may become *addicted* to the incitant, for exam-

TABLE 2-2. Stimulatory and Withdrawal Symptoms Associated with Exposure to Various Foods and Chemicals[a]

Behavior Classification	Response Classifications	Stimulatory Level	Response Characteristics	Typical Responses
Stimulated	Maladapted Cerebral and Behavioral Responses	+ + + +	MANIC, WITH OR WITHOUT CONVULSIONS	Distraught, excited, agitated, enraged, and panicky. Circuitous or one-track thoughts, muscle-twitching and jerking of extremities, convulsive seizures, and altered consciousness may develop.
		+ + +	HYPOMANIC, TOXIC, ANXIOUS AND EGOCENTRIC	Aggressive, loquacious, clumsy (ataxic), anxious, fearful, and apprehensive; alternating chills and flushing, ravenous hunger, excessive thirst. Giggling or pathological laughter may occur.
	Adapted Responses	+ +	HYPERACTIVE, IRRITABLE, HUNGRY, AND THIRSTY	Tense, jittery, "hopped-up," talkative, argumentive, sensitive, overly responsive, self-centered, hungry, and thirsty; flushing, sweating, and chilling may occur, as well as insomnia, alcoholism, and obesity.
		+	STIMULATED BUT RELATIVELY SYMPTOM-FREE	Active, alert, lively, responsive, and enthusiastic, with unimpaired ambition, energy, initiative, and wit. Considerate of the views and actions of others. This usually comes to be regarded as "normal" behavior.
"Normal"		0	BEHAVIOR ON AN EVEN KEEL, AS IN HOMEOSTASIS	Calm, balanced, level-headed reactions. Children expect this from their parents and teachers. Parents expect this from their children. We all expect this from our associates.

Withdrawal (maladapted) responses: loss of tolerance			
Maladapted Localized Responses	−	LOCALIZED ALLERGIC MANIFESTATIONS	Running or stuffy nose, clearing throat, coughing, wheezing. Asthma, itching (eczema and hives), gas, diarrhea, constipation (colitis), urgency and frequency of urination, and various eye and ear syndromes.
Maladapted Systemic Responses	− −	SYSTEMIC ALLERGIC MANIFESTATIONS	Tired, dopey, somnolent, mildly depressed, edematous with painful syndromes (headache, neckache, backache, neuralgia, myalgia, myositis, arthralgia, arthritis, arteritis, chest pain), and cardiovascular effects.[b]
Maladapted Advanced Stimulatory Responses	− − −	BRAIN-FAG, MILD DEPRESSION, AND DISTURBED THINKING	Confused, indecisive, moody, sad, sullen, withdrawn, or apathetic. Emotional instability and impaired attention, concentration, comprehension, and thought processes (aphasia, mental lapse, and blackouts).
	− − − −	SEVERE DEPRESSION, WITH OR WITHOUT ALTERED CONSCIOUSNESS	Unresponsive, lethargic, stuporous, disoriented, melancholic, incontinent, regressive thinking, paranoid orientation, delusions, hallucinations, sometimes amnesia and coma.

[a] Begin at 0 (normal behavior, feeling well), and follow the stimulated levels (+ up to + +, + + +, etc.) which result from exposure to a particular substance (tolerance or adaptation is occurring during these stages). With removal from exposure, the individual withdraws (+ +, down to + +, +, 0, −, − −, etc.) and experiences symptoms of withdrawal (loss of tolerance, or maladaptation).

[b] Cardiovascular manifestations, including rapid or irregular pulse, hypertension, phlebitis, anemia, and bleeding and bruising tendencies, may occur at any level.

Source: Randolph, T. and Moss, R., *An Alternative Approach to Allergies,* copyright J. B. Lippincott Co., Philadelphia, PA (1980).

ple, to some solvents. Addiction is most likely to be recognized for substances that have euphoric or other pleasant properties and less likely to be recognized for chemicals and foods without these properties.

By isolating his patients from their usual environments and then reexposing them to various foods and chemicals one by one, Randolph discerned that adaptation plays an important role in many common substances people eat, drink, or inhale. Virtually any food or chemical follows the same pattern: initially, the individual notes symptoms when the substance is first encountered; gradually, with continued exposure or multiple reexposures, tolerance or adaptation or acclimatization occurs. "Addiction" to commonly eaten foods, such as corn, wheat, milk, eggs, and citrus fruit, and to common chemical inhalants, for example, formaldehyde or gas combustion products, although generally not recognized by patients or their physicians, also occurs, according to ecologists.

As we indicated earlier, what Randolph contributed was a systematic approach to studying individual responses to foods and chemicals. By removing individuals from their total background of environmental incitants and exposing them to each food and each chemical individually, he was able to observe a biphasic response to some of these substances (Fig. 2-1). He noted that initially the individual might experience a stimulatory effect (adapted response; tolerance develops) lasting varying periods of time depending upon the incitant. However, this "up" phase was generally followed by a withdrawal phase (maladapted response; loss of tolerance). Upon beginning to experience unpleasant withdrawal symptoms, the individual would seek, consciously or unconsciously, more of the same substance. These ups and downs follow a sort of sinusoidal (biphasic) pattern, as depicted in Figure 2-1. On the graph, beginning at zero, the patient is free of symptoms and at baseline health status. Following a one-time or occasional exposure to a provoking substance, stimulatory effects result; after a period of time (minutes to hours to days, depending upon the nature of the incitant), the stimulatory effects subside and give way to withdrawal symptoms. The frequency of these up and down reactions depends upon the frequency of the person's contact with the incitant. The amplitude of the stimulatory and withdrawal portions of the reaction depends upon the substance and the individual's susceptibility (degree of adaptation or addiction) to it. The particularly sensitive person exhibits larger amplitudes than the normals. For example, an occasional drinker or a painter exposed to solvents the first few times might have a relatively pleasant up phase with relatively few withdrawal symptoms afterward. As exposures become more frequent, however, addiction may occur. A painter might visit other painters on his day off in order to "sniff" some solvent. (Ran-

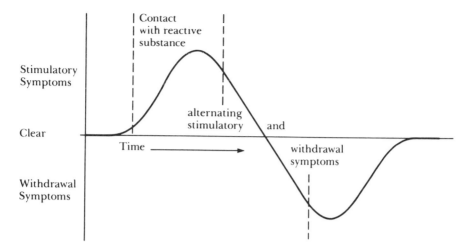

FIGURE 2-1. *Symptom progression of a single reaction to an incitant. During the early phases of exposure to a particular substance, stimulatory symptoms predominate ("up," "hyper," jittery). As exposure to the offending agent continues, adaptation occurs and fewer of these symptoms are experienced. With removal from (or discontinuance of) exposure, the individual experiences withdrawal symptoms ranging in intensity from mild to severe. (From O'Banion, D. R., Ecological and Nutritional Treatment of Health Disorders, 1981, p. 68. Courtesy of Charles C Thomas, Publisher, Springfield, Illinois.)*

dolph 1987, p. 109); perhaps drinking alcohol has a role in forestalling solvent withdrawal symptoms in some susceptible painters.

The key to understanding multiple chemical sensitivity may lie in recognizing these ups and downs that occur after exposure to many different substances. Table 2-2 illustrates the pattern of reaction Randolph claims he has observed in thousands of patients after exposure to an incitant. The amplitude of a reaction varies from person to person and incitant to incitant, but the pattern is quite constant. Beginning in the center of the table at zero, read upward for stimulatory effects and downward for withdrawal effects. Note that many of the stages, both stimulatory and withdrawal, are characterized by central nervous system (CNS) symptoms such as anxiety, confusion, depression, and irritability. Such symptoms are commonly noted by patients with multiple chemical sensitivities. The early stimulated (+ and + +) levels are adaptive responses by the body to an environmental incitant. Individuals at adaptation level + are stimulated but relatively free of symptoms. They may remain at this relatively desirable level (which often is confused with normalcy, level 0) indefinitely. According to Randolph, individuals rarely seek medical help at levels +, or + +. However, the onset of withdrawal (−) symptoms, whether systemic (fatigue, myalgia, or impaired concentration) or localized (rhinitis, asthma, or colitis), brings patients to the doctor. Often, the plus phase of any reaction is followed by a minus phase at least as intense, or perhaps one stage deeper. Thus a + + stimulatory phase may be followed by a − − or − − − withdrawal phase (see Table 2-2). In many individuals, every step of the entire sinusoidal progression of symptoms (for example, +, + +, +, 0, −, − −, − − − and finally back up to − −, −, and 0) can be observed. The most extreme case would be progression from mania to deep depression in a single patient, as in manic-depressive disease. Another interesting aspect is the tendency for psychotic (+ + + + or − − − −) and classical allergic (− and − −) manifestations to alternate in individuals. In the 1800s Savage described several cases in which insanity alternated with asthma; when one was present, the other disappeared. Old psychiatric texts refer to this vacillation between physical and mental manifestations as *alternation* (Randolph 1976b).

With long-term exposure to a given incitant (for instance, alcohol), especially in certain individuals, the degree and duration of stimulation may become less and less while the withdrawal or depressed phase becomes deeper and more prolonged. At face value, this sinusoidal reaction to a substance might seem a somewhat artificial construct, but Randolph asserts it is not. Randolph himself has hospitalized, fasted, and tested more than 10,000 people with many foods and chemicals since 1956, and his theories are distilled from observations of patients who have gone through an environmental unit (Randolph 1980, p. 169). For a patient with sensitivities involving foods alone, an elimination diet or fasting followed by reintroduction of single foods may be adequate for diagnosis. However, Randolph states that since World War II he has observed increasing numbers of individuals who respond adversely to the chemical environment. Subtle chemical sensitivities may be difficult to assess while a patient remains at home or even in most hospitals because these places generally contain background low levels of natural gas, disinfectants, perfumes, cleaners, tobacco smoke, paints, varnishes, adhesives, and other substances. According to Randolph, the patient's symptoms may be *masked* by the presence of these contaminants (more on this subject follows later).

With regard to chemical sensitivity (or *susceptibility*, a term Randolph prefers in order to avoid confusion with classical, IgE-mediated allergic sensitivity), Randolph (1987, p. 78) notes that, more than foods, chemical exposures are:

> associated with higher degrees of individual susceptibility and relatively greater persistence of susceptibility as well as more advanced clinical syndromes. Also, once individual susceptibility to one or a few environ-

mental chemical exposures has developed, it almost invariably tends to spread to involve other combustion products and derivatives of gas, oil and coal.

Randolph has referred to this as the "petrochemical problem." The stimulatory and withdrawal levels for foods and chemicals overlap each other (Fig. 2-2) so that in real life—outside an environmental unit—at any given moment what the organism is feeling is a summation of all effects, whether stimulatory or depressive, of all substances inhaled, contacted, or ingested. Figure 2-2 shows that attempts to identify the effects of single substances would be frustrated by the overlapping responses. Only by placing the individual in an environment devoid of chemical and food incitants is one able to determine whether the illness is alleviated. Assuming the patient improves (which occurs in the majority of cases, according to ecologists), the next step is to reexpose the person to individual substances in order to avoid overlapping responses and then to observe the result. According to Randolph, only in this way can the stimulatory and withdrawal phases associated with a given substance be discerned. If all possible food and chemical contributors are not removed, an effect may be missed. Hence, in order to rule out environmental illness definitively, an environmental unit would be required. An environmental illness could be ruled in on an outpatient basis, but not ruled out. However, environmental factors should be ruled out before psychiatric diagnoses and labels are applied to patients (see later discussion on psychogenic mechanisms).

In real life, following several days' avoidance of suspected incitants, a very robust response may occur with re-exposure, but not be recognized as such: An asthmatic might feel well after spending a week in a Caribbean island, breathing clean air and eating a diet devoid of usual foods, only to have a severe, life-threatening asthmatic response to exhaust from the engine of a boat taking the individual home. Once

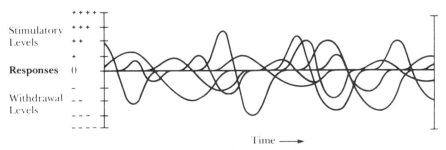

FIGURE 2 2. Overlapping of responses to food and chemical incitants in an individual with multiple exposures and multiple chemical sensitivities.

the asthmatic readapts, acclimatizes to auto exhaust combustion products and other air pollutants in the area, and experiences only chronic wheezing. Thus, following deadaptation (removal of incitants), the individual exhibits a more acute and convincing reaction upon reexposure. This appears to be what occurs in an environmental unit during testing. So acute and convincing are some of these reactions that patients themselves erroneously (at least in the eyes of some) surmise they must have an "allergy" to a particular substance. However, if the patient is not deadapted (unmasked) when tested, a reaction may not occur, convincing the physician that the "allergy" was all in the patient's mind. Many double-blind studies by allergists and others in the past have not taken this phenomenon of masking into account and therefore may be flawed. The sensitivity, if not tested for in the unmasked state, may easily be missed.

Occupational health has many widely recognized examples of adaptation that are analogous. They, too, fit a biphasic pattern. Industrial hygienists and occupational health physicians know that one of the most valuable clues to work-related illness is a history of intense symptoms following return to work *after* a vacation or weekend (leading to withdrawal and deadaptation). The following examples help underscore the existence of this phenomenon of adaptation. These particular examples may or may not also represent multiple chemical sensitivity. Individuals who are more sensitive or susceptible to the following substances may be the same individuals who are prone to developing multiple chemical sensitivities. In other words, multiple chemical sensitivities may reflect failure of adaptation in some sense. Failure may be the result of individual tendencies, an environmental insult, or some combination of the two. Bearing in mind that multiple chemical sensitivity might be what is actually occurring in the most sensitive subgroups exposed to these substances, let us now turn to some specific examples from occupational medicine:

1. Welding on galvanized metal causes evolution of zinc oxide fumes that, when inhaled, provoke an influenza-like syndrome with headaches, nausea, weakness, myalgia, coughing, dyspnea, and fever. The same symptoms may result from inhaling fumes of copper, magnesium, aluminum, and other metals. Hunter (1978, p. 407) writes:

   > The frequency and severity of the attacks are affected by the regularity of exposure for those who work continuously in the trade seem to acquire a tolerance which, however, is only transient, since it may be lost during a weekend away from work. In such cases the relapse of symptoms after working on a Monday gives rise to the name "Monday Fever."

Removal from exposure (deadaptation) for a couple of days is followed by an exacerbation of symptoms upon return to work.

2. Cotton (Hunter 1978, pp. 1043–1045), grain, and other organic dusts, as well as vapors from contaminated humidifiers, also produce an acute flulike illness, usually on the first workday after a period away from the job. In other words, after a period of deadaptation, reexposure to organic dust or vapors may provoke acute symptoms. In this example and the previous one, systemic symptoms (metal fume fever and flulike illness from organic dust exposure), appear, not just respiratory symptoms. Flulike symptoms involving multiple systems occur in a subgroup of exposed workers. Although the physiological responses to metal fumes and organic dusts have been characterized and differ markedly from that of other substances discussed here, nevertheless there is an adaptive component to these responses. Adaptation appears to be an attribute of many, if not all, physiological systems.

3. Nitroglycerin, used to manufacture gunpowder, rocket fuels, and dynamite, may cause severe headaches, breathing difficulty, weakness, drowsiness, nausea, and vomiting as a result of inhalation. Lesser symptoms may appear with oral administration (for example, as a drug for cardiac patients) or skin contact. Even wives of nitroglycerin workers who launder and iron their husbands' clothing may experience similar symptoms, headache being the most prominent.

> The headache may continue for one or two hours, or even for three or four days. The onset may be associated with exhilaration, but usually this passes and the victim becomes depressed. . . . Tolerance to exposure develops after three or four days of continued exposure, but is lost after two days away from work. Since the 19th century, workers have been known to avoid the Monday headache, once they become tolerant to nitroglycerin, by placing nitroglycerin under their hatbands over the weekend, or sucking occasionally on a piece of dynamite. Others inhaled the fumes from their work clothes over the weekend (Daum 1983, pp. 639, 648).

Stimulatory (exhilaration) and withdrawal (depression) symptoms occur in a biphasic manner. Alcohol may make dynamite headaches worse and cause confusion, extreme agitation, hallucinations, or violent behavior; individuals have even been known to commit murder (Daum 1983). Here then is an example of an individual exposed to a particular chemical who exhibits reduced tolerance for an ingestant.

Most studies have failed to show a difference in blood pressure between dynamite workers and controls; however, differences are noted between measurements taken in dynamite workers on Monday (lower) and those taken later in the week. Without insight into the principles of adaptation, one might overlook or discount such a phenomenon and examine only the average blood pressure during the week. Nitroglycerin has other effects on the central nervous system such as mania $(+ + + +)$, epilepsy $(+ + + +)$, depression $(- -,$ $- - -, - - - -)$, aphasia $(- - -)$, parasthesias $(- -)$, and headaches $(- -)$. Understanding adaptation, one could trace all of these symptoms through their stimulatory and withdrawal levels.

Noteworthy is a study by the Pennsylvania Department of Health of dynamite workers who suffered sudden death that was felt to be cardiovascular in origin (Carmichael et al. 1963). Almost all of them died after a period away from exposure on the job, that is, during the withdrawal phase (see footnote to Table 2-2 regarding cardiovascular manifestations). Most of these workers were 30 to 45 years old, and their physical exams were largely unremarkable. "Monday morning angina" has been described among dynamite workers; if their angina attack does not interfere, returning to work may cure it! The incidence of sudden death among workers at one dynamite plant was 15 times the expected rate. The mechanism is unknown; however, some have speculated that acclimatization followed by vasospasm on withdrawal from exposure suggests an increased sympathetic nervous system output compensatory mechanism (Daum 1983). William Rea, a cardiothoracic surgeon and ecologist, has written extensively concerning arrhythmias and coronary vasospasm—both of which he feels may contribute to sudden death—resulting from food and/or chemical sensitivities (Rea 1981, 1987b).

Hunter relates the story of a dynamite worker who had severe headaches and whose wife and children had terrible headaches as well. His physician recommended slipping two or three grains of powder under his hatband and taking a few more grains and hiding them around the house. This advice worked well; their headaches disappeared. However, anyone who came to the house as a visitor would get "a hell of a headache." Alfred Nobel, inventor of dynamite and founder of the Nobel prize, likewise suffered from dynamite headaches over many years. Hunter notes that individual susceptibility to nitroglycerin varies tremendously. In all but 2 to 3 percent of workers, tolerance is acquired within a few days of exposure (Hunter 1978, pp. 560–566). Conceivably, those who fail to develop tolerance to nitroglycerin may have the chemical susceptibility (multiple chemical sensitivity) problem.

4. Ozone, an air pollutant of special concern to residents of Los Angeles and other cities, has been the focus of considerable research relevant to adaptation. Intrigued by how little respiratory illness and death

occurred relative to the high levels of ozone in very polluted cities and suspecting adaptation might play a protective role, Hackney and associates (1977a) compared the responses of four Canadians (not adapted) and four Californians (adapted) to ozone challenges. Although reactivity varied greatly from individual to individual, Californians were only minimally reactive to levels that for the Canadians caused coughing, substernal discomfort and airway irritation, pulmonary function test decrements, and increased red blood cell fragility.

In another experiment, six volunteers with respiratory hyperreactivity were placed in an environmental chamber with ozone at 0.5 ppm (parts per million), typical of ambient levels, for 4 days (Hackney et al. 1977b). Five of six had decreased pulmonary function during days 1 to 3, but gradually improved, almost to baseline by day 4, suggesting adaptation had occurred. The authors note that not all adverse effects of ozone may be prevented by adaptation; for example, increased blood cell fragility may persist. Therefore, adaptation or masking of some symptoms may occur while other physiological alterations continue.

Individuals' abilities to adapt to ozone appear to depend upon their initial sensitivity to it. More sensitive persons adapt more slowly and cannot maintain the adaptation as long; they usually remain adapted less than 7 days following cessation of exposure (Horvath 1981).

Adaptation to ozone seems to be a concentration-dependent phenomenon. Chronic exposure to low levels of ozone in the air does not mitigate the effects of acute exposures to ozone at higher concentrations (Gliner 1983). These observations pertaining to ozone might also apply to the exposure experiences of chemically sensitive patients.

Because some tolerance to ozone can be induced in small rodents exposed to 1 ppm of ozone in as little as one hour, Stokinger (1965) speculated that the mechanism that seemed to explain observations in animals best was "either a cellular depletion phenomenon or enzyme stimulation." Mustafa and Tierney (1978), discussing oxygen (as opposed to ozone) toxicity caution, "the term 'tolerance' should not be considered to indicate absolute tolerance, because continuing injury does occur, and eventually, emphysema-like changes and fibrosis develop"; in the short term, tolerance may be protective, but "when exposures are continuous or intermittent for a period of weeks or even years, it is likely that unacceptable lung injury is necessary to keep the mechanism of tolerance or adaptation activated." With regard to ozone exposures among "adapted" Southern Californians, Mustafa and Tierney comment, "Whether or not they have continuing lung injury with increased probability of bronchitis, emphysema, or even malignancy is not known, although some reports have supported the concept."

Bell and King (1982), extending these ideas to chemically sensitive patients, speculate that their chronic symptoms "may reflect the more insidious, non-adapting changes induced by offending foods and chemicals. At the same time, the more obvious and dramatic adverse clinical effects may be masked or adapted."

5. Solvents are among the chemicals most frequently implicated by chemically sensitive patients who attribute the onset of their illness to a particular exposure (Terr 1989a; Cone et al. 1987; see Table 2-3). Vapors from various solvents are the most prevalent of indoor air contaminants (Molhave 1982). The volatile organic compounds (VOCs) associated with sick building syndrome are in large part solvent vapors. The sensory irritation, headache, drowsiness, and other symptoms noted by occupants of tight buildings are consistent with known effects of solvent vapors, albeit at much higher concentrations. *Solvent* is a very broad term encompassing a wide range of liquids that are capable of dissolving or dispersing other substances. They are found in many products commonly used at home and at work, for example, in paint, varnishes, adhesives, pesticides, and cleaning solutions.

The adverse short- and long-term effects of solvents are widely recognized by occupational health professionals. In recent years much effort has been directed toward reducing the solvent content of paints by using larger fractions of water as a vehicle for pigments and thereby reducing exposures for persons applying them. Nevertheless, exposures to solvents at home and at work may be significant, particularly in confined spaces or tightly sealed structures with inadequate fresh (outdoor) air for ventilation. Where solvents are applied over large surface areas, the opportunity for evaporation is increased, and the concentration of solvents in the air may reach high levels.

Those who have painted or used solvents to any major extent are well aware of the olfactory fatigue (nasal adaptation) that occurs and may have experienced the stimulatory and depressive properties of solvents. Alcoholic beverages contain the solvent ethanol, which has related and familiar stimulatory and withdrawal effects. A study of Finnish car painters demonstrated increased complaints of fatigue and nausea and reduced vigilance and concentration while at work (Husman 1980). Many (more than controls) reported they often felt excess tiredness after the workday (withdrawal). Inquiries to painters

TABLE 2-3. Causative Agent in Claims of Work-related Environmental Illness

Agent	Number	Agent	Number	Agent	Number
Paint, thinner	10	Carbon tetrachloride	2	Tar	1
Smoke	7	Paper	2	Wood dust	1
Organic solvents	7	Hair sprays	2	Beet sugar dust	1
Pesticides	6	Office machines	2	Detergent	1
Xylene	5	Photographic chemical	2	Nail polish	1
Phenol	5	New building	2	Polyurethane	1
Dust	5	Hydrogen	1	Soap	1
Formaldehyde	4	Helium	1	Germicide	1
"Chemicals"	4	Argon	1	Vehicle exhaust	1
New carpets	4	Nickel	1	Unspecified fumes	1
Ammonium compounds	4	Lead	1	Chairs	1
Freon	3	Ethanol	1	Heat at work	2
Solder flux	3	Nitric acid	1	Fall at work	2
Welding fumes	3	Sulfur dioxide	1	Stress at work	1
Fiberglas	3	Hydrochloric acid	1	Dry air	1
Perfumes	3	Polyvinyl chloride	1	Herpes zoster	1
Food	2	Toluene diisocyanate	1	Scabies	1
Hydrocarbons	2	Methylene chloride	1	Uncertain	2
Trichlorethane	2	Kerosene	1	None	5
Fuel	2	Glue	1		

Source: Terr, A., "Clinical Ecology in the Workplace," *Journal of Occupational Medicine* 31(3): 257–261, p. 258 (1989) (copyright American College of Occupational Medicine).

who had changed employment revealed that 52 of 101 had changed vocations and 26 of the 52, or half of those who had left car painting, mentioned health or other occupational hazards as reasons for leaving. This latter group reported having experienced a relatively high symptom frequency level while working at car painting. Thus workers who were sensitive to chemicals were more likely to migrate to other occupations, which could help to explain why chemically sensitive individuals are often seen among office workers or light industrial workers rather than among workers in heavy industry. Terr (1989a) summarized the occupations of 90 individuals referred to him for workers' compensation evaluation (Table 2-4). Migration of sensitive subgroups to other jobs or out of the workforce may contribute to the so-called healthy worker effect.

TABLE 2-4. Occupations of 90 Patients Claiming Work-related Environmental Illness

Occupation	Number	Occupation	Number
Office work		Education	
Clerical worker	12	Teacher	8
Telephone operator	4	Food industry	
Manager	2	Waitress	2
Salesperson	1	Baker	1
Transportation		Equipment mechanic	1
Flight attendant	6	Other manufacturing	
Mechanic	3	Lithographer	4
Pilot	2	Welder	1
Bus driver	1	Plastics worker	1
Train operator	1	Wood worker	1
Medicine and social work		Lumber worker	1
Nurse	4	Hairdresser	3
Social worker	2	Writer	2
X-ray technician	1	Engineer	2
Respiratory therapist	1	Dry cleaner	1
Psychotherapist	1	Carpet layer	1
Dental assistant	1	Security guard	1
Laboratory technician	1	Artist	1
Electronic manufacturing		Geologist	1
Assembler or technician	7	City dump operator	1
Solderer	3	Transformer repair	1
Tool and die maker	1	Fabric salesman	1
Research worker	1		

Source: Terr, A., "Clinical Ecology in the Workplace," *Journal of Occupational Medicine* 31(3): 257–261, p. 258 (1989) (copyright American College of Occupational Medicine).

Molhave (1982) identified chemicals emanating from 42 modern building materials, the ten most common of which were solvents (Table 2-5). Health effects most commonly involve the central nervous system. Molhave and associates (1986) exposed individuals who had previously complained of sick building syndrome symptoms to a mixture of 22 volatile organic compounds common in indoor air, predominantly solvents, for 2¾ hours. Levels were much lower than occupational health standards required and in the range of levels found in tight buildings. These healthy but sensitive subjects complained of nasal and throat irritation and inability to concentrate at levels of solvents far below permissible occupational exposure levels. (For a fuller discussion of Molhave's studies, see Chapter 3.) A similar study, using healthy subjects who had not previously complained of symptoms, showed no effect on the ability to concentrate (Otto et al. 1990).

Chemically sensitive patients commonly report central nervous system symptoms at solvent levels as low as those used by Molhave and lower. Their complaints are consistent with the recognized health effects of these substances, albeit the *levels* of exposure that trigger symptoms in these patients may be lower by orders of magnitude.

According to Randolph (1980), the stimulatory effects of solvents may be pleasant or unpleasant, depending upon the person and the exposure. At the + level (see Table 2-2), a normal individual exposed to solvents may experience being alert, enthusiastic, energetic, and witty

TABLE 2-5. Commonly Identified Chemicals Off-gassing from 42 Modern Building Materials

The 10 Most Frequently Identified Compounds	*The 10 Compounds in Highest Average Equilibrium Concentration*
Toluene	Toluene
n-Decane	3-Xylene
1,2,4-Trimethylbenzene	$C_{10}H_{16}$ (Terpene)
n-Undecane	n-Butyl acetate
3-Xylene	n-Butanol
2-Xylene	n-Hexane
n-Propyl benzene	4-Xylene
Ethyl benzene	Ethoxy ethyl acetate
n-Nonane	n-Heptane
1,3,5-Trimethyl benzene	2-Xylene

Source: Reprinted with permission from "Indoor Air Pollution Due to Organic Gases and Vapours of Solvents in Building Materials," in *Environmental International* 8:117–127, 122. Molhave, L., Moghisssi, A., and Moghissi, B. (eds.). Copyright 1982, Pergamon Press plc., Elmsford, NY.

and may sustain this level for long periods while a chemically sensitive person may progress quickly to + + and feel tense, jittery, argumentative, sensitive, and overly responsive and experience chills or flushing. Very sensitive individuals may become manic or develop seizures (+ + + +).

During withdrawal, symptoms for relatively healthy individuals may be only mild, localized symptoms (–) resembling an allergy, for example, a runny or stuffy nose, coughing, wheezing, eczema, hives, diarrhea, and eye and ear symptoms. However, the response is not actually an allergic one; there is no evidence that IgE plays a role. If withdrawal is more severe (– –), fatigue, depression, muscle and joint aches, headache, and a rapid or irregular heart rate may ensue. At more advanced stages in very sensitive individuals (– – –, – – – –), confusion, indecision, and apathy can occur with comprehension and concentration becoming impaired. The most severe stage (– – – –) may be attended by stupor, delusions, and hallucinations. From a clinical viewpoint, Randolph's model could serve as a useful framework for following the progression of these patients' symptoms. Many patients with chemical sensitivity seem to experience the more advanced stages of stimulation and depression in the model. According to Randolph, when they are deadapted and then exposed to a single incitant, the stepwise, sinusoidal progression up through stimulatory levels and back down through withdrawal levels can best be observed. In normal daily life, exposures overlap and discrete stages may not be discerned.

Studies of xylene, one of the most prevalent solvents in indoor air (see Table 2-5), conducted by Riihimaki and Savolainen (1980) demonstrate that its effects are attenuated as exposure continues, presumably due to adaptation. Their work is discussed in detail in Chapter 5.

We have mentioned a number of the exposures that are recognized as involving adaptation or addiction. No doubt the physiological events that allow us to adapt to ozone are quite different from those for nicotine or nitroglycerine. Nevertheless, it is clear that human beings adapt to a wide variety of substances in their environment. What is not clear is the specific role adaptation plays in the dramatic responses patients with food and chemical sensitivities have to low-level exposures that do not overtly affect others (for further discussion of possible mechanisms, see Chapter 4).

Without exception, all traditional allergists we interviewed recognized the phenomenon of acclimatization or adaptation and agreed that it was potentially a crucial variable that should be controlled in studies of low-level exposure to chemicals. These concepts are familiar to occupational health practitioners and industrial hygienists because they observe such effects firsthand among workers exposed to chemicals. Randolph (1962, p. 7) states that most physicians see patients long after adaptation has occurred and at the time when end organ damage is setting in: "It is much as if the physician arrived at the theatre sometime during the last scene of the second act of a three act play—puzzled by what may have happened previously to the principal actor, his patient." Through comprehensive environmental control (that is, an environmental unit), one can overcome the masking effect of adaptation and back up or reverse the exposure to allow monitoring of toxicity in progress. The environmental unit represents a kind of *dynamic toxicology;* traditional medical approaches provide only a snapshot of what is happening to the patient.

The sheer heterogeneity of substances that can evoke adverse reactions (those enumerated above and others) suggests a fundamental mechanism of adaptation to environmental substances. The mechanism may involve the nervous system or some other system(s) rather than the immune system. However, the intense symptoms that may occur with deadaptation and reexposure resemble a classic allergic response, that is, an untoward reaction to an incitant. This concept has clinical utility.

Because adaptation appears to be a generalized response (Selye 1946), a toxic insult to, for example, the sympathetic nervous system or enzyme detoxification pathways could cause a general loss of the ability to adapt to a wide variety of substances, including other chemicals and even foods (the spreading phenomenon). Knowing the mechanism by which this occurs would, of course, be ideal. Thus far, it has eluded clinical ecologists and placed them at a distinct disadvantage.

Observing a phenomenon and documenting its existence must, of necessity, precede knowledge of its mechanism. Of course, knowing the mechanism of a disease is not necessary in order to prevent it. A historic example occurred in 1854 when a London physician, John Snow, noted that individuals who developed cholera obtained their drinking water from the Broad Street pump. Medical folklore tells us that by ordering the removal of the pump handle, he stopped the epidemic (Snow 1936). Not until 1883, almost 30 years later, did Koch discover the bacterium responsible for cholera. Analogous to the current dilemma, understanding the mechanism for food and chemical sensitivities is not necessary in order to begin diagnosing and treating them. Eventually, knowledge of the mechanism may suggest better treatments.

With regard to patients with chemical sensitivities who also develop dietary intolerances, Bell (1982, pp. 35–36) notes that "foods are not only sources of nutrients, but also complex mixtures of organic chemicals. For instance, it is the unique pattern of chemical constituents that make a tomato a tomato rather than an apple." She provides a partial listing of chemical constituents of tomato, apple, milk, and orange (see

Table 4-2). Allergists Butcher, Salvaggio and associates (1982) reported an interesting case of a worker with toluene di-isocyanate (TDI) sensitivity who was also intolerant of radishes. Both TDI and radishes contain allyl isothiocyanate and benzyl isothiocyanate, but other foods containing these same chemicals did not provoke symptoms. The authors were unable to speculate as to the possible mechanism for this cross-sensitivity. Indoor air contains organic compounds also found in foods, such as the fragrances limonene and pinene (see Table 4-2).

McGovern and associates (1981–82) have also written about chemical and food cross-sensitivity and noted that many foods contain phenolic derivatives. Chemically sensitive patients also frequently react to phenolic inhalants. The McGovern group attempted to desensitize patients to particular phenolics and noted very robust reactions to such challenges. Rea (1988a) reports food sensitivities in 80 percent of his patients with chemical sensitivities. Like airborne pollutants, foods contain a wide range of chemical constituents and are in intimate contact with the organism for long periods of time. The surface area of the gastrointestinal tract is enormous, and the chemical load, in terms of both quantity and diversity of exposure, is huge. From a developmental perspective, the contents of the gastrointestinal tract can be thought of not simply as part of the organism, but as "an insinuation of the environment into the body" (Angyal 1981).

Those IgE-mediated allergic reactions affecting the skin or lungs are more accessible for study than those affecting the gastrointestinal tract. The skin can be viewed directly, and devices are available for measuring changes in pulmonary function. Dean Metcalfe of the National Institutes of Health (1986) comments:

> The situation is much different when it comes to allergic diseases that involve the gastrointestinal tract. This system is relatively inaccessible and difficult to study and, thus far, less information has accumulated relative to allergic reactions in its tissues.

Yet, the potential for allergic reactions in the gastrointestinal tract is present: mast cells, the cells that release chemical mediators that result in an allergic reaction, are more densely packed in the intestinal tract ($20,000/mm^3$) than in the skin ($7,000/mm^3$), another organ considered rich in mast cells (Barrett 1984).

The food intolerances of patients with multiple chemical sensitivities may or may not be IgE-mediated or involve mast cell release of histamine or other mediators. In any case, the difficulties in studying adverse reactions to foods are the same. The inaccessibility of the gastrointestinal tract and its enormous chemical and antigenic contents greatly encumber study of chemical mediators or subtle pathophysiological alterations. Whatever the mechanism, conducting blinded food challenges is difficult to accomplish without altering the food in some respect. Dehydrated foods administered in opaque capsules may not provoke the same response as larger quantities of fresh foods. Some individuals may react to the capsules themselves or fail to react if food does not contact the oral mucosa.

Whether in reference to foods, drugs, or other chemicals, adaptation may occur, altering the organism's later responses to and tolerance for other substances. The mechanisms are unknown, but this does not preclude our recognition of or intervention in this problem. The use of an environmental unit may provide a way of unmasking or backing up the experience or, as Randolph (1976a) states, provide "the means of reverting many chronic illnesses of unknown cause to acute illness in which specific etiology is readily demonstrated."

Summary of Adaptation Hypotheses. We acknowledge the complexity of the concept of adaptation. Here we summarize the salient points concerning this topic.

Symptoms of exposure to many chemicals, whether inhaled or ingested, appear to follow a biphasic pattern. Adaptation is characterized by acclimatization (habituation, tolerance) with repeated exposures that result in a masking of symptoms. Withdrawal occurs when exposure is discontinued. Once a person has adapted, then the experimental consequences are that further exposures have very little additional effect and therefore may not be observed. The observer may not be able to witness the stimulatory or reactive event because a kind of "saturation" effect has set in.

Adaptation and withdrawal occur for a wide variety of organic and inorganic substances in many physical forms, including various dusts and fumes, solvents, nitroglycerin, ozone, and foods.

An individual is exposed to a variety of substances at different times with varying frequency, duration, and intensity of exposure for each of these substances and with varying frequency and duration of reduction in or cessation of exposure for each substance. The individual may be in different stages (stimulatory or withdrawal) simultaneously for different substances. These stages may overlap (see Fig. 2-2) and interfere with attempts to observe cause-and-effect relationships.

Adaptation may mask some symptoms or effects while other physiological alterations may continue.

Comprehensive environmental control, that is, an environmental unit, may overcome the masking effect of adaptation and the problems of overlapping exposures that result in overlapping responses to multiple agents. The environmental unit may allow the investigator to back up or reverse the experience of adaptation and monitor toxicity in progress. The advantages dynamic toxicity of this nature may have over conventional methods for determining toxicity include facilitating detection of subclinical, prepathological effects of chemicals and providing more than just a snapshot of an individual's response to substances. Removing the person from interacting, time-dependent stimuli in this way may allow the unraveling of multiple causes. The environmental unit may be an essential tool. Many carefully conducted studies of chemical effects that have had negative or equivocal outcomes may be flawed by their failure to take adaptive mechanisms into account. The consequences of such an oversight could be major.

For further discussion of adaptation and related concepts, see "Experimental Considerations and Approaches to MCS" in Chapter 10.

The Environmental Unit

Adaptation and the use of an environmental unit are complex topics that do not lend themselves to the short presentations typical of most scientific forums. Yet, physicians must understand adaptation if progress is to be made in this field. Some of the allergists we spoke with recognize the pivotal role adaptation may play. Prominent among them is John Selner, an allergist who has long advocated that allergists take a more active role in understanding patients with alleged chemical sensitivities and who described in detail the design and operation of an environmental unit (Selner and Staudenmayer 1985a). Selner visited Rea's unit in Dallas and collaborated with Ken Gerdes, an ecologist who trained with Randolph, to establish a unit in Denver at Presbyterian–St. Luke Hospital in 1979. This unit, which operated for several years before closing for reasons unrelated to its utility as a diagnostic tool, incorporated many if not most of the features of existing clinical ecology units. Rea and Randolph, who had their own units, both visited Selner's unit when it opened and admired the care that had been exercised in its construction. Without exception, all allergists with whom we spoke agreed that an environmental unit like Selner's was an important tool for properly evaluating patients with alleged low-level chemical sensitivities. Few, however, appreciate the degree to which Selner patterned his approach after that developed by the clinical ecologists.

The clinical ecologists' environmental units and Selner's unit shared many of the same design and operational parameters (Table 2-6).

We are unable to discern any major differences in these two approaches. Even though Selner's unit is no longer in operation, he continues to employ some of the same principles, such as housing patients

TABLE 2-6. Features of Environmental Units[a]

Characteristics/Practices	Allergists' Unit[b]	Clinical Ecologists' Units[c]
Construction using materials that do not off-gas (primarily glass, steel, ceramic; cotton bedding and clothing). Avoidance of synthetic materials. No perfumes, cosmetics, odorous cleaners/soaps, etc.	Yes	Yes
Air supply filtered; patients' rooms under positive pressure to reduce contamination from adjacent areas; airlocks	Yes	Yes
Patients' medications discontinued insofar as possible; gradual withdrawal from steroids, etc.	Yes	Yes
Patients fasted for 4 to 8 days to clear symptoms.	Yes, if symptoms do not clear after several days in unit	Yes, at time of admission to unit
Organic foods used for food testing; commercial foods tested also	Yes	Yes
Patients tested for acceptable water	Yes	Yes
Challenges performed using single foods and chemicals after period of avoidance (to eliminate masking)	Yes	Yes

[a] None of the units described in this table is currently in operation.

[b] Selner in Denver (Selner and Staudenmayer, 1985a).

[c] Randolph in Chicago and Rea in Dallas.

in a relatively clean environment to try to avoid chemical exposure prior to testing. In Selner's view, the fundamental concept is still valid. He states, as do Rea and Randolph, that the majority of patients can be worked up as outpatients. However, a small percentage of patients are difficult to evaluate without such a facility.

Studies from the ecologists' units leave much to be desired in terms of study design. Unfortunately, no studies were ever published from the allergists' unit in Denver. *Every* traditional allergist we interviewed recognized that removal from exposure prior to testing might be a critical factor in studying reactions to low levels of chemicals. They felt that reestablishment of a unit would be an important step in understanding the problems of individuals who believe they are sensitive to low levels of chemicals. In fact, some of the allergists offered examples for which removal from exposure for several days prior to testing would be important, such as exposure to western red cedar or cigarette smoke. They were not at all opposed to the concept of an environmental unit and were aware of the considerable expense involved in establishing a well-designed and well-run environmental unit.

Some allergists we interviewed felt ecologists should be involved in the design of a study unit and appropriate protocols because of their experience in this area and so as to avoid later criticism from the ecologists regarding the protocols that are used. Two of the traditional allergists we interviewed praised Rea's engineering skills in designing and operating his unit, one saying, "No one does it as well as Rea does." Rea has, in fact, stated his willingness to cooperate with any impartial venture to undertake studies using his facilities or to design a model research unit elsewhere.

Although the detailed description of an environmental unit is beyond the limits of this discussion, some of the essentials are noted here.

First, by employing construction materials, furnishings, and clothing that are less likely to off-gas, very low levels of volatile organic compounds (for example, from synthetics) can be maintained inside the unit. To create and operate a unit that is as free as possible from chemical pollution requires knowledge, precision, and vigilance while working with architects, ventilation engineers, contractors and their suppliers, nurses, dieticians, food and water suppliers, and maintenance and custodial staffs. Obviously, "this is no trifling undertaking" (Selner and Staudenmeyer 1985a). Three basic scientific approaches for studying a disease are clinicopathological studies, animal experiments, and epidemiological investigations. To these, Randolph (1965) has added a fourth tool, comprehensive environmental control. Animal models, clinicopathological studies, and epidemiological investigations have certain important limitations for studying the phenomenon of multiple chemical sensitivities. None is as sensitive to low-level exposures and effects as the use of an environmental unit in which all exposures are controlled simultaneously and the individual is challenged with single substances while in the deadapted state. A theoretical, graphical representation of an individual's responses to environmental incitants before entering an environmental unit, after entering the unit, and during challenges to single incitants appears in Figure 2-3.

Animal models are best used to study relatively high doses of chemicals that result in distinctive physical or biochemical *pathology* that can be monitored. First, an appropriate animal must be found. Next are concerns about extrapolations to humans. More importantly here, rats, mice, and other animals are unable to tell researchers if they have headaches, feel depressed or anxious, or are nauseated. Thus the subtle effects of low-level chemical exposure may be missed entirely.

Epidemiology may have some utility with regard to tight buildings or community exposures to a toxic material. If everyone in the population responds with the same symptoms to the same agents, the task is relatively easy. However, if some people have headaches, others have muscle spasms, and still others are less able to concentrate, the results blur and may wash out entirely; that is, no single symptom has a statistically significant prevalence over controls. Thus, for multiple chemical sensitivities with multiple triggers and multiple health effects, epidemiology may be an insensitive tool. Further, although epidemiology can point to *associations* between events, other kinds of studies are needed to establish cause-and-effect relationships. In the study of chemical sensitivity, identification of an unaffected control group presents further difficulties.

Clinicopathological studies rely upon the presence of some clinical sign (for example, tachycardia or decreased reflexes), laboratory measurement, or tissue pathology. For meaningful data in humans, large numbers of similarly exposed individuals with similar end-organ effects must be examined. Again, multiple chemical sensitivities may involve multiple triggers and multiple effects. To date, no single laboratory test is abnormal in most, much less all, who are affected. At some point in the future, such a test or marker may be discovered, but for now no mass applicable clinical, laboratory, or pathological findings are available. Further, subjective complaints of patients may be overlooked particularly if they vary from one person to the next. Thus clinicopathological studies are not likely to be sensitive to the early effects of low-level exposures, that is, prior to end-organ damage.

What is needed is a sensitive tool that reliably detects symptoms of exposure to low levels of multiple chemicals in human beings, taking into account individual variability, a tool that will allow us to ascertain cause-and-effect relationships between exposures and symptoms. The environmental unit could be such a tool. Potentially, it may be the most useful of the four approaches for studying human response to environmental agents. For this reason, if for no other, the EPA and other governmental bodies concerned with regulating exposure to low levels of toxic environmental agents should take great interest in this approach. The individuals who might enter environmental units perhaps represent the most susceptible population. Their responses to chemical challenges while in a deadapted state in an environmental unit would further our understanding of low-level chemical sensitivity. Carefully designed and orchestrated studies with meticulous attention to the details of environmental control, as defined by those who have operated such units, are essential to resolving these issues to the satisfaction of all.

For further discussion of the environmental medical unit, see "Experimental Considerations and Approaches to MCS" in Chapter 10.

Graphical Representation of an Individual's Symptoms Before and After Entering an Environmental Unit

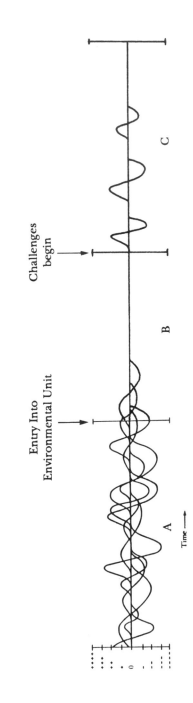

Figure 2-3. In time period A an individual is responding to multiple irritants encountered in normal daily living (chemicals and/or foods), with stimulatory and withdrawal effects that overlap in time. At any particular time, how the person feels is determined not only by ongoing exposures, but by previous exposures whose effects may still be waning. In time period B, the individual enters an environmentally controlled facility, fasting. With cessation of contributory exposures, withdrawal effects occur, for example, headache, fatigue, and myalgias. Symptoms continue for some time (typically 4-7 days) until the individual reaches "0" baseline. In time period C, single challenges to suspected irritants are administered. Symptoms, often robust, develop soon after challenges, allowing patient and physician to begin to observe the relationship between exposures and symptoms for that individual.

CHAPTER 3

Origins of Multiple Chemical Sensitivity and Effects on Health

Offending Substances

This chapter addresses the origins of multiple chemical sensitivity, that is, the offending substances that may induce the illness, as well as those that trigger symptoms once the problem has begun. In Chapter 2, adaptation was discussed in relation to several different materials, including ozone, nitroglycerin, cotton dust, metal fumes, solvents, alcohol, and tobacco smoke. Clinical ecologists and their patients have noted adaptation to an enormous range of substances that can be categorized as outdoor air pollutants; indoor air pollutants, both domestic and workplace; food contaminants and additives; water contaminants and additives; and drugs and consumer products.

In preparing this book, we considered referencing representative articles from toxicology that would show that substances in the above categories may be toxic to animals or humans. However, an encyclopedic listing would have little point. Certainly toxic substances in the environment have adverse health consequences. The question is whether certain persons develop heightened reactivity to chemicals and foods, and if so, why?

John Selner, an allergist critical of clinical ecology yet who advocates the use of an environmental unit for diagnosing certain patients, studied

59

these hyperreactive individuals but regrettably has published no studies of patients in his former environmental unit. His and others' criticisms and studies to disprove this field have not involved independent investigations in which food and chemical challenges on patients in a deadapted state were performed. Thus, Randolph's observations on adaptation have not yet been replicated, even though adaptation would seem an important hypothesis to test. Most ot the allergists' criticisms have been aimed at the efficacy of treatments, for example, provocation-neutralization, for chemically sensitive patients. These critiques of treatment modalitites and an absence of consistent laboratory abnormalities have been used as the basis for trying to disprove existence of the illness altogether.

Clinical ecologists, as well as some allergists (Selner 1988) with whom we spoke, invoke the concept of total body load or burden. To ecologists, this load is comprised of all incitants to which the body must respond (adapt) to maintain homeostasis. They may be chemical, biological (pollens, molds, bacteria, viruses), physical (heat, cold, radiation), or psychological. Notwithstanding its utility as a theoretical construct to help "explain" why this disorder occurs in a given individual, total load per se is not measurable. However, part of it can be quantified. For example, Laseter and co-workers (1983) measured levels of 16 chlorinated hydrocarbon pesticides in 200 chemically sensitive patients, 99 percent of whom had residues at or above 0.05 ppb in their blood, reflecting even higher tissue levels. The mean was 3.4 pesticides per patient. Volatile organic hydrocarbons (Rea 1987) and aliphatic hydrocarbon solvents (Pan et al. 1987–88) have also been measured in these patients, but their levels have not been compared with those of individuals who do not have such symptoms. Nevertheless, Rea and others feel that these substances are not normal constituents of the body and therefore represent a substantial burden for the individual. However, even if *no* differences in levels existed between chemically sensitive patients and so-called normals, these compounds still could be a source of their illness because chemically sensitive patients may be a subgroup of the population that is more susceptible to the effects of these chemicals. In addition, what may be most relevant is their *past* exposure that may have caused them to become sensitized in the first place.

Some authors have attempted to distinguish between those chemical exposures associated with the *onset* of multiple chemical sensitivity syndrome and those associated with *recurrence* of symptoms, that is, act as triggers once the syndrome has developed. Cone and associates (1987) studied workers with multiple chemical sensitivities, 11 of whom reported that solvent exposures of various types had caused their problem; three pointed to pesticide exposures; one, hydrogen sulfide; one, copy machines; one, new materials including carpets. Once the syndrome had developed, triggers for recurrence of symptoms in these same patients were far more diverse and included such common exposures as tobacco smoke, perfume, scented soap, car exhaust, copiers, gas stoves, tar, smog, newspapers, new clothing in stores, leather, printed books, office buildings, and glues.

In Terr's (1986) review of 50 cases seen by clinical ecologists, 43 cases of which were referred to him by workers' compensation carriers for independent evaluation, the patients also attributed their illness to a wide array of exposures: 16 complained of acute exposure to a chemical, three from pesticides, two from phenol; 34 patients felt their illness resulted from chronic exposures, six from unspecified chemicals in their homes, four from office machines, three from organic solvents, three from smoke, three from foods, two from formaldehyde, two from the hospital environment, and two from airplanes. These reports reflect fairly well the variety of exposures that clinical ecologists allege precede their patients' illness.

Subsequently Terr (1989a) compiled findings from 40 of these 50 cases plus an additional 50 cases; all patients had previously seen an ecologist and had applied for workers' compensation. Of these 90 patients, 28 had symptoms corresponding to a single organ system and 62 had multisystem polysymptomatic complaints. Table 2-3 summarizes the exposures these patients felt had caused their condition: 83 identified one or more (up to six) causative agents. Exposure durations ranged from a few seconds to 20 years. Table 2-4 lists their occupations: 19 were engaged in office work, 13 in transportation, 12 in electronics manufacturing, 11 in medicine or social work, eight in education, eight in other manufacturing, and the remainder distributed among a variety of other occupations.

It is conceivable that chemical sensitivity involves a two-step process: Certain exposures may induce the illness, whereas others may simply trigger symptoms once the syndrome has developed. We next discuss in more detail the range and nature of exposures that are thought to contribute to this problem and explore the five subgroups of exposures mentioned at the beginning of this chapter.

Health effects data on chemicals are notoriously inadequate. A 1984 study by the National Research Council attempted to assess the testing needs for various industrial and consumer chemicals (National Research Council 1984). Figure 3-1 shows existing needs for health hazard assessment and toxicity data. No toxicity data or minimal data are available for 66 percent of pesticides and their supposedly inert ingredients, 84 percent of cosmetic ingredients, 64 percent of drugs, 81 percent of food additives, and 88 to 90 percent of chemicals in commerce.

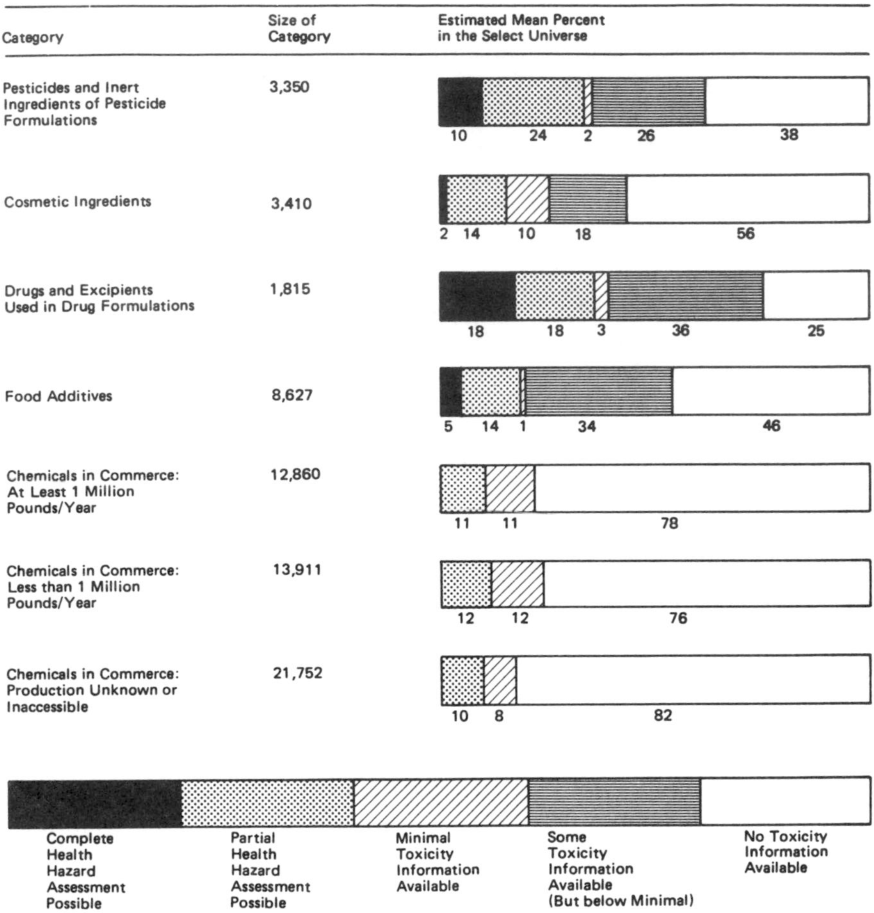
Category
Size of Category
Estimated Mean Percent in the Select Universe
Pesticides and Inert Ingredients of Pesticide Formulations
3,350
10 24 2 26 38
Cosmetic Ingredients
3,410
2 14 10 18 56
Drugs and Excipients Used in Drug Formulations
1,815
18 18 3 36 25
Food Additives
8,627
5 14 1 34 46
Chemicals in Commerce: At Least 1 Million Pounds/Year
12,860
11 11 78
Chemicals in Commerce: Less than 1 Million Pounds/Year
13,911
12 12 76
Chemicals in Commerce: Production Unknown or Inaccessible
21,752
10 8 82
Complete Health Hazard Assessment Possible
Partial Health Hazard Assessment Possible
Minimal Toxicity Information Available
Some Toxicity Information Available (But below Minimal)
No Toxicity Information Available

FIGURE 3-1. Ability to conduct health-hazard assessment of substances in seven categories of a select universe of chemicals.
Source: National Research Council 1984.

Thus, scientific data concerning health effects of the vast majority of chemicals are woefully lacking. Chemically sensitive patients may fill in the gaps long before toxicologists do.

Outdoor Air Pollutants

Among the most hazardous exposures for patients seem to be pesticides sprayed either outdoors or indoors. Alone, pesticides have accounted for some of the most advanced and persistent cases of chemical sensitivity known to clinical ecologists. As early as 1966, occupational health practitioners observed that certain persons who had "recovered" from acute organophosphate pesticide poisoning experienced protracted symptoms of nausea, headache, irritability, insomnia, inability to concentrate, blurred vision, or shakiness (Tabershaw and Cooper 1966). Twenty of 114 individuals stated they could no longer tolerate smelling or contact with pesticides. Depression and schizophrenia occurred in others (Gershon and Shaw 1961). Neuropsychiatric, cardiopulmonary, and gastrointestinal symptoms may persist long after exposure to organophosphate insecticides (Namba et al. 1971), which are widely used by exterminators indoors and out-of-doors. Other outdoor exposures presenting problems for the chemically susceptible patient include vapors from solvents and fuels, combustion products, tar fumes, paint vapors, diesel and auto exhaust, and industrial air pollution (Randolph 1962).

The adverse effects of air pollution upon individuals with respiratory or cardiac compromise are widely acknowledged. Less well known, but increasingly studied, have been associations between outdoor air pollutant levels and psychiatric emergency room visits (Briere et al. 1983; Strahilevitz et al. 1979), psychiatric hospital admissions (Strahilevitz et al. 1979), family disturbances (Rotton and Frey 1985), and anxiety symptoms (Evans et al. 1988).

Randolph described a woman who became ill each time she journeyed through the industrial pollution of northwestern Indiana and the south side of Chicago (Randolph 1987, pp. 73–76). Other patients note difficulty in any large metropolis, in the vicinity of airports, at bus or train stations, or in heavy traffic.

Diesel exhaust is a particular problem for many patients. In an EPA review of the toxicology of diesel exhaust, Nelson projected, "I think we can conclude quite straightforwardly that a major increase in the Diesel fleet is not going to produce a disastrous epidemic of lung cancer," but "risk assessment should be the ultimate goal and should be given the highest priority" (Nelson 1982). Many chemically sensitive patients experience severe symptoms with exposure to diesel exhaust. Interestingly, a Japanese study suggests that the striking increase in allergic rhinitis triggered by pollens that has occurred in that country over the past 30 years may be in part the result of lenient regulation of diesel exhaust and increased numbers of diesel vehicles (Muranaka et al. 1986). The authors note that the Japanese cedar, a tree indigenous to Japan for at least a million years, was never known to cause allergic rhinitis until 1964; and that before 1950 allergic rhinitis was virtually unknown in their country, although even then it had been recognized among Japanese living in the United States. Muranaka points to diesel exhaust as a possible cause for Japan's increasing allergic rhinitis.

In guinea pigs, short-term exposure to the ubiquitous air pollutant sulfur dioxide (SO_2), even at levels below national ambient air quality standards, augments subsequent allergic sensitization of the airways (Riedel 1988). Under the microscope, the bronchi (large airways) in SO_2-exposed guinea pigs appear inflamed and thus may be more permeable, facilitating access of antigens to the immune system. When provocation with inhaled antigen (ovalbumin) is performed repeatedly over a 3-week period, most of the SO_2-exposed animals show maximal bronchial reactions during the first few challenges. Continued provocation leads to a decrease in obstructive reactions that the authors describe as "allergen tachyphylaxis" but that might also fit the model of adaptation.

Chemical waste disposal sites may contaminate the air and groundwater of nearby communities. David Ozonoff of the Boston University School of Public Health and his co-workers (1987) surveyed households surrounding an odorous chemical waste disposal site and found that exposed individuals more often complained of respiratory symptoms (wheezing, shortness of breath, chest discomfort, persistent colds, coughs) and constitutional symptoms (chronic fatigue, bowel dysfunction, and irregular heartbeat) than did controls. Levels of air contaminants that were detected were exceedingly low and led the authors to conclude that the general population may react to chemicals at concentrations much lower than previously thought. Most studies of populations near hazardous waste facilities have focused on serious but low-prevalence diseases such as cancer. In small populations, such outcomes are difficult to measure because of low statistical power. The authors suggest that future investigations should concentrate instead upon common medical complaints: "Not only are such outcomes more amenable to study because of their higher prevalence, they may have considerable importance because of their impact on the efficiency, well-being, comfort, and productivity of a community."

As is discussed in the following section, indoor air pollution rather than outdoor air pollution accounts for the greatest number of and most intense exposures (Nero 1988). Most people spend the majority of their time indoors, either at work or home. Moreover, the levels of exposure to many contaminants, particularly volatile organic compounds (many of which are uncharacterized and whose health effects are unknown) are much higher indoors than out-of-doors.

Indoor Air Pollutants, Domestic and Workplace

The scope of indoor air pollutants has been reviewed by others (Spengler and Sexton 1983b; Nero 1988; Cone and Hodgson 1989). John Spengler of the Harvard School of Public Health, an authority on indoor air, predicts that the problem of buildings with unhealthy air is likely to continue for years because another generation of "pathologic" buildings is already on the drawing boards (Spengler 1989). He describes the "space flu" experienced by astronauts before NASA recognized that construction materials and supplies off-gassing inside a tightly sealed spacecraft were the source of the illness. Subsequently, NASA has compiled a detailed inventory of all materials used in their vehicles and the types and amounts of chemicals they release into the air and has designed spacecrafts to minimize exposures.

The range of indoor air pollutants affecting industrial workers is enormous. Seemingly, almost any process involving chemicals appears to have the potential for initiating chemical hyperreactivity via long- or short-term exposure. Two general types of exposures seem particularly apt to initiate hypersusceptibility:

1. A massive, overwhelming exposure, such as a chemical spill, a fire involving synthetic materials, pesticide spraying, or working with chemicals in a confined, unventilated space.

2. Repeated, low-level exposure to a complex array of synthetic organic compounds, as occurs with combustion products (such as diesel exhaust), tight buildings, and soldering (Miller 1979).

Gas chromatographic analysis of air samples from problem buildings or homes typically reveals the presence of multiple spiked peaks, each representing a particular organic compound (Fig. 3-2). Additional sampling and analytical approaches are needed to measure oxides of nitrogen from gas combustion, ozone, pesticide residues, and other air contaminants that may be present in the same environment.

At home, troublesome exposures for the chemically sensitive patient include the gas stove, one of the most commonly identified triggers of symptoms in these patients; combustion products from gas- or oil-fired furnaces and space heaters, water heaters, and central air heating systems; sponge rubber bedding, padding, and upholstery; plastics (especially pliable odorous plastics such as shower curtains); insecticides; perfumes; paints and decorating materials; fireplaces; cleaning agents; disinfectants; deodorizers; mothballs; cedar closets; newsprint and other printed materials; fabrics in clothing, bedding, and window coverings, especially synthetics or coated fabrics; particleboard; gasoline vapors from attached garages; and carpeting and carpet padding. Disinfectant liquids and sprays containing phenolics frequently provoke symptoms in these patients. Interestingly, researchers first became concerned about orthophenyl phenol when they noticed that mice housed in cages washed with this common institutional and household disinfectant showed markedly depressed immune responses after 4 to 6 weeks of exposure (La Via 1979).

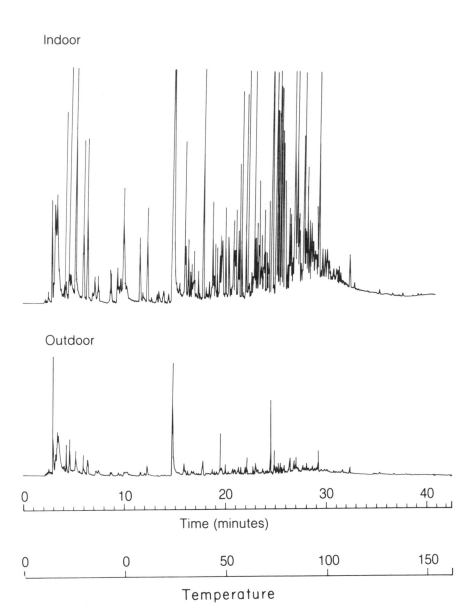

FIGURE 3-2. *A comparison of gas chromatographs of equal volume air samples taken indoors and outdoors near a complaint office building.*
Source: Miksch, R. R., Hollowell, C. D., Schmidt, H. E. "Trace Organic Chemical Contaminants in Office Spaces," Environ Int *8:129–138, 1982. Courtesy of Lawrence Berkeley Laboratory, University of California.*

The Environmental Protection Agency recently conducted air sampling for pesticides inside homes in Jacksonville, Florida, and Springfield and Chicopee, Massachusetts (Immerman 1990). (See Fig. 1-7). Analysis revealed a profile of multiple agents that had been applied over the life of each home. Pesticides applied years before and subsequently banned could be detected. Approximately eight different pesticides were detected in the average home in Florida, compared to four in Massachusetts (out of a total of 32 pesticides for which testing was done). Patients with chemical sensitivities sometimes associate onset of their illness with a particular move, and the potential role of previously applied, long-lived chemicals merits thoughtful scientific exploration.

Several guides for constructing homes from "safer" materials that are less toxic and/or do not off-gas have been written (Rousseau et al. 1988; Good and Dadd 1988; Zamm and Gannon 1980). Urea formaldehyde foam insulation, which may have provoked this illness in many in the past, has ceased being used by insulators since a flurry of successful lawsuits. As part of its research and information mandate under Title IV of the Superfund Amendments and Reauthorization Act of 1986, the EPA Indoor Air Division is currently developing several guidance documents on designing and operating buildings to ensure indoor air quality. These documents address the design and construction of residential and public and commercial buildings, the operation and maintenance of public and commercial buildings, and indoor air quality management in schools.

Mobile homes and automobile interiors present their own special problems. Indoor air pollutants in other settings may present problems: shopping malls, perfume counters, detergent and insecticide aisles, fabric stores, dry cleaners, deodorizers and hairspray in public rest rooms, tobacco smoke, incense, Sterno used in buffets, gas cooking combustion products in restaurants, and perfume, cologne, or mothball odors on garments worn in theaters, churches, and public transport commonly cause difficulty for these patients. "Odors" of virtually any desription, especially petrochemical odors but also "natural" odors from cedar or pine terpenes or cooking foods, may provoke symptoms; the presence of an odor implies that the substance in question has a vapor pressure and that molecules of it are present in the air. The subject of odors and their role in this syndrome is discussed in Chapter 4.

Historical Notes.   Randolph first discussed the topic of indoor air pollution in a series of articles published in 1954 and subsequently in *Human Ecology and Susceptibility to the Chemical Environment* in 1962. That same year, President Kennedy called the first national conference on air pollution in Washington, D.C. It was a 3-day program with only an hour and a half at the end for open discussion. During the discussion, Randolph remarked that in 3 days of presentations not a single reference had been made to *indoor* air pollution. In his clinical experience, he said, indoor air pollution was 8 to 10 times more important as a source of illness in susceptible individuals than outdoor air pollution. Whereas outdoor air pollution tended to be intermittent and variable, indoor air pollution was much more constant. Further, individuals spend the majority of their time indoors. He also noted that more than 800 gas stoves had been removed from the homes of his highly susceptible patients.

Twenty years later, pollution from the same sources Randolph had identified as triggers of his patients' symptoms was documented by advanced and sensitive analytical techniques such as gas chromatography with mass spectrophotometry, which was not available at the time of Randolph's original writings. Between 1979 and 1985, the EPA undertook an extensive study of exposures to volatile organic compounds of 400 residents in three states (Wallace 1987). The TEAM study (total exposure assessment methodology) employed state-of-the-art monitoring and analytical methods. Each subject wore a personal air sampler for 24 hours and provided a breath sample at the end of the day. Personal exposures were consistently greater than outdoor levels, sometimes by factors of 10 or more (closely approximating Randolph's estimates 20 years earlier), implying important indoor sources of exposure. Smoking, visiting the dry cleaner or gas station, and certain occupations resulted in very elevated exposures. Breath levels were 30 to 40 percent of personal air concentrations for 9 of 11 compounds but ranged as high as 75 to 80 percent for benzene (from gasoline) and 90 percent for tetrachloroethylene (Wallace et al. 1985). Summarizing data from nine separate studies involving more than 1,000 homes, Wallace reported agreement on these points:

1. Essentially every one of the 40 or so organics studied has higher indoor levels than outdoor.

2. Sources are numerous, including building materials, furnishings, dry-cleaned clothes, cigarettes, gasoline, cleansers, moth crystals, hot showers, printed material, etc.

3. Ranges of concentrations are great, often 2 or more orders of magnitude.

Clearly, exposures in most indoor situations occur at levels well below current OSHA or EPA standards. At a given moment, several hundred different chemicals may be present in air samples from a home or office. One question that arises is whether the summation of all of these chemicals' effects could be responsible for symptoms, even though no single constituent accounts for them. To this end, Molhave (1986) of Denmark exposed 62 individuals to a mixture of 22 volatile organic compounds (VOCs) that commonly occur as indoor air pollutants. Three concentrations of total pollutants were used: 0 mg/m³ (control), 5 mg/m³, and 25 mg/m³ of the same mixture of 22 compounds. Using "healthy" subjects (both male and female) who had previously complained about symptoms of sick building syndrome, Molhave exposed them to each concentration for 2.75-hour periods. As the air concentration increased, complaints of nasal and throat irritation and inability to concentrate (measured by digit span memory performance) rose. Thus, as the dose increased, these more "susceptible" individuals who otherwise appeared healthy were significantly affected by the indoor air pollutants, to the point of having difficulty with tasks requiring concentration.

In a follow-up study, undertaken to confirm and extend the 1986 Molhave study, Otto, Molhave, and co-workers (1989) exposed 66 normal, healthy males (with no history of chemical sensitivity) for 2.75 hours to a complex VOC mixture at 0 and 25 mg/m³. The study confirmed perceptions of odor unpleasantness found in the 1986 study. However, VOC exposure did not affect performance on any behavioral tests. Certainly, the 2.75-hour exposure time used in both studies does not compare to the day in, day out exposures of occupants of tight buildings. Nevertheless, comparison of the two studies suggests differences in central nervous system effects between healthy individuals and those who have complained of indoor air problems previously. Unfortunately, because of the irritant effects from exposure to higher levels of VOCs, blinding is exceedingly difficult to achieve in the design of such studies.

Other Scandinavian researchers have found total volatile organic compound concentrations in homes with complaints by occupants to average 1.3 mg/m³ (range 0.092 to 13 mg/m³), whereas the concentrations in houses where there were no complaints averaged 0.36 mg/m³ (range 0.02 to 1.7 mg/m³) (Molhave 1986–87). Thus, levels of volatile organic compounds were generally higher in problem houses.

Studying the white-collar work environment, Robertson et al. (1985) compared health problems in two office buildings, one fully air-conditioned and the other naturally ventilated. Sickness was significantly increased in the air-conditioned building versus the naturally ventilated building, particularly rhinitis (28 percent versus 5 percent), nasal blockage and dry throat (35 percent versus 9 percent), lethargy (36 percent versus 13 percent), and headache (31 percent versus 15 percent). That temperature, humidity, air velocity, and other such factors did not differ between the two buildings suggests that the sickness was caused by indoor air pollutants. Similarly, Finnegan and co-workers (1984) found significant excesses of eye, nose, and mucous membrane symptoms, as well as lethargy, dry skin, and headaches among workers in air-conditioned versus naturally ventilated offices.

An unanticipated and unwelcome opportunity for the EPA to study the effects of indoor air pollution firsthand arose when 27,000 square yards of new carpeting were installed in the agency's headquarters in Washington, D.C., in 1987 and 1988 (Hirzy and Morison 1989a, 1989b). An estimated 124 of 2,000 employees exposed to volatile off-gassing from the carpet became ill, exhibiting symptoms ranging from eye, nose, and throat irritation and breathing problems to nausea, headache, dizziness, difficulty in thinking, fatigue, and increased susceptibility to many exposures formerly tolerated. At least two employees quit their jobs as a result of illness. Seventeen were unable to work in their assigned spaces. Some now work at home or in other locations. Eight report new sensitivities to common substances, including perfumes, auto exhaust, and tobacco smoke. Symptoms of the 20 or so most severely affected individuals appear identical to those of patients seen by clinical ecologists. Agency scientists in the employee's union who analyzed air samples felt the culprit might be 4-phenylcyclohexene (4-PC), which is used to bind carpet fabric to its backing. Estimates of the exposures that initiated illness in the susceptible subgroup range from 5 to 15 ppb of 4-PC (Hirzy and Morison 1989b). These same persons, now "sensitized," experience symptoms upon reexposure to less than 1 ppb of the substance; symptoms include respiratory difficulty, dizziness, "spacey" feelings, and general malaise. The EPA problem lends further credence to Randolph's and other clinical ecologists' observations with respect to: (1) diverse symptoms occurring in different individuals even with the same exposure, (2) "spreading" of sensitivities to other low-level chemical exposures and to foods that formerly had been tolerated, and (3) adaptation, that is, the less severely affected employees noted improvement in symptoms while away from work with marked increase upon return and gradual subsidence during the workweek as tolerance developed (Hirzy 1989).

Foods, Food Additives, and Contaminants

Rea (1988a) estimates that food sensitivity occurs in about 80 percent of his patients with chemical sensitivities. Ecologists observe that excessive chemical exposure may result in loss of tolerance to foods, sometimes to every food in the diet, and that removing the individual from such exposures and rotating foods so that no food is eaten more than once every 4 days may restore dietary tolerance. Pesticide residues, can linings (the gold-brown lining of cans may contain a phenolic resin), fumigants, fungicides, sulfur treatment, artificial colors, sweeteners, preservatives, ripening procedures (such as ethylene gas), protective waxes, and packaging materials, especially plastics, may trigger symptoms in patients. When patients are challenged with foods in an environmental unit, at first they are given chemically less contaminated foods such as organic meats and produce wherever possible. Once a variety of "safe" foods has been determined and prior to discharge from the unit, patients may be given several consecutive meals of commercial preparations of their safe foods. These meals might include commercial apples that have been sprayed (consider the recent concern over spraying apples with Alar, causing possible long-range effects in children who consume these apples or apple juice), canned foods, and nonorganic meats. After 2 days of such feedings, many patients reportedly experience fatigue, headache, myalgia, arthralgia, arthritis, depression, and other muscular, skeletal, and/or neurological symptoms (Randolph 1987).

Water Contaminants and Additives

According to Rea (1988a) as many as 90 percent of his patients with chemical sensitivities may have reactions to contaminants in drinking water. While fasting in an environmental unit, patients test waters from a variety of sources including tap water, specially distilled or filtered water, and various spring or well waters until they find one that does not evoke symptoms. Drinking water may be contaminated by leaching of chemicals from plastic storage containers, rubber hoses or connectors in distilling apparatus, or plastic or rubber fittings in drinking water dispensers. Uncontaminated well or spring water in glass bottles may be preferable for particularly sensitive individuals.

Chemical contamination of groundwater is a growing national concern. Aldicarb, a carbamate insecticide and nematocide used extensively since the 1960s, was first noted as a groundwater contaminant in the late 1970s, when more than 1,100 wells in New York's Suffolk County, a potato-farming region on Long Island, tested positive for aldicarb (levels greater than 7 ppb). Since that time, aldicarb has been found in groundwater in Maine, Florida, California, Arizona, North Carolina, Virginia, and Wisconsin. In Wisconsin, 23 apparently healthy women who consumed groundwater with detectable aldicarb were found to have altered T-cell subsets with a decreased T_4:T_8 ratio of 1.88 versus 2.54 in an unexposed control group ($p < 0.05$) (Fiore et al. 1986). Unlike AIDS, in which the T_4:T_8 ratio is decreased primarily because T_4 (helper) cells are destroyed by the virus, these women had an increase in T_8 (suppressor) cells. In addition, lymphocyte proliferation in response to *Candida* allergen was increased ($p < 0.02$) versus controls. Likewise, among residents of Woburn, Massachusetts, who drank water that had been contaminated with industrial solvents, excess leukemias (12 versus 5.3 expected), immunological abnormalities including decreased T-cell ratios (p <0.01), increased autoantibodies and infections, and neurological, cardiac, and skin abnormalities were noted (Byers et al. 1988). (See also Chapter 4.)

Ingestion is not the sole route of exposure to contaminants in water. Brown and associates (1984) reported that skin absorption (for example, from bathing or showering) may be a significant portal of entry for water contaminants accounting for 29 to 91 percent (average 64 percent) of the total daily dose of these substances. Aside from skin contact, showering volatilizes contaminants in water and leads to inhalation of chlorine, chloroform, and organic compounds (Bailey and Vanderslice 1987; Foster and Chrostowski 1987). Water contaminated with organic material and subsequently chlorinated contains chlorinated hydrocarbons that are potentially carcinogenic. Interestingly, chemically sensitive individuals frequently note symptoms while bathing or showering, and some claim they must use specially filtered water or at least water treated to remove the chlorine.

Drugs and Consumer Products

Physicians recognize that a person who has an adverse reaction to one drug is more likely to react to other drugs. The allergist Sullivan (1989) reported that individuals who experienced an adverse reaction to penicillin are much more likely to react adversely to other drugs, in particular, other antibiotics. Interestingly, he calls this phenomenon, which occurs in a very small percentage of the population, the "multiple drug allergy syndrome." The mechanism is unclear but he postulates it may be related to faulty regulation of antihapten immune responses. Meggs (1989a) compiled a list of symptoms that have been reported for seven well-known pharmaceuticals (indomethacin, propranolol, azatadine, pseudoephedrine, captopril, diazepam, and reserpine); these symptoms reproduced about 80 percent of the symptoms and complaints Terr (1989a) reported in patients exposed to various organic chemicals. Meggs comments, "Perhaps there is a similarity between adverse reactions to pharmaceuticals and volatile organic compounds found in the workplace. Again we are dealing with low molecular weight carbon-based compounds of similar structure in the two cases."

Randolph (1962, pp. 85–87) surveyed one series of 80 and another series of 250 of his chemically sensitive patients who had "known" reactions to some facet of their chemical environment and found that an extraordinarily high percentage had reacted adversely to one or more medications. One quarter to one half claimed to have reacted to aspirin, barbiturates, or sulfonamides. According to Randolph, because of this proneness to drug reactions and because many physicians do not understand this problem, many individuals with chemical sensitivities are reluctant to seek health care. Thus there seems to be an important overlap between individuals who react badly to medications and chemically sensitive patients. Investigating whether a disproportionate number of the idiosyncratic reactions listed in the *Physicians' Desk Reference* occur in the same subgroup of patients would be worthwhile. The psychiatrist Schottenfeld (1987) confirms that many individuals with multiple chemical sensitivities appear unusually sensitive to the anti cholinergic and sedative effects of tricyclic antidepressants.

Drugs, of course, contain much more than active ingredients. They also contain excipients (for example, cornstarch or lactose in tablets), diluents, coloring agents, flavorings, various coatings, and/or preservatives (as in allergy shots, which often contain about 0.4 percent phenol). Mineral oils, petroleum jelly, ointment, lotions, laxatives, synthetic vitamins, and adhesive tape cause problems for many patients. Most cosmetics, scented soaps, shampoos, hand lotions, personal hygiene products, perfumes, colognes, deodorants, hairsprays, hair dyes, mouthwashes, denture adhesives, and bath salts and oils have been reported to provoke reactions in individual patients. Many patients do better by substituting "natural" products for petrochemical ones (Ziem 1989).

In addition, permanent press finishes (especially during ironing); synthetic textiles; clothes that have been dry-cleaned; residues of detergents and fabric softeners; electric blankets (the plastic coatings over the wires off-gas when heated); waterbeds; mattresses treated with flame retardants; felt-tip pens; odorous books, magazines, and newsprint; polishes, cleaners, and bleaches; and chlorinated swimming pools and even bath and shower water have also been associated with intolerance (Randolph 1962, pp. 112–114).

The very ill patient may be sensitive to most if not all of these substances and products and has difficulty avoiding them and finding suitable substitutes. Mail-order services, often begun by patients, have developed to help sensitive individuals find products that are better tolerated. Exposures may be very subtle. For example, individuals may find themselves irritable or anxious when talking on the telephone, but if they substitute Bakelite phones for their new colored plastic ones or use speakerphones instead, their problem resolves (Randolph 1962). Clothing that was stored in particleboard drawers may emit formaldehyde and trigger symptoms. Synthetic fabrics have been implicated in elevated blood pressure, increased heart rate, arrhythmias, and angina

(Seyal et al. 1986a–d). Acrylic dentures may provoke headache, joint pain, fatigue, and rashes (Kroker et al. 1982).

The process of discovering the limits of one's tolerance may be long and tedious, with many setbacks. The setbacks can be so painful and disabling that patients go to great lengths to educate themselves about chemicals and avoid them. Very sensitive patients may react adversely to contact lenses, dental materials, medical implants or prostheses, local anesthetics, plasticizers leaching from plastic IV or oxygen lines, lubricating jelly applied during an examination, or alcohol evaporating on the skin when blood is drawn. Such patients view any encounter with an unknowing or disbelieving dentist or physician with great trepidation. Radio-contrast dyes may be of special concern. If surgery is planned, patients may inquire what intravenous solutions will be used (D_5 is 5 percent dextrose and thus contains corn sugar; corn is the most common food that provokes symptoms in these patients [Randolph and Moss 1980, p. 109]), what anesthetic drugs will be used, and the like, so as to prepare themselves and their doctors for any adverse reaction and attempt to avert it. Many practitioners find such inquiry intimidating or view the patient as demanding or hypochondriacal, when in fact the patient, in need of an operation or special procedure, only wishes to avoid an adverse reaction. Practitioners need to understand these patients' concerns and realize that the patients' fears may be well founded in prior experiences with very painful or embarrassing reactions. When they must place themselves in the hands of the medical establishment, chemically sensitive patients feel a lack of control and a vulnerability most would not understand.

Health Effects

According to the clinical ecologist, the symptoms and diseases caused by food and chemical exposures involve any and every system of the body and are so diverse that many traditional practitioners find them unbelievable. Some of the physicians we interviewed recalled being told as medical students that the more symptoms a patient complained of, the less validity any of them had. Clearly, such a belief by physicians could pose an obstacle in that the average patient with food and chemical sensitivities who enters an environmental unit has five symptoms, many of them neurological (Johnson and Rea 1989).

To many nonecologists, a troublesome aspect of the provocative food and chemical challenges performed by clinical ecologists has been the differences seen in symptoms on challenge versus those that were part of the patient's chief complaint at presentation. In his critique of 50 patients who previously had been seen by clinical ecologists, Terr (1986), an allergist, noted that 30 of the patients (60 percent) developed one or more *new* symptoms during their diagnostic and therapeutic experience with the clinical ecologist. Of these, the most frequent were headache (30 percent), fatigue (23 percent), confusion or loss of memory (20 percent), swelling (20 percent), dizziness (17 percent), depression (17 percent), nausea (17 percent), and rash, drowsiness, anxiety, and abdominal pain (13 percent). Forty-three of Terr's 50 patients were referred for workers' compensation evaluation and thus represent the "worst" cases. In comparison, less than 5 percent of the patients seen by Randolph and about 20 percent of those seen by Rea (who sees sicker patients referred by other ecologists) apply for disability.

In a more recent paper, Terr (1989a) compiled data from 40 of his original 50 patients plus 50 others claiming disability because of chemical sensitivities. Patients' symptoms were again diverse: 28 had symptoms referable to a single organ system, whereas 62 had "multisystem, polysymptomatology." After diagnosis of environmental illness by ecologists, 75 of these 90 workers reported one or more new triggers for their symptoms, including foods. These triggers were associated with one or more *new symptoms* in 34 patients.

The frequent emergence of new symptoms during deadaptation and reexposure is well known to clinical ecologists. Unfortunately, Terr does not offer any of the patients' commentary about their illness. However, 62 percent had a long history of multiple symptoms involving many systems and parts of the body and had been examined, tested, and treated unsuccessfully for years by many physicians prior to seeing a clinical ecologist. Perhaps some of their new symptoms did occur in the past but were transient in nature and forgotten. Perhaps adaptation masked certain symptoms. To the clinical ecologist, patients with very advanced environmental illness are manifesting the most extreme, overlapping stimulatory and withdrawal reactions to multiple substances. Chronic disability may ensue. However, the patient who is withdrawn from inciting chemical exposures and placed on a "safe" diet may be able to reverse the experience and begin to associate cause and effect. Specific symptoms can then be attributed to identifiable chemicals or foods in a reproducible way. Following deadaptation, symptoms that have not been experienced for decades may be manifested; that is, unmasking takes place. As long as multiple exposures causing multiple effects are overlapping, symptoms are masked and the person may experience chronic disability or ultimately end organ failure of some type.

The acute symptoms experienced by a patient when a clinical challenge is performed must be differentiated (Table 3-1) from the chronic disease states that are purported to result from chronic exposure to

TABLE 3-1. Possible Acute Reactions to Incitants during Provocation

Nasal	Throat, mouth	Ears	Lungs, heart, blood vessels	Joints	Muscles
Urge to sneeze	Itching, sore, tight, swollen	Itching	Coughing	Ache, pain	Tight, stiff
Itching, rubbing	Dysphagia, difficulty in swallowing, choking	Full, blocked	Sneezing	Stiff	Aches, soreness, pain
Obstruction	Weak voice, hoarse	Erythema of pinna (reddening)	Reduced air flow	Swelling	Neck
Discharge	Salivation, mucus	Tinnitus (ringing in ears)	Retracting, shortness of breath	Erythema, warmth, redness	Upper, lower back
Postnasal drip	Bad or metallic taste	Earache	Heavy, tight chest		Upper, lower extremities
Sinus discomfort		Hearing loss	Not enough air		
Stuffy feeling		Hyperacusis (abnormal sensitivity to sound)	Hyperventilation, rapid breathing		
			Chest pain		
			Tachycardia (rapid pulse)		
			Palpitations (rapid, violent or throbbing pulses; extra or skipped beats)		
			Blood vessels— spontaneous bruising and petechiae, cold sensitivity, swelling, acneform lesions		

Skin	Eyes	Vision	Cerebral, head	Genitourinary	Gastrointestinal
Itching, local or general	Itching, burning, pain	Blurring	Headache, mild–moderate; migraine	Voided, mild urge	Nausea
Scratching	Lacrimation (tearing)	Acuity decreased	Ache, pressure; tight, exploding feelings	Frequency in voiding	Belching
Moist, sweating	Injected light sensitive	Spots, flashes	Throbbing, stabbing	Urgency, pressure	Full, bloated
Flushing, hives	Sensitivity (allergic) shiners	Darker, vision loss	Fainting	Dysuria (painful or difficult urination)	Vomiting
Pallor (white or ghostly)	Feel heavy	Photophobia (brighter)	Depression	Genital itch	Pressure, pain, cramps
		Diplopia (double vision)	Mood swings	Vaginal discharge	Flatus, rumbling, gas
		Dyslexia–difficulty reading, transposition of letters; letters or words becoming small or large; words moving around	Hallucinations	Yeast infection	Diarrhea
			Hyperactivity		Gallbladder symptoms
			Irritability		Hunger, thirst
			Fatigue		Hyperacidity
			Apathy		
			Confusion		
			Lethargy		
			Blackouts		
			Insomnia		
			Somnolence		

Source: Rea, W., *Outpatient Information Manual* (1984a; 1988 revision), Environmental Health Center, Dallas.

incompatible foods and chemicals. The latter, according to clinical ecologists, include a wide range of diseases and disorders. Traditional practitioners consider many of these disorders idiopathic or essential (as in "essential hypertension") or give them names that are descriptive (as "asthma" or "urticaria") and are not revealing about possible causes. Clinical ecologists claim that many of these conditions are caused by environmental (food or chemical) incitants.

Perhaps the definitive test for chemical sensitivity would be to have the patient fast in an environmental unit. The resolution of chronic, debilitating symptoms then might suggest an environmental cause. Proof of environmental causation involves rechallenge with single foods and chemicals while noting the effects. If an effect were reproducible, causation would be inferred. Confidence regarding causation would be strengthened by double-blind, placebo-controlled challenges.

Scientists at Research Triangle Institute have been seeking better methods to assess health effects associated with complex mixtures of chemicals present in indoor air. A number of chemical compounds were selected as representative of a particular newly renovated office building; for example, certain chemicals that were measured outgassing from carpet samples and office partitions. Health effects reported in the TOXNET database for these individual chemicals were compared with complaints by office workers in the building (Pierson et al. 1990). These are shown in Table 3.2. Health effects reported by the employees were similar to those found in TOXNET, although the literature documents effects at much higher levels of exposure.

Appendix A contains an annotated bibliography for health effects that may be related to foods and chemicals. Many, but not all, sources listed were written by clinical ecologists. For certain diseases such as migraine and atopic dermatitis (eczema), traditional practitioners as well are coming to accept that foods may play an important role in certain patients.

Appendix A is not intended to be encyclopedic. Rather, it is an attempt to present the range and diversity of diseases for which environmental (food or chemical) origins have been proven or proposed. It is also designed to help the reader identify key articles on particular disorders because many of these articles have appeared in older or less widely circulated journals and would otherwise be difficult to locate.

We have highlighted studies with positive outcomes, that is, those in which a relationship between symptoms and food or chemical exposures was confirmed. Many studies that fail to show an association between symptoms and exposure exclude only a single food from the diet. Those studies of hyperactivity (Kaplan 1989), seizures (Egger et al. 1989), headaches (Egger et al. 1983), and rheumatoid arthritis (Kroker et al.

1984; Marshall et al. 1984) that do show an association tend to use simultaneous avoidance of multiple incitants. Their design should serve as models for future studies in this field.

The majority of the articles discuss foods rather than chemicals as potential factors in disease. Nevertheless, most of the diseases listed here have also been attributed to chemical exposures by some observers. Randolph's observations of patients worked up in an environmental unit are summarized in his books and papers and provide interesting anecdotal accounts of the role chemicals might play in particular conditions.

By presenting this material we are not affirming an environmental cause for these diseases but hoping to alert the reader to that possibility and the need for evaluating such patients in an environmental unit when more traditional approaches have failed. What might seem obvious—that foods and chemicals are not significant factors in most of these disorders—could change if one were to eliminate masking and control for the effects of adaptation.

We cannot overemphasize the importance of avoiding the frequent error of confusing diagnosis and etiology. Terr (1986, 1989a) in his reviews of ecology patients criticizes ecologists for attributing illness in these patients to environmental factors where they clearly had other well-defined clinical diseases such as depression. In a critique of Terr's most recent review of ecology patients, William Meggs (1989b) of East Carolina School of Medicine asserts:

> First, both Dr. Terr and the clinical ecologists consistently confuse diagnosis and etiology. Environmental illness, ecological illness, or similar terms should not be used as a diagnosis, which is the error of the clinical ecologists. Dr. Terr's error is to state that since a patient has another diagnosis, the diagnosis of environmental illness is wrong, and therefore there is no environmental cause of the illness. . . . In his methods section Dr. Terr does not discuss how he determined that the patients' symptoms were not triggered by environmental exposures. Many of the symptoms he lists in Table 4 of his article such as asthma, rhinitis, and dermatitis are known to have an environmental etiology in some patients, and generally accepted methods are available for verification. . . . Correctly diagnosing an autoimmune condition in a patient claiming environmental illness, rather than disproving an environmental etiology, should alert the physician to look for an environmental cause. The claim that psychiatric disorders can be triggered by chemical exposures is worthy of serious scientific study, particularly with increasing rates of depression. . . . Cases of depression related to exposure to furnace fumes were described by Randolph thirty years ago.

The study regarding furnace emissions as a cause of depression is by Randolph (1955).

Table 3-2. Summary of Reported Effects of Indoor Air Pollutants[a]

Body System Effect	Compound[b]								
	AcA	ACE	CUM	DCB	EtB	FOR	STY	TOL	XYL
Eyes									
Irritation[c]	×	×		×	×	×	×	×	×
Irrit. mucous membranes						×	×	×	
Conjunctivitis	×	×					×		×
Lacrimation	×	×			×	×			
Diplopia (double vision)[c]									
Photophobia	×								
Nose									
Irritation[c]	×	×		×	×	×	×		×
Irrit. mucous membrane						×	×	×	
Runny nose[c]				×					
Respiratory									
Irritation		×			×	×	×	×	×
Pharyngitis		×						×	
Throat irritation[c]			×	×		×	×		×
Bronchitis	×	×				×			×
Coughing[c]	×					×			
Shortness of breath						×			
Asthmatic reaction						×			
Pulmonary edema	×					×			
Central Nervous									
Tinnitus							×		
Headache				×				×	
Dizziness[c]		×		×				×	
Depression[c]							×		
Fatigue[c]							×	×	
Confusion[c]							×		×

	AcA	ACE	CUM	DCB	EtB	FOR	STY	TOL	XYL
Drowsiness	X								
Vertigo			X				X	X	
Slowed reaction time									X
Intoxication: Euphoria, exhileration, boastfulness, talkative			X	X				X	X
Incoordination (ataxia)			X				X	X	X
Anasthesia[c]							X	X	
Edema							X		
Weakness			X		X				
Skin									
Erythema, irritation					X			X	
Dermatitis	X					X	X		
Blood									
Leukopenia					X			X	
Leukocyctosis					X			X	
Macrocytosis								X	
Reduced erythrocytes						X		X	
Liver injury				X				X	
Miscellaneous									
Gastritis								X	
Nausea and vomiting								X	
Dysphagia					X				
Menstrual disorders[c]								X	X
Weight loss			X	X					X

[a] Absence of symptoms does not inherently mean that these do not exist for a given compound—only that they were not reported.

[b] **Key:** AcA = acetaldehyde; ACE = acetone; CUM = cumene (isopropylbenzene); DCB = dichlorobenzene; EtB = ethylbenzene; FOR = formaldehyde; STY = styrene; TOL = toluene; XYL = xylene.

[c] Effects reported associated with the example complex mixture.

Source: Pierson et al. 1990.

Allergists accuse ecologists of overzealously diagnosing environmental illness and overlooking other important medical conditions (Bardana and Montanaro 1989; Terr 1986, 1989a); however, some allergists may have wrongly assumed that a patient's condition that has an accepted medical label cannot have an environmental etiology. Physicians need to be aware of the wide variety of medical conditions for which environmental (either food or chemical) etiologies are being considered. We have attempted to pull together some of the most pertinent articles for Appendix A.

If foods and chemicals are responsible for even a modest percentage of the diseases listed in Appendix A, the implications are staggering. The trend toward recognition of chemical and food factors in many diseases is growing. Caution must be exercised, however. Perhaps only a subset of patients with a particular illness (for example, rheumatoid arthritis) responds to environmental manipulation. Even one individual who responds positively while fasting in an environmental unit can be an important finding if that finding is reproducible. Thus responses may occur only in a subgroup of sensitive patients. Unless objective testing is limited to that sensitive subgroup, positive results in a few may be diluted by nonresponders, and prevalence studies will not be statistically significant. Clearly, studies of these patients must be carefully constructed if health effects are to be discerned.

How the disorders that have just been discussed relate to the concept of adaptation is unclear at present. Bell and King (1982a) propose that "the chronic symptoms which ecology patients reportedly have with repeated exposures to offending agents may reflect the more insidious, non-adapting changes induced by offending foods and chemicals." For example, individuals may adapt to the *acute* effects of ozone on their upper and lower respiratory tracts, but red blood cell fragility may persist. Others may adapt to the stimulatory effects of caffeine, only to develop fibrocystic disease (Russell 1989; Hindi-Alexander et al. 1985; Boyle et al. 1984) or urticaria (Pola et al. 1988).

Randolph depicts the development of chronic illnesses as in Table 3-3. In the left column are intermittent (acute) responses, and on the right are chronic responses. At the top are stimulatory levels, as discussed in Chapter 2. With continued or repetitive exposures to an incitant, the course of the reactions moves from left to right. Over time, if no intervention occurs, advanced sustained stimulatory responses (upper right) ultimately move toward sustained withdrawal responses (lower right) (Randolph 1987, pp. 248–249). Rea (1988c) speculates that adaptation may indirectly contribute to total body load by covering up (masking) acute reactions with chronic exposure responses so that affected individuals are unaware of the relationship between exposures and their ill-

TABLE 3-3. Environmental Personal Interrelationships

Intermittent Responses	Levels	Sustained Responses
Mania[a] (agitation, excitement, blackouts, with or without convulsions)	+ + + +	Drug addiction (both natural and synthetic)
Hypomania[a] (hyperresponsiveness, anxiety, panic reactions, mental lapses)	+ + +	Alcoholism (addictive, drinking)
Hyperactivity[a] (restless legs, insomnia, aggressive forceful behavior)	+ +	Obesity (addictive eating)
Stimulation[a] (active, self centered with suppressed symptoms)	+	Absent complaints (the desired way to feel)
Behavior on an even keel, as in homeostasis	0	Behavior on an even keel
Localized physical ecologic manifestations[b] (rhinitis, bronchitis, asthma, dermatitis, gastrointestinal, genitourinary syndromes)	–	Impaired senses of taste and smell, Meniere's syndrome
Systemic physical ecologic manifestations[b] (fatigue, headache, myalgia, arthralgia, arthritis, edema, tachycardia, arrythmia)	– –	Small vessel vasculitis, hypertension, collagen diseases
Brain fag—moderately advanced cerebral syndromes[b] (mood changes, irritability, impaired thinking, reading ability and memory)	– – –	Mental confusion and obfuscation, morose inebriation
Depression—advanced cerebral and behavioral syndromes[b] (confabulation, hallucinosis, obsessions, delusions and temporary amnesia)	– – – –	Dementia, stupor, coma, catatonia, residual amnesia

[a] Specifically adapted stimulatory levels.

[b] Specifically maladapted withdrawal levels.

Source: Randolph, T. and Moss, R., *An Alternative Approach to Allergies*, copyright J. B. Lippincott Co., Philadelphia, PA (1989).

nesses. Smokers, for example, may learn to tolerate the irritating properties of tobacco smoke, but their adaptation only allows them to continue the habit more comfortably, oblivious to the fact that continued exposure may lead to emphysema, cancer, and cardiovascular disease. Patients with multiple chemical sensitivities, on the other hand, frequently experience intense discomfort whenever smoke is present, even in low concentrations. Interestingly, many smokers who quit (deadapt to tobacco) also report acute symptoms with minimal exposure, not unlike those reported by chemically sensitive individuals. This underscores an important point: it is impossible to know how sensitive individuals are to an environmental agent until they are deadapted. Those who appear least sensitive (smokers, for instance) may in fact be most sensitive with their addiction only masking that sensitivity.

Thus adaptation could play a central role in the development of many medical disorders. Perhaps the best evidence thus far for the existence of adaptation in humans comes from clinical observations of withdrawal symptoms when individuals are removed from their usual exposures and subsequent resolution of formerly chronic symptoms occurs. According to some physicians, this process is viewed optimally in the setting of an environmental unit while patients fast.

In the next chapter we discuss some of the mechanisms that have been proposed for multiple chemical sensitivities.

For insights gained about the origins of chemical sensitivity since the first edition, see the section on "Origins of Chemical Sensitivity" in Chapter 8.

PART · II

Mechanisms, Diagnosis, and Treatment

CHAPTER 4

Mechanisms of Multiple Chemical Sensitivities

Possible Physiological Mechanisms

A useful review of this topic can be found in Bell (1987b). The limited data available at this time suggest that any mechanism or model that would purport to explain the syndrome of multiple chemical sensitivities would need to address the features most closely associated with this illness:

1. Symptoms involving virtually any system in the body or several systems simultaneously

2. Differing symptoms and severity in different individuals, even those with the same exposure

3. Induction (that is, sensitization) by a wide range of environmental agents

4. Subsequent triggering by lower levels of exposure than those involved in initial induction of the illness

5. Concomitant food intolerances, estimated to occur in a sizable percentage of those with chemical sensitivities

6. "Spreading" of sensitivity to other, often chemically dissimilar substances; each substance may trigger a different constellation of symptoms

7. Adaptation (masking), that is, acclimatization to environmental incitants, both chemical and food, with continued exposure; loss of this tolerance with removal from the incitant(s); and augmented response with reexposure after an appropriate interval (for example, 4 to 7 days)

8. An apparent threshold effect referred to by some (including certain traditional allergists we interviewed) as the patient's *total load*. Total load is a theoretical construct that has been invoked by clinical ecologists to help explain why an individual develops this syndrome at a particular time. Illness is said to occur when the total load of biological, chemical, physical, and psychological stressors exceeds some threshold for the patient. This concept has emerged from clinical observations; no direct experiments have been done to test its validity in humans; however, animal models do exist. The concept aligns with Selye's (1946) work on the general adaptation syndrome.

Randolph knew Selye and was intrigued by his ideas but failed to see their clinical utility. Hans Selye (1977) was concerned with the general response of an organism to stressors. He observed that animals treated with a variety of toxic substances (indeed, Selye's supervisor accused him of spending his entire life studying the pharmacology of dirt) reacted in the same way: all had a generalized response involving increased endocrine activity and adrenal size, reduction of lymphatic tissue including the thymus (where T lymphocytes undergo maturation), and peptic ulcers in the stomach and duodenum). Selye defined stress as the nonspecific response of the body to any demand. Three phases occur:

1. An alarm reaction accompanied by increased adrenocorticotrophic hormone (ACTH) release

2. Resistance, during which demands upon the organism are met with little increase in ACTH or steroid hormone production; this stage corresponds with Randolph's adapted stage

3. Exhaustion, when animals succumb to stress, having expended their adaptive energy

The most characteristic feature of the stress syndrome, according to Selye, is its nonspecificity. All stresses (typhoid, cold, ecstasy, malnutrition) have their own features and causes, but all require the body to adjust to the demand for adaptation. When organisms succumb to excessive stress, their individual manifestations may differ somewhat, but the result is the same: as with a chain, there is always a weakest link, a point at which things are most likely to break down.

Randolph's concept of specific adaptation parallels Selye's, but differs in one key respect: adaptation in a given individual is specific to the incitant. Selye recognized this possibility but chose to focus on the more global aspects of adaptation. In Randolph's view, the accumulation of multiple stressors could overwhelm the organism's ability to adapt. In the exhaustion phase, Selye's general adaptation syndrome and Randolph's concept of specific adaptation coincide. From these ideas emerged the ecologists' concept of total load of stressors or incitants as the determinant of an individual's ability to adapt or failure to adapt. Ecologists consider Randolph's substance-specific and individual-specific view of adaptation to be the clinical counterpart of Selye's general adaptation syndrome (Randolph 1962, pp. 6–8; 1976a).

Rea probably has performed more clinical laboratory tests on chemically sensitive patients than any other clinical ecologist. When we asked him what mechanism he thought was responsible for these patients' illness, he responded, "Which one?" In his view, many interacting factors may be present. No single biochemical or immunological abnormality appears consistently in every patient. Some may have abnormal levels of immunoglobulins, complement, immune complexes, T-cells, B-cells, prostaglandins, kinins, serotonin, histamine, acetylcholine, vitamins, minerals, or detoxification enzymes (such as glutathione peroxidase) (Johnson and Rea 1989). Rea sees a great diversity of patients because he receives referrals of more difficult cases from other physicians. More clearly defined, homogeneous patient groups, such as those from a specific workplace, contaminated community, or tight building, might very well exhibit less variation in their laboratory profiles. (See discussion in Chapter 6.)

Before we examine some of the specific theories that have been proposed to explain multiple chemical sensitivity, three important points must be recognized:

1. The human body is an integrated system that traditionally has been separated into its component parts or systems to facilitate study. Interactions of these parts are necessarily more complex. For example, multiple chemical sensitivities conceivably could involve the entire neuroimmunoendocrine axis. Teasing out the subtle biochemical interactions involved in adaptation to the plethora of substances in the environment may be extremely difficult.

2. Traditional allergists who have studied sensitivity to industrial chemicals have been as baffled as the ecologists in trying to discern a mechanism for hyperreactivity. Butcher and co-workers (1982) remarked upon the continuing controversy over the mechanism for isocyanate hyperreactivity. Although an immunological theory has

been proposed, specific antibody is demonstrable in only 15 to 20 percent of reactive individuals. That antibodies also may persist beyond loss of reactivity casts doubt upon their role. More recently, Stankus, Salvaggio and associates (1988) demonstrated airway hyperreactivity to cigarette smoke among asthmatics who lacked specific IgE to tobacco smoke components. They report that the mechanism(s) behind this hyperreactivity remain unclear. These observations concerning the effects of cigarette smoke on some individuals parallel similar observations by clinical ecologists.

3. Although knowledge of the mechanism of a disease may be useful for developing better therapies, such knowledge is not a prerequisite for intervention. Preventing the development of multiple chemical sensitivities in those not yet afflicted may be possible by controlling environmental exposures that cause the initial sensitization.

The most frequently cited theories to explain chemical sensitivity involve the nervous system, the immune system, or the interaction between them because these two systems most clearly link the external environment and the internal milieu (Bell 1982). The rapid responsiveness of these systems also makes them attractive candidates because symptoms of food or chemical sensitivity have been reported to develop within seconds of exposure. As early as the 1940s and 1950s, the allergist Coca recommended sympathectomies (surgical interruption of certain sympathetic nerve pathways) in some cases of multiple food sensitivities, but benefits were often short-lived (Randolph 1987).

David Ozonoff (1989), professor of medicine and chief, Environmental Health Section, Boston University School of Public Health, suggests that because low levels of exposure do not trigger symptoms in everyone, perhaps a small initiating stimulus occurs, which the body of the chemically sensitive patient then amplifies or magnifies. In the case of multiple chemical sensitivities, the nervous system, the immune system, or both might amplify an external signal. Many chemicals, such as polybrominated biphenyls (PBBs) and trichloroethylene, affect both the nervous system and the immune system. Until 1980, the idea of a possible direct communication between the nervous and immune systems was widely debated. Subsequently, the existence of a neuroimmunoendocrine axis has been increasingly realized. Payan (1989, Payan et al. 1986) cites several discoveries that have helped to confirm the presence of two-way communication between the nervous and immune systems.

1. Studies show that neuropeptides (for example, substance P and somatostatin) and the nerve ganglia from which they arise project into immunological tissues.

2. Receptors for these neuropeptides occur on immunologically active leukocytes.

3. Certain immunologically active substances such as the interleukins can activate or be activated by cells in the nervous system.

4. Electrolytic lesions in the hypothalamus of animals produce distinct alterations in antibody production as well as abnormalities in the number and role of natural killer cells and T-lymphocytes. These alterations occur because of the interruption of a network of noradrenergic and peptidergic fibers that project into lymphoid tissues, including the thymus, spleen, Peyer's patches of the intestine, and bone marrow.

Therefore, the endocrine, immune, and nervous systems, once perceived as separate compartments, are increasingly recognized as interconnected.

Mechanisms Involving the Limbic System

The hypothalamus (part of the limbic system) has attracted considerable attention because it is the focal point in the brain where the immune, nervous, and endocrine systems interact (Bell 1982). Bell notes that assuming a direct cause-and-effect relationship would be premature, but that the hypothalamus could mediate food and chemical addictions in patients with multiple chemical sensitivities. The olfactory system has known links to the hypothalamus and other parts of the limbic system, which has led Bell (1982) to speculate that "the olfactory system, hypothalamus and limbic system pathways would provide the neural circuitry by which adverse food and chemical reactions could trigger certain neural, psychological and psychiatric abnormalities." Many different chemicals have been reported by clinical ecologists to trigger food cravings, binges, violence, or hypersexual activity. A model involving the hypothalamus could help to explain such behavioral changes in response to chemical exposures.

Some authors have alleged that psychological conditioning to odors is responsible for patients' reactions to chemicals. Of course, odor conditioning may occur in selected cases. (See the next section for a discussion of psychogenic mechanisms.) However, physiological mechanisms involving the limbic system may be at work. A direct pathway from the oropharynx to the brain and hypothalamic and limbic region has been demonstrated in rats (Kare 1968; Maller et al. 1967). Substances placed in the oropharynx migrated to the brain in minutes via a pathway other than the blood stream and in higher concentrations than if administered via the gastrointestinal tract, suggesting a direct route from mouth (or nose) to brain. Similarly, Shipley (1985) showed that inhaled substances that contact the nasal epithelium may cross into the brain and be distributed widely via transneuronal (through the nerve cell) transport. Thus, molecules that are inhaled and contact the olfactory apparatus could influence functions in other parts of the brain.

Bell (1990), a psychiatrist, notes that individuals who are shy tend to be more atopic (have a greater tendency to manifest allergies). She hypothesizes that chemical mediators, for example, histamine, VIP (vasoactive intestinal peptide), and prostaglandins, released in the nose in response to environmental agents may undergo transneuronal transport to the limbic system, temporal lobe, and other parts of the brain and there influence thought, mood, and personality traits such as shyness.

Ryan and co-workers (1988) studied 17 workers who attributed changes in thought processes, particularly memory and concentration difficulties, or changes in mood to their exposure to solvents. Those workers with "cacosmia" (a heightened sensitivity to odors) performed most poorly on neurobehavioral tests requiring verbal learning or visual memory. Although olfactory functioning was not tested objectively in this study, the authors felt their findings supported a hypothesis that chronic solvent exposure may affect the "rhinencephalic structures" (the primitive "smell" brain), the evolutionary precursor of the limbic system.

This phylogenetically ancient part of the brain (Fig. 4-1) is present in all mammals. It influences the organism's interaction with its environment in many subtle ways essential for preservation of the individual, its offspring, and the species. *Limbus* (Latin for "margin" or "rim") refers to its appearing like a rim around the edge of the cerebral hemispheres. Figure 4-2 shows its component parts. Note the close anatomical relationship to the olfactory bulb. Strong odors and even milder ones may provoke increased electrical activity in the amygdala and hippocampal areas of the limbic system (Monroe 1986). Subsensory exposure to chemicals can cause protracted, if not permanent, alterations in the electrical activity of the brain, beginning first with the most sensitive structures, particularly that portion of the amygdala that analyzes odors (Bokina 1976).

The amygdala is involved in feelings and activities related to self-preservation, such as searching for food, feeding, fighting, and self-protection (MacLean 1986). The cingulate gyrus appears to influence maternal care and nursing, separation cries between mother and offspring, and playful behavior, including wit and humor (MacLean 1986). The septum involves feeling and expression relating to procreation. Lesions in the septal area may cause hyperresponsiveness to physical stimuli (such as touching, sounds, or temperature changes), hyperemotionality, loss of motivation, excessive sugar and water intake, and fear of unfamiliar situations (Isaacson 1982).

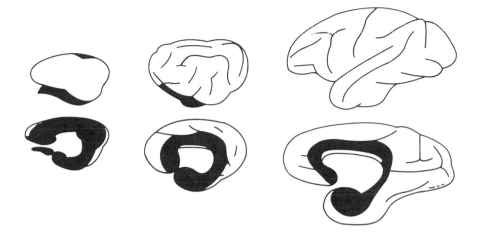

FIGURE 4-1. The cortex of the paleomammalian brain (limbic system) is contained in the great limbic lobe surrounding the brain stem. Shown in black is the location and relative size of the limbic lobe in the brains of the rabbit, cat, and monkey; the ring of limbic cortex is found as a common denominator in the brains of all mammals. The surrounding cortex of the neomammalian brain which undergoes a rapid expansion in evolution is shown in white.
Source: MacLean, P. D., "The Brain in Relation to Empathy and Medical Education," Journal of Nervous and Mental Disease (1967) 144(5):374–382: Williams & Wilkins, Baltimore, Maryland, p. 377.

The hippocampus appears important for laying down new memories and thus is essential for learning (Gilman 1982). Hippocampal lesions may cause difficulty in retaining recent memories (Isaacson 1982). The hippocampus, at the intersection of numerous neural pathways and in a critical position to affect the transfer of information from one brain region to another, acts as an information switching center. Learning and memory decrements are a frequent consequence of exposure to toxic substances, and some researchers view the hippocampus as a prime target for such toxins (NAS 1990, Walsh 1988). Damage to the hippocampus itself, or to nerves leading to or from it, may adversely affect the synthesis, storage, release, or inactivation of the excitatory and inhibitory amino acids that serve as neurotransmitters in this region of the brain. Toxins may disrupt the delicate balance of these amino acids, perhaps leading to the release of a flood of excitatory neurotransmitters that damage neighboring cells, a phenomenon that has been called *excitotoxicity* (U.S. Congress 1990). Relatively small perturbations of hippocampal function may have large and long-lasting effects upon behavior and cognition (Walsh 1988).

The most vital component of the limbic system, the hypothalamus,

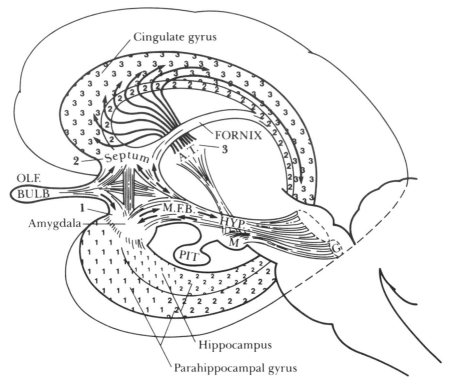

FIGURE 4-2. Three major subdivisions of the limbic system. The small numerals 1, 2, and 3 overlie, respectively, the amygdaloid, septal, and thalamocingulate divisions. The corresponding large numerals identify connecting nuclei in the amygdala, septum, and anterior thalamus. Abbreviations: AT, anterior thalamic nuclei; G, tegmental nuclei of Gudden; HYP, hypothalamus; M, mammillary bodies; MFB, median forebrain bundle; PIT, pituitary; OLF, olfactory.
Source: MacLean, P. D., "A Triune Concept of the Brain and Behavior," in Boag, T., and Campbell, D., The Hincks Memorial Lectures (1973), University of Toronto Press, Toronto, Ontario, p. 15.

governs: (1) body temperature via vasoconstriction, shivering, vasodilation, sweating, fever, and behaviors such as moving to a cooler or warmer environment or putting on or taking off clothing; (2) reproductive physiology and behavior; (3) feeding, drinking, digestive, and metabolic activities, including water balance, addictive eating leading to obesity, and complete refusal of food and water leading to death; (4) aggressive behavior, including such physical manifestations of emotion as increased heart rate, elevated blood pressure, dry mouth, and gastrointestinal responses (Gilman 1982).

The hypothalamus is also the locus at which sympathetic and parasympathetic nervous systems converge. Many symptoms experienced by patients with food and chemical sensitivities relate to the autonomic (sympathetic and parasympathetic) nervous systems; for example, altered smooth muscle tone produces Raynaud's phenomenon, diarrhea, constipation, and other symptoms. Recently, Rea and his co-workers acquired an iriscorder (to be distinguished from iridology) from the Japanese who have used this instrument to measure pupillary reactivity in persons with organophosphate pesticide toxicity. A brief pulse of intense light evokes pupillary constriction (a parasympathetic response) followed by dilation (a sympathetic response). With the aid of this device, they are attempting to monitor objectively sympathetic and parasympathetic nerve function in persons with chemical sensitivities.

The hypothalamus appears to influence anaphylaxis and other aspects of immunity (Stein 1981). Likewise, antigens may affect electrical activity in the hypothalamus (Besedovsky 1977).

Thoughts arising in the cerebral cortex that have strong emotional overtones can trigger hypothalamic responses and recreate the physical effects associated with intense anger, fear, and other feelings. To implement its effects, the hypothalamus not only has a direct electrical output to the nervous system but also produces its own hormones, many of which stimulate or inhibit the pituitary's production of hormones (Gilman 1982).

Corwin (1978) describes the complexity and delicacy of hypothalamic function.

> Imagine, if you will, a chemical laboratory set up to monitor a stream of fluid continuously. This laboratory is equipped to perform analyses for simple substances such as acids, bases, and salts, ions, such as sodium, potassium, calcium, and chloride, and more complex substances such as glucose and cholesterol, simple hormones such as adrenaline and thyroxine, more complex hormones such as the peptide hormones. This laboratory has a built-in computer which evaluates the balance between all these substances and controls this balance so that in case one is formed in excess, the valves can be tightened electrically to decrease the production or to generate an antagonist. In addition, this laboratory has the means for the synthesis of organic chemicals which can be released into the flowing stream of fluid at appropriate points to alter the action of the way stations producing the desired or undesired materials. . . . The human body has its analytical laboratory, computer controller, and hormone factory compressed into a few grams of hypothalamus, a true marvel of microminiaturization.

Most of the neural input to the hypothalamus comes from the nearby limbic and olfactory areas (Isaacson 1982). Lesions in the limbic region may be associated with irrational fears, feelings of strangeness or unreality, wishing to be alone, and sadness (MacLean 1967). A feeling of being out of touch with or out of control of one's feelings and thoughts, not unlike that described by many patients with chemical sensitivity, may be perceived.

Doane (1986) describes potential difficulties for patients with limbic dysfunction.

> Activity controlled by the limbic system may seem largely irrational and often is not perceived within one's self in ways that are easily understood or communicated in verbal language. These observations do not detract, however, from the reality of limbic determinism in human psychic functions and psychiatric disorders.

The dynamic involvement of the hypothalamus and limbic system in virtually every aspect of human physiology and behavior makes injury to these structures an intriguing hypothesis to explain chemical sensitivity's myriad manifestations. Rich neural connections lie between the olfactory system and the limbic and temporal regions of the brain. Surgical or epileptic patients with damage to the limbic or medial temporal portions of the brain may have persistent alterations in odor perception (for example, an unusual smell that characteristically precedes seizure activity) as well as learning and memory difficulties (Ryan 1988).

Bell (1990) hypothesizes that chemically sensitive patients may have olfactory-limbic-temporal pathways that are more easily "kindled." In other words, a small signal or insult would more readily trigger firing of nerve cells in brain regions where kindling was present. Kindling might be enhanced by genetic endowment, prior environmental exposures, psychological stress, hormonal variations, or other factors. Unlike surgical ablation, which destroys a brain area, kindling is a kind of stimulatory lesion (Girgis 1986). Stimulation of the amygdala with electrodes may elicit rage or loss of control of emotions, a phenomenon frequently reported by patients with multiple chemical sensitivities.

Kindling has been described previously in the context of seizures. The amygdala, for example, which is particularly susceptible to electrical discharge following either electrical (Girgis 1986) or chemical provocation (Bokina 1976), is subject to long-lasting alteration when given repeated stimuli. Very potent or repeated stimuli, whether electrical or chemical, may permanently augment the tendency for neurons to fire in the presence of future stimuli, even when challenged with much lower levels than those originally involved. Girgis (1986) reports a decrease in acetylcholinesterase (AChE), an enzyme that breaks down the neurotransmitter acetylcholine in junctions between nerve cells, that parallels the increase in supersensitivity to stimuli. The limbic system is especially rich in AChE, which is strongly bound to the nerve cell membranes and very stable. The AChE may play a protective role by enzymatically maintaining acetylcholine concentrations at nerve junctions within safe bounds and protecting susceptible cells in the limbic system from developing "bizarre sensitivity" (Girgis 1986). Interestingly, physicians who treat patients with multiple chemical sensitivities have noted some of the most severe and debilitating exposures for these patients have been to organophosphate pesticides, which inhibit AChE.

Bokina (1976) found impaired speed of execution and coordination of complex motor processes in humans repeatedly exposed to carbon disulfide for 10- to 15-minute intervals at subsensory levels. Animals primed by high concentrations of various chemicals (such as formaldehyde and ozone) and subsequently reexposed to low concentrations of the same chemicals showed an increased tendency toward paroxysmal electrical discharge in the amygdala (Bokina 1976). Bokina observed that although the chemicals he used to sensitize the animals were *different in terms of their structure and physical and chemical properties, their effects upon the limbic system were remarkably similar.*

Kindling could help to explain the apparent loss of adaptive capacity in multiple chemical sensitivity. Formerly well-tolerated low-level exposures to, for example, tobacco smoke or perfume might trigger symptoms in individuals whose limbic areas have been kindled by a prior exposure. Likewise, spreading of sensitivities to chemically unrelated substances might be understood on this basis.

One intriguing aspect of the limbic system as a mechanism for multiple chemical sensitivities is its responsiveness to both chemical and cortical stimuli. Therefore, conscious thought processes and emotional states influence limbic activity just as chemical or physical stimuli can. The former may be under more or less conscious control of the individual, whereas the latter are almost entirely unconscious and automatic. However, conscious efforts that play into the delicate circuitry of the limbic system may be able to alter or suppress concurrent electrical activity evoked by environmental agents. Nevertheless, very potent exposures may not be susceptible to conscious will. Some patients with chemical sensitivities report being able to "will" their way out of a mild reaction to a food or chemical and attempt to control their symptoms in this manner. Most say such efforts do not work for their most problematic incitants. In fact, the ability to exercise any conscious effort, even that of simply getting away from the exposure or taking alkali salts or a neutralizing dose, may be lost during a reaction. Monroe (1986) reported the case of a man for whom exposure to the odor of stale beer caused greatly increased electrical activity in the limbic system (amygdala and hippocampal areas). Various memories, some associated with beer, also increased electrical activity in the same region. However, simple

arithmetic computations would immediately stop such activity. Therefore, conscious thought processes could override preexisting activity in
the limbic system.

An intriguing example of the competing effects of exposure and psychological state has been reported by Sanderson (1989; for more detail
see in Appendix A "Neurobehavioral and Psychiatric Manifestations").
Carbon dioxide at levels greater than 5 percent in the air has been
shown to induce panic attacks ("fight or flight" responses depend upon
limbic activity) in patients suffering from panic disorder. While this effect
is not chemical sensitivity, the fact that patients in this study who believed
that they had control over the carbon dioxide level to which they were
exposed had fewer and less intense panic disorder symptoms suggests that
psychological factors (the illusion of control) can indeed mitigate the
biological response to an environmental stressor.

Thus, experimental evidence suggests a delicate interplay occurring
in the limbic region. Conceivably, chemicals contacting olfactory nerve
projections in the nose could either be transported into or relay electrical
signals to the limbic region, leading to a vast array of symptoms. Likewise, thought processes and mood states trigger limbic activity or may,
in some cases, interrupt preexisting limbic activity. At present, however,
no evidence suggests that limbic activity triggered by environmental
exposures can be entirely overcome by psychological interventions.

Immunologic Mechanisms

Immunological alterations are another possible explanation for multiple
chemical sensitivities. Most of us have been taught that the immune
system evolved to defend our bodies against microorganisms and other
"foreign invaders." In truth, the immune system may have evolved to
help control the body's internal milieu. Thus, its purpose is not simply to
ward off infections but also to carry out precise regulatory interactions
between the immune system, endocrine system, and nervous system.

Scientists are only beginning to learn which chemicals affect our
immune system and what those effects mean in terms of our health. Animal experiments demonstrate immunotoxicity from halogenated
aromatics, heavy metals, and organochlorine pesticides (Cone et al.
1987). Accidental human exposures to aldicarb, polybrominated biphenyls (PBBs), dioxin, and other toxins also provide data that chemicals can impact the immune system. Volumes, for example, Sharma
(1981), have been written on the subject of immunotoxicology. Descotes
(1986) has attempted to catalog the extensive published literature on
the immunomodulatory action of chemicals and drugs. Of special concern to allergists and some clinical ecologists have been Levin and

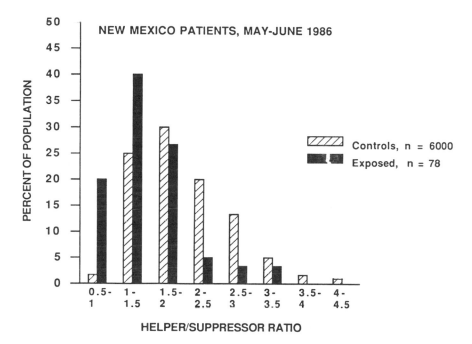

FIGURE 4-3. *New Mexico patients, May–June 1986. Helper/suppressor ratios obtained by standard clinical laboratory procedures in 78 injured workers from a computer chip manufacturing plant in Albuquerque, New Mexico, compared with the standard laboratory control population of 6,000 randomly selected asymptomatic people. The exposed population is statistically significantly different from the controls (chi-square = 39.34063: p = 2.62 × 10< − 6>).* Reprinted with permission from Levin, A. and V. Byers, "Environmental Illness: A Disorder of Immune Regulation," in Workers with Multiple Chemical Sensitivities, M. Cullen, Ed. *(copyright 1987, Hanley & Belfus, Inc., Philadelphia, PA), p. 672.*

Byers's (1987) assertions that environmental illness is a disorder of immune regulation. These authors point to decreased T-lymphocyte helper-suppressor ratios in four different populations exposed to environmental toxins. Figures 4-3 through 4-6 depict helper-suppressor ratios for these four groups.

Note that all four figures show a shift to the left (decrease) of the ratio of helper to suppressor T-lymphocytes.

The Woburn, Massachusetts, data (Fig. 4-5) is taken from 25 surviving family members of leukemia patients, all of whom drank water from wells contaminated with industrial solvents. Not only did these individuals have a statistically significant reduction in their T-cell helper-to-suppressor ratios (1.49 versus 1.94 in age- and sex-matched asymptomatic controls, $p < 0.01$) but also 48 percent (11 of 23) tested positive for autoantibodies. In addition, 88 percent (22) had frequent or chronic sinusitis or rhinitis, and 52 percent (13) had gastrointestinal complaints

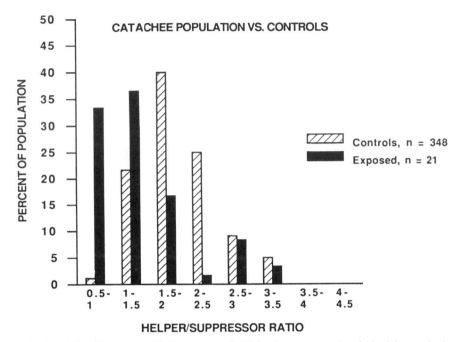

FIGURE 4-4.   Catachee population vs. controls. Helper/suppressor ratios obtained by standard clinical laboratory procedures on 21 environmentally ill patients who were domestically exposed to high levels of polychlorinated biphenyls (PCBs) over a period of 5 to 10 years in Catachee, South Carolina, compared with the standard laboratory control population of 348 asymptomatic individuals. The exposed population is statistically significantly different from the controls (chi-square = 63.48208: p = 1.37 × 10<—6>).   Reprinted with permission from Levin, A. and V. Byers, "Environmental Illness: A Disorder of Immune Regulation," in Workers with Multiple Chemical Sensitivities, *M. Cullen, Ed. (copyright 1987, Hanley & Belfus, Inc., Philadelphia, PA), p. 673.*

often described as irritable bowel syndrome. Rashes were frequent. Fourteen adults complained of rapid heart rate at rest, palpitation, or near syncope; of 11 who underwent cardiac workups, eight had multifocal premature ventricular contractions, and six were felt to need cardiac medications (Byers et al. 1988).

What is remarkable are the many similarities between the Woburn data and data gathered by Johnson and Rea (1989) on patients who have been worked up in the ecologists' environmental unit in Dallas. Of 150 ecology patients, 19 percent were positive for antinuclear antibodies. Many others had antithyroglobulin or other autoantibodies. In addition, the polysymptomatic complaints of the Woburn study group resemble those of the ecology patients. However, differences are present too. In 70 ecology patients with vascular dysfunction, the T_4-T_8 (helper-suppressor) ratio was *increased* (2.20) versus 60 controls (1.70) ($p = 0.001$).

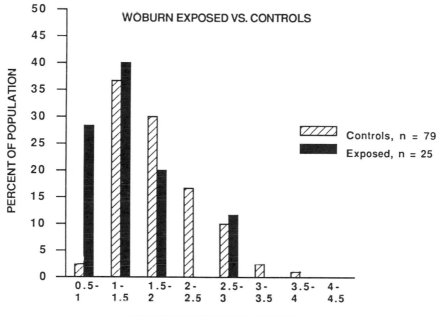

FIGURE 4-5. *Woburn exposed population vs. controls. Helper/suppressor ratios obtained by standard clinical laboratory procedures on 25 environmentally ill patients from Woburn, Massachusetts, who were domestically exposed to trichloroethylene (TCE) over a period of 5 to 10 years compared with age and sex matched asymptomatic controls. The exposed population was statistically significantly different from the controls (chi-square = 42.18912: p = <1 × 10 <—8>). This control population is not significantly different from the standard laboratory controls used in the other studies Reprinted with permission from Levin, A. and V. Byers, "Environmental Illness: A Disorder of Immune Regulation,"* in Workers with Multiple Chemical Sensitivities, M. Cullen, Ed. *(copyright 1987, Hanley & Belfus, Inc., Philadelphia, PA), p. 674.*

Seven rheumatoid arthritis patients showed similar increases in T_4-T_8, whereas 27 asthmatics showed no significant differences from controls. Why certain individuals have increased T_4-T_8 ratios while others have decreased ratios is unclear. Perhaps differences exist in the kinds of patients in these studies, the exact nature of their exposures, or the time elapsed since exposure. Interestingly, cigarette smoking, which is well recognized for its long-term adverse health consequences, recently has been linked to an increased number of T_4 (helper) cells (p = 0.002) and an increased T_4-T_8 ratio (p = 0.02) (Tollerud et al. 1989).

Such data warrant further studies employing carefully matched controls. Levin's work has stirred considerable controversy among allergists and clinical ecologists. His focus on the immune system has drawn allergists and immunologists into the fray because it is their area of

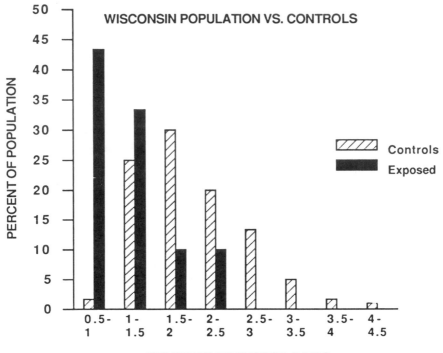

FIGURE 4-6.   *Wisconsin population vs. controls. Helper/suppressor ratios obtained by standard clinical laboratory procedures on 10 environmentally ill patients from rural Wisconsin who were domestically exposed to a variety of industrial dyes, solvents, and pesticides over a 5 to 10 year period compared to the standard laboratory control of 6,000 randomly selected asymptomatic people. The exposed are significantly different from the controls (chi-square = 73.58482: p = 4.77 × 10 <— 6>). Reprinted with permission from Levin, A. and V. Byers,* "Environmental Illness: A Disorder of Immune Regulation," *in* Workers with Multiple Chemical Sensitivities, M. Cullen, Ed. *(copyright 1987, Hanley & Belfus, Inc., Philadelphia, PA), p. 675.*

specialization. Levin and traditional allergists often serve as expert witnesses on opposing sides in lawsuits and disability evaluations. Terr (1986) asserts that immune parameters of patients who have seen clinical ecologists fall within expected normal ranges "except for several patients who had immunoglobulin (IgA) and lymphocyte levels above the normal range, reflecting a history of infections." Levin (1989) counters by arguing that only 2.5 percent of the population would be expected to fall outside the normal range, whereas in Terr's data 20 to 30 percent of these patients fall outside the normal range.

As can be seen from Figures 4-3 through 4-6, individual data points (a single individual's helper-suppressor ratio) may be difficult to interpret as normal or abnormal because the ranges for normal data are quite wide. However, when one looks at an entire exposed population, the data appear to be skewed to the left, that is, toward a *reduced* helper-suppressor ratio. Levin (1989) suggested about some recent data that the reduced helper-suppressor T-lymphocyte ratios seen in many chemically exposed individuals may be the result of increased numbers of cytotoxic lymphocytes in the suppressor cell population of these patients. Such cytotoxic lymphocytes could be "reactively cloned in response to a somatically transformed cell," that is, a cell somehow transformed by a chemical agent.

Similarly, Broughton and associates (1988) report T-lymphocyte activation, particularly increases in Ta1 positive T-lymphocytes, in persons exposed to formaldehyde, even after exposure has ceased. The Ta1 positive cells are elevated in certain autoimmune disorders including multiple sclerosis (Hafler 1985) and juvenile-onset diabetes (Jackson 1982) and after immunization (Yu 1980). These activated T-lymphocytes may serve as an index of immunological stimulation (Yu 1980). Broughton and co-workers (1988) propose that lymphocyte activation may continue in spite of formaldehyde avoidance by these chemically sensitive individuals as a result of the emergence of cross-sensitivities to other environmental chemicals that are not being avoided.

The idea that relatively low-molecular-weight chemicals can somehow alter native protein, perhaps by acting as haptens, and elicit a sort of autoimmune response to that altered protein is gaining support. Formaldehyde (Thrasher et al. 1987), trimellitic anhydride (Akiyama et al. 1984), isocyanate (Butcher et al. 1982), hydantoins (Kammuller et al. 1988), which are present in many drugs and foods, and hydrazine (Reidenberg et al. 1983), which occurs in mushrooms, plastics, pesticides, tobacco smoke, and various drugs, have all been reputed to cause immune derangement, possibly by such a mechanism.

Broughton, and Thrasher (1988) have studied more than 200 cases involving formaldehyde exposure and reported the development of antibodies to formaldehyde-albumin conjugates, evidence of immune system activation (activation marker Ta1 on T-lymphocytes), low titers of a variety of autoantibodies, and altered IL-1 (interleukin) production in these individuals, suggestive of "subtle but chronic activation of the immune system" (Broughton et al. 1990). They state that similar alterations occur in patients exposed to chlordane (a termiticide) and solvents in drinking water. Testing blood from symptomatic patients with a history of chemical exposure, Broughton and Thrasher noted a much higher incidence of autoantibodies among these patients than among controls (Broughton 1990). Individuals exposed to formaldehyde, trichloroethylene, and chlordane were tested for antinuclear antibody

(ANA) and autoantibodies to parietal cells, smooth muscle, brush border, mitochondria, and myelin. The clinical significance of some of these antibodies, particularly in low titers, is not known, but the differences between exposed and unexposed groups are striking and warrant further investigation (Table 4-1). Exposed groups were three to four times more likely than controls to have one or more autoantibodies present. In addition, the incidence of autoantibodies appeared higher when exposure was ongoing than when some time had elapsed. Many such chemicals are metabolized in the liver via the cytochrome P450 enzyme system and excreted in the urine. However, these authors speculate, highly reactive intermediate compounds, such as epoxides, formed during processing may damage liver cells and signal immune system cells to enter the area and clean up the debris. Macrophages (the "trash collector" cells of the immune system) produce interleukin-1, which can evoke flulike symptoms. Cell fragments, previously hidden to the immune system, may also trigger autoantibody production.

Antibodies to albumin conjugates of formaldehyde, tolulene diisocyanate, and trimellitic anhydride were reported in symptomatic workers in a newly remodeled building (Thrasher et al. 1989). Patterson and co-workers (1989) at Northwestern University Medical School analyzed blood from symptomatic individuals exposed to formaldehyde via inhalation and from hemodialysis patients who had intravenous exposure to formaldehyde. In striking contrast to Broughton and associates, they found no correlation between the presence of IgG antibodies to either formaldehyde or formaldehyde conjugated to albumin and the symptoms from formaldehyde exposure.

Some individuals exposed to toluene diisocyanate (TDI), used in the manufacture of urethane foams and plastics, develop respiratory difficulties and may experience symptoms upon reexposure to even very low concentrations of TDI. Yet, the majority of TDI-sensitive individuals do not have antibodies to TDI in their blood. Conversely, the presence of antibodies to TDI in the blood is felt to reflect prior exposure, not illness resulting from that exposure. Some, however, argue that antibodies to foreign chemicals do not occur naturally and that subclinical effects— effects not detectable using current diagnostic tools but nevertheless real —may be occurring.

Future governmental and scientific investigations must include measurement of T- and B-cell numbers, lymphocyte activation, and other relevant immune parameters as possible indices of toxicity. Levin and others have helped draw attention to the need for these data.

Clinical ecologists have examined other indicators of immune system function for their relevance. Commonly, IgE levels are normal or even low in their patients, but some are elevated. According to ecologists,

TABLE 4-1. Autoantibody Presence Among Chemically Exposed Groups

			% of These Groups Reporting Symptoms						
Chemical	Symptoms Reported	Autoantibodies Present[a]	Controls (n=28)	Mobile-home Dwellers (n=19)	Office Workers (n=20)	Occupationally exposed (n=6) Workers	Currently Exposed (n=39)	Removed from Exposure (2 years) (n=39)	Exposed (n=20)
Formaldehyde	Mucosal irritation, fatigue, flu-like syndrome	1 or more	21	89	80	66			
		2 or more	7	58	45	33			
		3 or more	0	37	15	33			
		4 or more	0	0	0	0			
Trichloroethylene[b]	Fatigue, flu-like syndrome, cognitive difficulties	1 or more	20				91	41.2	
		2 or more	0				43	26	
		3 or more	0				30	12.8	
		4 or more	0				0	0	
Chlordane[c]	Fatigue, flu-like syndrome, cognitive difficulties	1 or more	21						95
		2 or more	7						60
		3 or more	0						35
		4 or more	0						10

[a] Autoantibodies measured for the formaldehyde and chlordane-exposed control groups were ANA (antinuclear antibody) and antiparietal cell, brush order, mitochondria, and smooth muscle. For the trichloroethylene-exposed and control groups, antimyelin antibodies were also measured.

[b] Exposed through drinking water.

[c] Exposed in individuals' homes.

Sources: Broughton, A., Thrasher, J. D. Personal communication, 1990. Broughton A., Thrasher, J. D., Madison, R., "Chronic Health Effects and Immunological Alterations Associated with Exposure to Pesticides," *Comments Toxicology* (in press), 1990. Thrasher, J. D., Broughton, A., Madison, R., "Immune Activation and Autoantibodies in Humans with Long-term Inhalation Exposure to Formaldehyde," *Archives of Environmental Health* (in press, July/August), 1990.

abnormal activation of the complement system may occur; increased autoantibodies may be present. Lymphokines, prostaglandins, kinins, and a host of other mediators may be affected, but, again, none of these applies for all patients (Johnson and Rea 1989).

McGovern (1983), a clinical ecologist, challenged six normal controls and six patients who had multisystem clinical syndromes with foods or chemicals to which they were sensitive and monitored prechallenge and postchallenge blood levels of serotonin, histamine, epinephrine, norepinephrine, dopamine, immunoglobulins, immune complexes, complement, and prostaglandins. Patients included five females and one male, 25 to 75 years of age. Absolute lymphocyte counts for all patients were low or at the low end of normal ranges, that is, 700 to 1100 (normal range: 900 to 2900). Four patients were tested by feeding them a single food to which they were sensitive; one was exposed to 1 ppm of phenol for 5 minutes, and another was exposed to the emissions of a photocopy machine for 5 minutes. Controls underwent identical challenges but reportedly had no symptoms. Unfortunately, challenges were not blinded. Within 15 to 30 minutes after challenge, all patients developed some abnormal *physical findings,* for example, asthma, tachycardia, ataxia, ophthalmoplegia, finger swelling, cough, rhinitis, or shaking chills. Symptoms persisted for 2 to 8 hours after challenge. The results are shown in Figure 4-7.

From the graphs, levels of serotonin, histamine, complement, and immune complexes following provocative challenge appear to be more stable in controls, varying no more than about 10 percent from baseline, except for an early rise in serotonin in controls. In contrast, patients' responses appear far more variable.

The authors speculate that the patients appeared to be having their reactions via various immunological pathways, some type I (IgE-mediated), others type III (IgG-mediated), and some both. Drawing conclusions regarding mechanisms from this paper is difficult, but certainly future challenge studies, preferably blinded, will need to investigate alterations in a panoply of biochemical and immunological markers. Studies of this kind require an enormous amount of preparation and involve costly laboratory analyses, but they are needed to document reactions and elucidate mechanisms.

An intriguing paper concerning the effects of aldicarb (a widely used carbamate insecticide and nematocide) on the immune system of mice demonstrated that aldicarb in the drinking water suppressed the immune response (to sheep red blood cells) more at 1 ppb than at 1,000 ppb (Olson et al. 1987)! This result is a surprising departure from classical toxicological dose-response curves, where dose and toxicity increase together (see Chapter 1). The experiment was carried out several times with two mouse strains and two sources of aldicarb, with the same result. The animals did not die or develop the opportunistic infections usually associated with immune deficiency; however, the authors comment that "such animals will usually not survive a frank challenge with a virulent microorganism." They speculate on the reason for the inverse dose-response curve for aldicarb:

> This phenomenon may be associated with dose related detoxification/ elimination in the intestinal tract or body, differential rates of clearance by the kidneys, or possibly the clearance of antigen aided by antibody (induced through conjugation of the chemical to naturally occurring proteins and followed by elicitation of specific antibodies (Olson et al. 1987).

Conceivably, then, lower levels of toxic substances could be more damaging than higher levels, perhaps because damage from the former is so slight that usual cell repair mechanisms are not triggered and the damage becomes permanent.

Biochemical Mechanisms

Rea and other ecologists have noted vitamin and mineral abnormalities in many of their patients (Johnson and Rea 1989; Rogers 1990). Their detractors argue that these patients are often sick, debilitated, and malnourished, and therefore such findings are not surprising. Such a contention is difficult to disprove, even if it were incorrect.

Individuals who have defective enzyme detoxification systems may be more susceptible to low level exposures. Ecologist Rogers reasons that chemically sensitive individuals must have defective detoxification pathways, because others in the same environment tolerate the same exposures without symptoms. Rea has noted that many of his chemically sensitive patients have decreased levels of detoxifying enzymes, such as glutathione peroxidase. This possibility is particularly intriguing because such enzyme systems are inducible (that is, can be stimulated) and thus might conform to an adaptation hypothesis. Scadding and associates (1988) noted poor sulfoxidation ability in 58 of 74 patients with well-defined reactions to foods versus 67 of 200 normal controls ($p < 0.005$). Similarly, Reidenberg and co-workers (1983) reported the case of a laboratory technician who developed a lupuslike disease in response to hydrazine. She was genetically a slow acetylator, which, they felt, might have predisposed her to developing a lupuslike disorder after sufficient exposure to an inciting chemical. A deficiency of one or more particular enzymes could help to explain why some

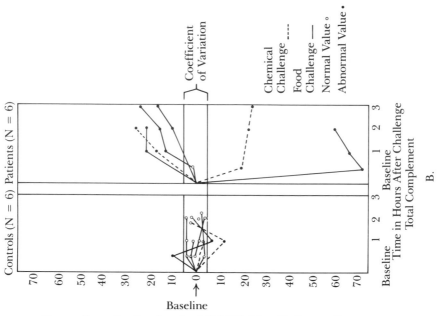

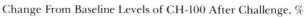

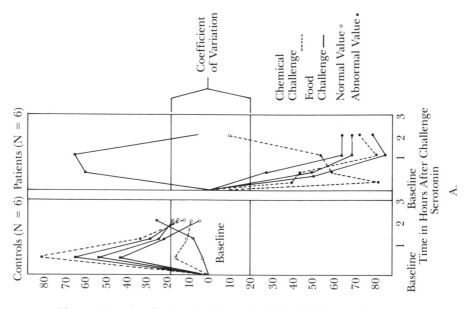

108

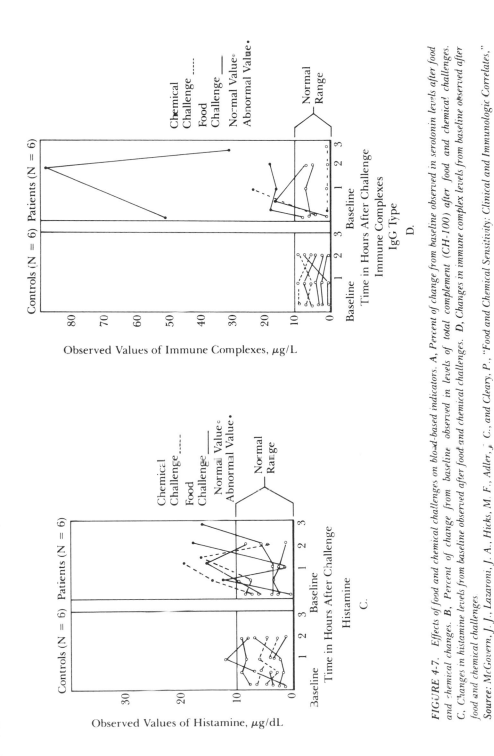

FIGURE 4-7. *Effects of food and chemical challenges on blood-based indicators. A, Percent of change from baseline observed in serotonin levels after food and chemical changes. B, Percent of change from baseline observed in levels of total complement (CH-100) after food and chemical challenges. C, Changes in histamine levels from baseline observed after food and chemical challenges. D, Changes in immune complex levels from baseline observed after food and chemical challenges*

Source: McGovern, J. J., Lazaroni, J. A., Hicks, M. F., Adler, J. C., and Cleary, P., "Food and Chemical Sensitivity: Clinical and Immunologic Correlates," Archives of Otolaryngology 109:292–297, p. 296 (copyright 1983, American Medical Association).

patients are more susceptible to foods and chemicals than others. Further, damage by a toxin might compromise detoxification pathways so that other substances formerly metabolized by this pathway could not be degraded properly and thus might provoke symptoms at low exposure levels (a hypothetical basis for the spreading phenomenon).

Levine (1983) has proposed that environmental sensitivities are the result of toxic chemicals reacting with cell constituents to create free radicals (which are formed when a molecule loses an electron). He hypothesizes that if an antioxidant molecule (such as vitamins A, C, E, and selenium) is not present nearby to supply the missing electron, then an electron may be removed from an unsaturated lipid (lipid peroxidation) in a cell membrane, leading to membrane damage, release of prostaglandins and other inflammatory mediators, and formation of antibodies to chemically altered tissue macromolecules.

In 1950, Randolph, collaborating with a surgeon patient of his, Harry G. Clark, published an abstract on the "acid-anoxia-endocrine theory of allergy." Clark, who had food sensitivities, felt that in view of the speed of acute food reactions, changes in electrolytes must be involved (Randolph 1987). Clark knew that allergy was often associated with edema and that one-celled marine organisms swell when acidified; from this he reasoned that because the end products of digestion are acids, perhaps in food-sensitive individuals these acid products of catabolism accumulate intracellularly more rapidly than they can be neutralized by the more alkaline extracellular fluid (including pancreatic bicarbonate). He thus surmised that treatment with alkali salts (that is, bicarbonate salts of sodium and potassium) might be helpful. Indeed, Randolph and Clark found that if alkali salts were administered shortly after an acute food reaction, symptoms were dramatically relieved for many patients. Almost 30 years later, this form of treatment is still used for acute food reactions by clinical ecologists because of its efficacy in many patients. (See further discussion in Chapter 5.)

Vascular Mechanisms

Rea, who began his medical career as a cardiothoracic surgeon, hypothesizes that blood vessel constriction, inflammation, or leakage in multiple organ systems may explain the bizarre combinations of symptoms in these patients. In his view, particular complaints may simply mirror the site and size of affected blood vessels. Spasms in large-caliber arteries, either acutely or chronically, could reduce blood supply to an organ or limb and result in dysfunction, pain, or even necrosis (Rea 1975). Chemical injury to the fragile walls of smaller vessels, however, would be more likely to cause hemorrhage (resulting in petechiae and bruises) or edema

(Rea 1979a). The walls of blood vessels contain smooth muscle. Rea notes that if a patient's symptoms are not explainable by vascular involvement, then other tissues containing smooth muscle such as the respiratory, gastrointestinal, and genitourinary systems are frequently implicated (Rea 1977). Impaired blood vessels or altered smooth muscle function are attractive hypotheses that may explain the diverse and seemingly unrelated symptoms occurring in patients with multiple chemical sensitivities. In the case of either blood vessel or smooth muscle dysfunction, clearly, neurological and immune alterations may play primary roles. A vascular hypothesis might also explain why patients may experience increased pain or other symptoms at the site of an earlier injury or surgery, where blood flow may be relatively compromised.

A final comment regarding the association of food sensitivities with chemical sensitivities is that foods are aggregates of chemicals (Bell 1982; Kammuller et al. 1988), as Table 4-2 demonstrates. The human diet is an important source of exposure to both low- and high-molecular weight compounds. The antibodies to foods that are present in the blood of many individuals attest to the fact that molecules from foods do leave the gut and enter the bloodstream. Thus, any mechanism for the development of sensitivities that might be proposed for chemicals could pertain to foods as well. Butcher and associates (1982) evaluated a worker with TDI sensitivity who could not eat radishes. One bite of a small radish caused severe, immediate bronchoconstriction with a 75 percent decrease in FEV_1, 5 minutes after challenge and necessitated epinephrine treatment. When 26 months later this individual was again able to tolerate isocyanates, he was challenged with 14 grams of radish with no ill effects. The authors note that radishes contain allyl isothiocyanate and benzyl isothiocyanate. However, these chemicals are also present in other foods that the patient was able to eat without adverse effects. This example illustrates a possible connection between sensitivities to environmental chemicals and sensitivity to particular foods. Many similar cases of coexisting food and chemical intolerance have been cited by clinical ecologists. Although their work is often dismissed as "anecdotal," only through observations like these can patterns be discovered, which in turn suggest a hypothesis, which then leads to experiments to prove or disprove that hypothesis. We are currently at the pattern-recognition stage with regard to multiple chemical sensitivities. Finding a mechanism to explain these patterns lies down the road.

TABLE 4-2. Chemical Constituents of Tomato, Apple, Milk, and Orange

Food	Component	Olfactory Threshold (Parts per Billion)
Tomato[a]	Hex-cis-3-enal	0.25
	Deca-trans, trans-2, 4-dienal	0.07
	Dimethylsulfide	0.33
	β-ionone	0.007
	Linalool	6
	Guaiacol	3
	Methyl salicylate	40
	2-Isobutyl thiazole	3.5
Apple[b]	Ethanol	100
	Hexanol	0.5
	Hexanal	0.005
	2-Hexanal	0.017
	Butyl Acetate	0.066
	2-Methylbutylacetate	0.005
	Ethyl 2-methylbutyrate	0.0001
	Hexyl acetate	0.002
Milk[c]	p-Cresol	
	4-Ethylphenol	
	3-n-Propylphenol	
	Phenylacetic acid	
	Hippuric acid	
	Caprylic acid	
	Palmitic acid	
Orange[d]	α-pinene	
	Myrcene	
	Limonene	
	Linalool	
	Cis-2, 8-p-menthadien-l-ol	
	Decanal	
	Carvone	
	Valencene	

[a] From Buttery, R. G., Seilfert, R. M., Guadagni, D. G., and Ling, L. C. Characterization of additional volatile components of tomato. *J Agr Food Chem.* 19(3):524–29, 1971.

[b] From Flath, R. A., Black, D. R., Guadagni, D. G., McFadden, W. H., and Schultz, T. H. Identification and organoleptic evaluation of compounds in Delicious apple sauce. *J Agr Food Chem.* 15(1)29–35, 1967.

[c] From Brewington, C. R., Parks, O. W., and Schwartz, D. P. Conjugated compounds in cow's milk-II. *J Agr Food Chem.* 22:293–94, 1974

[d] From Moshonas, M. G., and Shaw, P. E. Composition of essence oil from overripe oranges. *J Agr Food Chem.* 27(6)1337–39, 1979.

Source: Bell, Iris R., *Clinical Ecology* (Common Knowledge Press, Bolinas, CA, 1982), p. 36.

Possible Psychogenic Mechanisms

Only a decade ago, when news of the first cases of tight building syndrome reached the public, some psychiatrists and psychologists were quick to attribute the subjective complaints of individuals exposed in these buildings to "mass hysteria" or "mass psychogenic illness." In 1979 NIOSH held a symposium entitled "The Diagnosis and Amelioration of Mass Psychogenic Illness," at which one of the authors of this book presented the paper "Mass Psychogenic Illness or Chemically Induced Hypersusceptibility?" (Miller 1979). The presentation was devoted to a discussion of subjective symptoms provoked by exposure to low levels of chemicals, and it sparked a great deal of controversy. The same conference today would more likely be entitled "Indoor Air Pollution" because this phenomenon is now widely recognized. For the most part, mass psychogenic illness has faded from view (Kreiss 1989). However, the possible psychological causes of environmental illness remain a controversial area. To some, the distinction between mind and body is artificial: They are simply two different ways of viewing the same event. In fact, a single process is transpiring. A comprehensive, biopsychosocial approach to patients' problems avoids the pitfalls of reductionistic viewpoints and is thus preferred by many psychologists and psychiatrists (Lipowski 1989).

Two separate issues arise in the context of possible psychiatric origins or contributions to chemical sensitivity. The first relates to the plausibility, nature, and extent of these contributions; the second concerns the most prudent approach toward diagnosis and treatment of the patient when both physiological and psychological factors might be involved. We address the second issue in Chapter 5.

Psychological symptoms are not necessarily psychological in origin. Advances in biological psychiatry focus on genetic and biochemical factors as contributors to central nervous system dysfunction and behavioral disturbance. Environmental exposures can also have psychological sequelae.

The symptoms of low-level chemical exposure may include depression, difficulty concentrating, anxiety, peculiar bodily sensations, headaches, and other subjective symptoms. Patients with multiple chemical sensitivities often report that before they understood what was causing their symptoms they felt as if they could not trust their own bodies or feelings. At any moment, they might feel fine, making plans and commitments for the future; then, the next day or even later the same day, they might feel lethargic, unmotivated, headachy, sleepy, and depressed, as if they had flu. Suddenly, they are unable to fulfill commitments made when they felt energetic. These ups and downs are frequently interpreted by psychiatrists as responses to psychosocial stresses, and patients may be willing to accept such insights because they lack a better explanation. In contrast, patients who have been worked up in an environmental unit often say they are amazed to find direct, clear-cut, cause-and-effect relationships between their symptoms and various foods and chemicals. For the first time, they say they are able to discriminate between their real feelings and those triggered by chemicals. They report that their emotions are appropriate to the situation thereafter, unless, of course, they are having a reaction to a chemical. Not infrequently, such patients feel hostility toward the physicians and psychiatrists who for so long overlooked the chemical basis for their symptoms and instead attributed them to psychiatric causes. These patients wonder how psychiatrists, who routinely use minute doses of chemicals called *drugs* to effectuate behavior, fail to recognize that chemicals in the air or foods can impact the brain or cause marked behavioral changes.

Many of the chemicals these patients implicate as triggering their symptoms are solvents, pesticides, and other substances whose primary target organ, in terms of classical toxicity, is the brain. Interestingly, these individuals who "react" to levels well below those heretofore considered toxic also complain of central nervous system symptoms. Therefore, their complaints are in many respects consistent with known toxic actions of these substances, albeit the *levels* of exposure triggering their reactions are quite different.

Unquestionably, enzymes and various nutrients such as vitamins and minerals act as biological catalysts and regulators. Clearly, any disruption of their function by environmental chemicals might have diverse and far-reaching effects. The limitations of medicine's ability to understand the health complaints of these patients must be honestly and fully acknowledged so that these patients' symptoms are not dismissed as psychiatric when in fact physicians are just beginning to understand the gamut of effects of chemicals on the brain and central nervous system.

Some patients report adverse reactions to particular chemical or food odors, for example, when they smell nail polish remover, cigarette smoke, or popcorn popping. Whether such reactions are classically conditioned responses (Bolla-Wilson 1988) or actual effects of only a few molecules perhaps entrained in the nasopharynx and rapidly transported to the brain is unclear. These responses to odors must be differentiated from those in which mere sight of a food or thought of it produces symptoms.

Researchers at McMaster University in Ontario used classical Pavlovian conditioning to demonstrate that rats sensitized to egg albumen injections in the presence of strobe lights and a humming fan release mediators from their mast cells (IgE mediated) either when injections are given or when the lights and sound are used as the sole stimuli (MacQueen et al. 1989). The authors comment that their results support a role for the nervous system as a regulator of immune function, and they suggest that the nervous system may play a role in colitis, irritable bowel syndrome, and food sensitivities. However, others feel the evidence at present does not support a major role for classical conditioning in human allergic reactions (Metcalfe 1989).

John W. Crayton (1986), a psychiatrist at the University of Chicago, cautions that conditioned responses are more likely to occur in individuals who have true intolerances.

> Undoubtedly, subjects can learn adverse reactions to foods in a classic conditioning sense. However, the presence of these learned reactions may not be helpful in determining whether an individual has "true" adverse reactions to foods. Indeed, it is probably more likely that learned reactions co-exist with true ones. A similar situation occurs in "hysterical epilepsy" in which the patient has gained some conscious or unconscious control over the induction of seizures.

Thus, possible conditioned responses occurring in patients with multiple chemical sensitivities may draw attention away from the true adverse reactions they experience.

That odor conditioning may occur in selected cases is clear. However, patients experience reproducible symptoms to specific chemical exposures (1) often before the odor is perceived (for example, some patients experience symptoms in a particular room or building without detecting an odor, only to learn later that the facility had recently been sprayed with pesticides; Ziem 1989), (2) with their noses clamped during provocative testing, and (3) when anosmia is present (Shim and Williams 1986). These observations weigh heavily against classical conditioning as any more than a partial explanation in certain patients.

It has been suggested that inhaled chemicals may irritate trigeminal free nerve endings in the nose (Doty 1988). Irritation causes a reflexive, involuntary disruption of breathing, a phenomenon familiar to anyone who has inhaled smelling salts. Although trigeminal irritation could explain certain symptoms, such as breathlessness or faintness, in response to pungent olfactory stimuli, it does not explain the entire range of symptoms reported by these patients, often in response to odors that are barely discernible. In addition, patients describe feeling symptoms moments after exposure via intradermal or sublingual routes of administration. Likewise, suppositories placed in contact with mucosal surfaces may provoke onset of symptoms in these patients. Thus, many avenues of exposure may lead to reactions in these patients.

Ecologists observe that patients with chemical sensitivities experience very specific and reproducible constellations of symptoms in response to particular exposures. For example, in one individual, diesel exhaust might trigger sleepiness, mental confusion, and ravenous hunger; rest room deodorizers might be associated with nervousness and irritability; and exposure to a pesticide might be associated with rage followed by uncontrollable crying. Although conditioning could play a role, the fact that responses are *specific* for particular exposures suggests otherwise. Unfortunately, theories of causation that do not fit the reported experiences of these patients, many of whom are highly educated (Doty 1988), abound.

Doty and associates (1988) at the University of Pennsylvania tried to determine whether patients with multiple chemical sensitivities are more sensitive to smells. They exposed 18 patients and 18 healthy control subjects to very low levels (less than 1 percent of permissible occupational levels) of phenyl ethyl alcohol (the principal component of rose oil) and methyl ethyl ketone (a common solvent). Although no difference in ability to detect odors was found between the two groups, nasal resistance was 2 to 3 times higher in the chemically sensitive group than in controls both prior to and following exposure to the odors. Respiratory rates were also higher in subjects than in controls, which the authors felt could be due to nasal constriction because the "nasal airway represents the single largest component of man's total airway resistance and significantly influences tidal volume, respiratory frequency and expiratory time" (Doty 1988). Patients' subjective feelings of not getting enough air or of having labored breathing may reflect increased nasal airway resistance in some patients. Bascom (1990) has noted similar increases in nasal airway resistance among persons claiming sensitivity to tobacco smoke when they are exposed to relatively high concentrations. Doty's patients with chemical sensitivities were no more capable of detecting low-level odors than controls, but they had significantly greater nasal resistances than controls, both at baseline and after challenge. Moreover, 89 percent of the patients had central nervous system symptoms, 67 percent had respiratory symptoms, and 67 percent had gastrointestinal symptoms. At least some of their respiratory symptoms may have been related to their nasal resistance. In addition, these patients had significantly ($p < 0.005$) higher depression scores than did controls.

Urich and co-workers (1988) at the University of Toronto investigated the role of psychological suggestibility in individuals who react to passive smoke. Asthmatics viewed a bank of burning cigarettes during each of their exposures to cigarette smoke but were found not to be particularly suggestible. As the smoke they inhaled increased from zero to moderate to heavy concentrations, there was a progressive dose-response increase in symptoms, deterioration of pulmonary function, increased carboxy-hemoglobin in blood, and increased nasal air flow resistance in subjects that corresponded to exposure levels.

Results of psychological tests also may be misleading in these patients. For example, the Minnesota Multiphasic Personality Inventory (MMPI), a widely used psychological instrument, includes questions concerning peculiar bodily sensations, feelings of inappropriateness, depressed feelings, and many other symptoms, any of which could result from chemical exposures. The chemically sensitive patient who has such symptoms may "read out" as depressed, hypochondriacal, or hysterical on MMPI scales when, in fact, depression, hypochondriasis and hysteria might not be the cause but rather the result of their food and chemical sensitivities.

Food intolerance is a component of several different psychological syndromes involving multiple somatic complaints. Crayton notes overlapping symptoms among patients diagnosed with neurasthenia, allergic tension fatigue syndrome, and somatoform disorder (Table 4-3), any of whom may complain of food intolerance.

In 1880, Charles Beard published a monograph on neurasthenia that described fatigue, irritability, mental confusion, food intolerances, and numerous other complaints of these patients. He felt diet played an important role and noted that fasting for 4 or 5 days resulted in rapid improvement in some patients. Neurasthenia was regarded (even by Freud) as a predominantly physiological rather than psychological illness (Crayton 1986). The term *neurasthenia,* still commonly used in Europe and the Soviet Union, is no longer favored in the United States. Other diagnostic labels such as *somatization disorder, conversion disorder, dysthymic disorder,* and *neurosis* may be applied to these patients (Crayton 1986).

The diagnostic manual published by the American Psychiatric Association and used by psychiatrists in the United States categorizes a number of disorders characterized by physical complaints under the rubric *somatoform disorder* (DSMIII 1980), which includes somatization disorder, conversion disorder, psychogenic pain disorder, and hypochondriasis.

The syndromes listed in Table 4-3 share many common features. One must wonder whether they might not also share the same etiology, that is, food and chemical incitants. Certainly, the incidence of atopic disorders such as asthma and hay fever is significantly higher among patients with affective disorders (for example, depressives) and their first-degree relatives than among schizophrenics (p <0.005) (Nasr 1981). Undoubtedly, most individuals with food or chemical intolerance who have nasal

TABLE 4-3. Syndromes of Multiple Somatic Complaints

Symptoms	Neurasthenia	Allergic Tension Fatigue	Somatoform Disorder
Fatigue, sickly	+	+	+
Food intolerance	+	+	+
Gastrointestinal symptoms (pain, nausea, vomiting, etc.)	+	+	+
Arthralgias	+	+	+
Myalgias	+	+	+
Cognitive deficits (memory, concentration, etc.)	+	+	+
Palpitations	+	+	+
Insomnia	+	+	0
Headache	+	+	0
Depression	+	+	0
Irritability	+	+	0
Nasal symptoms	0	+	0
Hives	0	+	0
Eczema	0	+	0
Deafness	0	0	+
Blindness	0	0	+
Loss of voice	0	0	+
Convulsions	0	0	+
Sexual indifference	0	0	+

Source: Crayton, J., "Adverse Reactions to Foods: Relevance to Psychiatric Disorders," *Journal of Allergy and Clinical Immunology* (1986) 78(1):243–250, p. 246 (Copyright C. V. Mosby Co., St. Louis).

or skin manifestations would prefer to be seen by an allergist even though they may also experience fatigue, headaches, and memory or concentration problems. Theoretically, different medical specialists such as neurologists, psychiatrists, allergists, gastroenterologists, and rheumatologists may see patients with chemical or food sensitivities in whom varying complaints predominate that drive them to select one specialist over another. Each specialist could be viewing the same problem from a different perspective. Indeed, patients with chemical and food sensitivities may have seen physicians decades ago but were given diagnostic labels such as neurasthenia or the "vapors," an archaic term for a depressive or hysterical neurological condition. Perhaps the latter term will come into vogue again if a chemical etiology for these conditions is affirmed.

A study of 42 patients admitted to the Dallas environmental unit employed a battery of clinical instruments including the MMPI and the Weschler Adult Intelligence Scale–Revised (WAIS-R). Analysis of test results, before and after entering the unit and being on safe foods, showed "statistically significant and clinically meaningful" improvement in five factors: "alienated depression, ineffectiveness, effortful processing, vigilance and effective energy." According to Bertschler et al. (1985), depression lifted, mental acuity improved, feelings of despondency and hopelessness resolved, concentration and short-term memory increased, and energy returned.

Kaye Kilburn (1989a) of the University of Southern California School of Medicine proposes that the human nervous system, because it is so highly evolved, may be most susceptible to environmental agents.

> Sensitivity may be its undoing. The intuitive hypothesis is advanced that the nervous system is the most liable of the body's systems to damage from environmental toxins. Appreciation of damage may be masked because subtle dysfunction is concealed by the nervous system's remarkable redundancy and substitution of functions, or it is overlooked in clinical evaluations which are usually only qualitative.

Physicians who see patients complaining of concentration or memory problems find objective assessment of these complaints difficult. Without careful, quantitative testing and precise knowledge of the patient's abilities prior to the exposure, physicians may erroneously attribute the patient's complaints to anxiety, lack of intelligence, or aging.

A model study of 14 firefighters exposed to polychlorinated biphenyls (PCBs are used to insulate electrical transformers) and their combustion products in a transformer room fire showed significant impairment of memory, cognitive function, and perceptual motor speed compared to unexposed firefighters from the same department (Kilburn 1989). Two days to 3 months after the fire, all 14 of those exposed noted symptoms such as extreme fatigue (8), headaches (7), muscle weakness (9), joint aches (5), memory loss (8), and impaired concentration (6). Only by employing an extensive battery of neurobehavioral tests and comparing scores with controls were their physicians able to detect these alterations, which were very apparent to the firemen themselves. Patients with multiple chemical sensitivities are unlikely to receive such a careful evaluation routinely, nor are preexposure test results or appropriate controls generally available.

Evidence suggests that psychosocial events, such as the death of a spouse or divorce, can suppress immune system function and may predispose certain people to being more sensitive to chemicals at low levels. Certainly, the relationship between psychological and physiological systems is an intricate one.

Selye in his theory of general adaptation and the ecologists both view psychosocial stressors as part of the organism's total stress load. Dantzer and Kelley (1989) review animal and human data suggesting that stress impacts the immune system. The authors caution against the temptation to interpret such data in the context of the psychosomatic model of disease, that is, that one's psychological traits somehow cause immune-related disease via neurological and hormonal correlates. However, because of the feedback loops between the brain and the immune system, an "immunoneuropsychological" interpretation may be closer to the biological truth. Presumably, undetectable changes in the immune system may alter central nervous system function and produce psychological and emotional manifestations. For example, in the early stages of cancer, products released by immune or tumor cells may produce the helpless, hopeless feelings that have been associated with progression of the disease.

Analogously many of the psychological and emotional problems experienced by patients with multiple chemical sensitivities may be the result of currently undetectable alterations in their immune or nervous systems, rather than the result of personality problems or a belief system.

Two of the most vocal critics of clinical ecology, allergists John Selner and Abba Terr, are of the opinion that multiple chemical sensitivity patients adhere to a "belief system" that chemicals are the cause of their health problems. Staudenmayer and Selner (1987) describe what they term "an irrational belief system":

> The ecology belief system usually is deeply entrenched and its logic well developed by intricate rationalizations and indoctrination. Social factors feed on the primary and secondary gain of the victim. "True believers" are more than willing to present their testimonials, seeking and affording mutual assurance. The social and psychological dynamics of the cult apply. In addition, there exists a plethora of "health publications" that provide the authority of print, while an impulsive media, eager for news, often is duped by unsubstantiated and unscientific claims of so-called ecology authorities.

Terr (1989c) believes that no psychotherapeutic intervention will help these patients, whereas Selner (1988, p. 51) advocates systematic deprogramming of the patients to purge them of their beliefs and believes that 50 to 75 percent of receptive patients can be deprogrammed. Staudenmayer and Selner emphasize that those patients with chemical sensitivity who adhere to a belief system, particularly so-called universal reactors, must be separated from those who are truly sensitive to specific, identifiable chemicals.

Some psychiatrists strongly feel that individuals with multiple chemi-

or combinations of these (Schottenfeld 1987; Brodsky 1987). According to Schottenfeld (1987), "the early childhood history of individuals with M.C.S. [Multiple Chemical Sensitivities] is often notable for the presence of physical or sexual abuse, severe medical illness during childhood, death of one or both parents, or other severe disturbances of early caregiving relationships."

According to Staudenmayer (1989), at first many patients will not reveal problems of childhood abuse, but when trust is established in a therapeutic relationship, they will. Whether persons with chemical sensitivities experience as youths more psychological trauma than the "average" has not been determined. Knowing what percentage of "normal" individuals undergoing the same degree of intensive psychological inquiry would confess to similar difficulties is important. Otherwise, this particular approach to the problem suffers from the same flaws the clinical ecologists have been accused of with regard to study design.

Staudenmayer and Selner (1990) compared 58 patients with multiple chemical sensitivity, 89 patients from a psychology practice (diagnoses included depression, anxiety, mood swings, phobia, panic, and insomnia), and 55 controls reported not to have had psychological symptoms for at least one year (diagnoses included asthma, gastrointestinal problems, headaches, skin problems, hypertension, and menstrual pain). All patients underwent a battery of neuropsychophysiologic tests. Electroencephalograms (EEGs) for the psychology patients and for the multiple chemical sensitivity patients differed significantly from those of the controls (p <0.001). However, the authors conclude that "the universal reactor group was not statistically different from the psychologic group" (p = 0.22). Patients with multiple chemical sensitivity had significantly higher scalp electromyographic (EMG) activity than did the other groups (p <0.001 in both cases). The authors felt their data confirmed their hypotheses "that the group of universal reactors would not be significantly different from a group of outpatients with overt psychologic disorder who did not project them onto the environment" and "that universal reactors manifest psychosomatic illness rather than true environmental disease." While these hypotheses could explain their data, their psychology practice population might also have chemical sensitivities, accounting for the similarities in EEGs. The statistically significant difference in EMGs suggests real differences between universal reactors and psychologic patients. Unfortunately, this study does little to clarify the etiology of multiple chemical sensitivity. In a published critique, Davidoff et al. (1991) also question the authors' interpretation of their data and suggest that "Psychological and psychiatric disturbances could simply be consequences, rather than causes, of MCS [Multiple Chemical Sensitivities]."

Even if chemically sensitive patients have had more early trauma, two other questions must be asked. First, could major psychological trauma somehow predispose individuals to developing bona fide sensitivities to chemicals? Ecology patients sometimes report a major life event coincided with the onset of their difficulties; however, such life changes are frequently associated with changes in *exposures,* for example, a move to a different home following a divorce or spouse's death, or taking medications during stressful times. Consider chemical sensitivities that might arise during exposures from remodeling that could be manifested as irritability or depression. If a divorce ensues, the development of chemical sensitivities might be attributed to the "stress" of the divorce, when in fact the sensitivities may have contributed to both the stress and the divorce. Staudenmayer and Selner (1987; 1990) assert that they have performed blinded chemical testing using sham challenges as controls with patients who claim to have this condition and that these challenges result in both false positives and false negatives. In their view, these alleged erroneous reactions by patients confirm the lack of true sensitivities and provide a point of departure for the psychologist to explore with patients their "belief system" about having chemical sensitivities. In examining the experimental design for these challenges, crucial questions that have not yet been addressed in published studies are:

1. Are subjects in a deadapted state prior to the challenge so that extraneous exposures during and prior to the challenge (up to several days before) do not interfere with testing?
2. Are open challenges performed first to confirm that the placebo (a masking odor such as peppermint) is in fact a placebo and that the "active" challenge is something to which the patient has had demonstrable reactions?
3. What is the recency and latency of the patient's exposure to the substance being tested? In other words, has enough time elapsed (about a week or so) that the person is no longer adapted or reacting to the last exposure but not so much time that the sensitivity has waned? Recency of exposure is recognized as a crucial variable in conducting challenges in patients with occupational asthma, for example.

Finally, with regard to the issue of childhood abuse or childhood illness, one must ask whether the parents and families of chemically sensitive patients (patients who often have psychological manifestations) might not also have such problems. Ecologists suggest some genetic predisposition to this problem. Abusive or alcoholic parents of chemically sensitive patients may have suffered from unrecognized environmental sensitivities themselves (see the appendix regarding alcoholism and drug abuse and neurobehavioral and psychiatric manifestations).

Ecologists argue that major illness during childhood may have been the result of undiagnosed chemical sensitivities or that sensitivities may have been triggered by infection or medications that were administered, rather than viewing these events as a disruption in the care-giving relationship or the beginnings of secondary gain, that is, seeking attention or nurturing via illness. Therefore, even if one could prove that childhood trauma were more prevalent among patients, such a finding neither proves psychological interpretations nor disproves chemical causes.

Schottenfeld (1987) offers advice to physicians who work with these patients: "Regardless of the original etiology of symptoms, these individuals tend to amplify their symptoms and to develop the mistaken belief that the symptoms are indicative of severe disease."

> Changes in the workplace that reduce toxic exposure and the risk of exposure may provide the most reassurance—the installation of a new exhaust system in Mrs. A.'s workplace was an extremely effective psychotherapeutic intervention in addition to its obvious benefit in the prevention of occupational respiratory disease (Schottenfeld 1987).

Another psychiatrist (Brodsky 1987) writes:

> A review of medical history and literature that reflects on medical cultures reveals that there have always been people who have had unpleasant physical and emotional symptoms and experiences for which they sought explanations. . . . In the culture of 20th century medicine, a disorder of the immune system would represent a sophisticated and acceptable explanation, because the immune system is demonstrably complex and is interrelated with all other systems, and no one would disagree that many of its mechanisms and manifestations are still unknown.

Recently, Black et al. (1990) reported that 15 of 23 subjects (65%) diagnosed with environmental illness by clinical ecologists had at one time met the criteria for a mood, anxiety or somatoform disorder versus 13 of 46 matched healthy controls (28%). These authors conclude that such patients may have one or more commonly recognized psychiatric conditions that could explain some or all of their symptoms. Critics of Black's study draw attention to the fact that more than a third of the patients claiming environmental illness had no history of significant psychopathology in their lifetimes, therefore mental illness did not explain their problems (Galland 1991). The study was further criticized for failing to consider that individuals who have quit working or been hospitalized for *any* illness will exhibit more psychological symptomatology than healthy controls (Galland 1991). Black responds that while he believes that psychiatric conditions may explain most, or all, symptoms in some individuals, he suspects that other patients diagnosed as having environmental illness may have a "verifiable physical disorder that would explain their symptoms equally well" (Black 1991).

The issue of whether chemical sensitivity is a bona fide physical entity, or, an "irrational belief system" that may be "systematically deprogrammed," or a form of psychopathology amenable to psychotherapeutic interventions is a critical one, one that merits thoughtful consideration and rigorous scientific inquiry.

Conclusions

Perhaps the mechanism for multiple chemical sensitivities is not identifiable; that is, after all avenues of biochemical and immunological inquiry have been exhausted, no single explanation for this disorder is forthcoming. The theory of substance-specific adaptation is based upon observations of the responses of patients in a deadapted state who are evaluated in an environmental unit. Adaptation is *only* an observation at this time, not a mechanism. However, biological *limits* might regulate how much an organism can adapt, limits that could be highly individual and vary by orders of magnitude. Certainly adaptation occurs at all levels of biological systems, from enzyme systems to cells, tissues, organs, and even behavior (Fregly 1969). Theoretically, a major insult or the accumulation of lower-level injuries within these systems could lead to a kind of "overload" or "saturation" effect with respect to adaptive capacity that would cause an individual to have environmental responses, which, instead of being flexible and fluid, are now fragile and overly responsive. Many patients we interviewed for this book told us that even years and in some cases decades following the onset of their problems they had recovered only a portion of their former energies and tolerance for their environment. Their descriptions seem to suggest the loss of an intangible capacity to adapt, parts of which may be temporary and recoverable and other parts of which may not. We are reminded here of the teaching: "Listen to the patient. He is telling you the diagnosis." Perhaps they are telling us the mechanism as well.

For insights gained since the first edition about possible mechanisms underlying chemical sensitivity, see the section on "Mechanisms" in Chapter 8.

CHAPTER 5

Diagnosis and Treatment

Diagnostic Approaches

As with most fields in medicine, meticulous history taking is the most important element in making a diagnosis. However, history taking for multiple chemical sensitivities involves obtaining a chronology not only of illness but of exposures as well. "Although the physical examination is an integral part of all medical investigation, 'examination of the environment' of a patient tends to be relatively more rewarding" (Randolph 1987, p. 274). Physicians today must ask their patients what kind of work they do and inquire about specific chemical exposures on and off the job and changes in symptoms at work, on weekends, and during vacations. Ramazzini, the father of occupational medicine, instructed physicians to ask, "Of what trade are you?" On the whole, occupational health practitioners today, more than other medical specialists, take the most comprehensive exposure histories. Thus, for patients who may have multiple chemical sensitivities resulting from industrial, tight building, or community exposures, the physician group most attuned to and therefore likely to discover the potential link between the patient's illness and a chemical exposure is the occupational physician. The "new generation" of occupational health physicians is well informed about chemicals, various processes, and associated exposures, as well as signs and symptoms resulting from chemical exposure (Rosenstock 1984). They are familiar with the industrial hygienists' measurements of chemical exposure. However, before they can help chemically sensitive patients, they will require instruction in the particular symptoms, provoking exposures, and special problems of these patients. Occupational health physicians and clinical ecologists have overlapping interests and would benefit from information exchange and cross-training. Likewise, allergists, by learning more about chemicals and toxicology as recommended by some of their spokesmen (Selner and Staudenmayer 1985b; Bardana and Montanaro 1989) and by taking exposure histories that go beyond the confines of IgE-mediated disease, could emerge as a major physician group specializing in the problems of these patients in the future.

An adequate exposure history with attention given to pinpointing when symptoms began in relation to other factors (for example, drugs, chemical exposures, job changes, household moves, operations and hobbies) is essential. Concurrent illness in other household members, co-workers, and even pets may provide clues. This time-consuming detective work is the *sine qua non* for discovering an inciting exposure. Properly designed patient questionnaires may facilitate the process by enabling the patient to engage in the detective work as well. Questions concerning the patients' likes or dislikes for certain odors may be revealing because aversion to particular odors has been noted commonly among patients who have multiple chemical sensitivities (Randolph 1980). Over the years, Randolph has noted that the more odors checked off on his questionnaire as "strongly like" or "strongly dislike," the more likely the patient is to have chemical sensitivities. In addition, patients can prepare a chronology or time line of major events such as household moves, job changes, surgeries, and pregnancies and the onset and duration of symptoms or illness. Similarly, a daily log of activities or exposures that notes any symptoms may facilitate recognition of patterns.

Of course, the more symptoms and the more systems of the body affected, especially the nervous system, and the more these symptoms fluctuate in intensity, the stronger should be the physician's suspicion of multiple chemical sensitivities. Johnson and Rea (1989) report that the average patient entering the Dallas environmental unit has five symptom complaints, many of which are neurological. Industrial workers with multiple chemical sensitivities exhibit similar constellations of symptoms (Cone et al. 1987). A history of multiple "idiosyncratic" drug reactions or alcohol or food intolerance or cravings may also be suggestive (see Chapter 3).

The physical exam, traditionally an important diagnostic tool, disappointingly is often normal in these patients. Symptoms or signs may occur only with exposure. Injury may be subclinical and prepathological. Laboratory findings may be normal or, if abnormal, provide no pattern or clue that seems to have clinical relevance. Altered helper-suppressor T-lymphocyte ratios and the presence of autoimmune antibodies may be suggestive (see Chapter 4), but are not diagnostic. Subtle signs of vasculitis can be noted in some; spontaneous bruising, petechiae, and cold or blue extremities (Raynaud's phenomenon) may occur.

Rea has noted yellowish skin discoloration with normal liver function tests in some patients and refers to this as the "chemical yellows." Allergic shiners (dark circles under the eyes, or "racoon eyes," from renous congestion), Dennie's lines (creases under the eyes), reddening of the ears, and the "allergic salute" (nose rubbing) might provide clues in some children (Rapp and Bamberg 1986); these facial features are recognized by allergists in children with both classical IgE-mediated allergy and nonallergic, non-IgE rhinitis (etiology of the latter is unknown but conceivably could be related to food or chemical exposures).

Rea and associates are exploring more sophisticated and objective ways of measuring changes in their patients, such as monitoring sympathetic and parasympathetic nervous system activity by recording pupillary reactions to a light stimulus (discussed in Chapter 4) and using a balance recorder, a platform on which the patient stands and attempts to maintain balance. Movements of the patient are recorded and reflect disturbances in one or more of the three physiological inputs that regulate balance: visual input, the inner ear, and proprioceptive signals. PET (positron emission tomography) or other brain-scanning techniques may prove helpful in the future (Morrow 1990).

As discussed in detail in Chapter 2, the gold standard for diagnosing chemical hyperreactivity in a patient is the environmental unit, coupled with fasting. Although this approach may be too costly and time-consuming for the average patient, for the very ill, it may be the *only* way to unravel this multifactorial, polysymptomatic illness. Eventually, biomarkers may be developed for chemical sensitivity, especially if the mechanisms of the disease are biochemical or immunological. However, if the nervous or limbic system is key, it may not be possible to identify biomarkers.

Short of a several-week stay in an environmental unit, might any other approach be used to diagnose chemical and food sensitivities in patients who do not require hospitalization or who may wish to be worked up as outpatients? Certainly an elimination diet could be attempted to identify food incitants. Patients may have difficulty fasting or avoiding common incitants, rotating their foods, or obtaining chemically less contaminated foods. (Even so-called organic foods may not be entirely free of pesticides and other contaminants.) Detecting subtle chemical sensitivities while at home could be quite difficult. Masking or adaptation to chemicals in one's home environment (such as gas furnace emissions) might go unrecognized. If comprehensive environmental control were attempted at home, major remodeling or overhauling of furniture, heating systems, wardrobe, and other major changes may be required in order to achieve a chemically less contaminated environment for the patient. Such interventions done in a hit-or-miss fashion could be costly. Far better would be residences that are relatively "safe" habitats, such as specially constructed trailers or homes in which patients could reside temporarily while they sort out their sensitivities and undergo food and chemical testing. We inspected trailers lined with porcelain that have been specially made for chemically sensitive individuals and visited a small community outside Dallas, where specially constructed homes are occupied by chemically sensitive patients.

Clearly, the cost and trouble of such a rigorous diagnostic approach may be prohibitive for the average patient with chemical sensitivities. For this reason, provocation-neutralization has been promoted by ecologists as a way to diagnose and treat at least some of their patients' sensitivities to biological inhalants, foods, and chemicals. This procedure is considered in detail in the next section, which discusses therapies. Less widely accepted and far more controversial diagnostic approaches used by the minority of clinical ecologists include electroacupuncture and kinesiology. The basis for these procedures is speculative at best; they are not addressed in this book.

Therapies

To the traditional practitioner, perhaps the most disturbing feature of clinical ecology is the wide range of therapeutic modalities used by various practitioners and the lack of proof for many of them. Many allergists with whom we spoke expressed frustration with an attitude among certain clinical ecologists that they do not need science because they are right. Allergists are critical of clinical ecology's lack of randomized, double-blind clinical trials. Randolph (1987, p. 220) and other ecologists feel this criticism is "overdrawn." They emphasize the clinical nature of the field: its concepts and techniques are inductively derived from careful clinical observation. As rigorous and cautious as Randolph's use of an environmental unit might have been, the same cannot be said for other treatment approaches used by clinical ecologists. We discuss here some of the more frequently used ecologists' therapies, including provocation-neutralization, nutritional supplementation, detoxification, and the treatment of acute reactions to foods and chemicals. Fundamentally, clinical ecologists agree that avoidance of incitants, both food and chemical, is the treatment of choice and allows the best possibility for recovery. Clearly, however, this treatment is not entirely satisfactory.

Avoidance can lead to an ascetic life-style that is unacceptable to many patients. Some who are very ill feel they have no choice. A number of treatment modalities have been employed by ecologists in an attempt to speed their patients' recoveries; however, no study has been done to demonstrate whether patients receiving these treatments recuperated any faster than if they had practiced avoidance alone.

Provocation-Neutralization

The majority of clinical ecologists use provocation-neutralization to a greater or lesser extent. This technique involves provoking a patient's symptoms by injecting under the skin or administering sublingually a small dose of an inhalant, food, or chemical while observing the patient for symptoms and/or increase in wheal size if given via a cutaneous route. This diagnostic test is used to identify incitants for a particular patient. Subsequently, various dilutions of the same substance that produced symptoms or a wheal are injected or given sublingually until one dilution is found that turns off the patient's symptoms or that results in no increase in wheal size following intradermal injection. This dose is called the *neutralizing* dose.

A lengthy review of all studies of provocation-neutralization done to date is beyond the scope of this book. Further, we feel strongly that too much emphasis has been placed upon trying to disprove this method as if the existence of the problem of multiple chemical sensitivities depended on provocation-neutralization. The existence of multiple chemical sensitivities and the efficacy of provocation-neutralization are independent issues and ought to be treated as such.

Suffice it to say that provocation-neutralization may be an evolving technique, just as classical allergy testing is still evolving. Salvaggio, an allergist, has remarked upon the paucity of evidence to support the efficacy of mold immunotherapy that is used by classical allergists (Salvaggio and Aukrust 1981). Others offer similar views:

> Immunotherapy has been used empirically over the past 70 years, primarily because the actual immunologic mechanism has continued to elude investigators (Gurka and Rocklin 1988).

> The mechanisms by which hyposensitization is achieved are not completely understood. . . . While statistically controlled blinded studies on the efficacy of allergens and immunotherapy have been made, for the most part, only for some pollens, extension of these results to other allergens and certain conditions is generally considered acceptable (package insert from an allergenic extract).

Immunotherapy was used by allergists for decades before the discovery of IgE by Ishizaka in 1969. Controlled trials demonstrating its effectiveness have been available only since the 1950s. Van Metre and Adkinson (1988, p. 1329) describe the difficulties faced by investigators who wish to design controlled trials for testing the efficacy of immunotherapy:

> Design requirements are complex and difficult to accomplish in any one single trial. These difficulties can be addressed by developing a model of specific aeroallergen disease with which multiple groups of investigators can work over a relatively long period of time. Methods and reagents are refined until consistent, accurate results are achieved.

Such trials are difficult and costly to conduct. Large, *homogeneous* groups of patients must be recruited, for example, a large number of patients with seasonal hay fever triggered by ragweed pollen. Here each patient has the same symptom resulting from the same exposure. The added complexity of multiple symptoms resulting from many divergent exposures (as occurs in multiple chemical sensitivities) is obvious. The same authors recognize that trials using provocation-neutralization have had major problems with reproducibility, nonstandardized extracts, nonhomogeneous patient populations, and disparate methods of measuring outcome; yet they comment favorably on studies by Boris and others (1985a; see also Boris et al. 1988) using provocation-neutralization in two randomized, placebo-controlled, double-blind, crossover studies for cat and dog extract causing asthma and state that "the work deserves careful study and attempts at replication." Their comments contrast sharply with the position paper by the American Academy of Allergy (1981) on this subject: "Subcutaneous provocation and neutralization as a method for the treatment and diagnosis of allergic disease has no plausible rationale or immunologic basis."

A recent and comprehensive study of provocation-neutralization was supported by the American Academy of Otolaryngic Allergists (AAOA) and reported in *Otolaryngology: Head and Neck Surgery*, a leading ENT journal. Approximately 1,800 members of AAOA use these methods, which are endorsed by the 8,000 members of the American Academy of Otolaryngology Head and Neck Surgery, the largest group of ENT physicians in the country. William King and co-workers' studies (1988a, 1988b, 1989), sponsored by the AAOA, reported that provocation-neutralization had a sensitivity of 79.7 percent and a specificity of 72.4 percent, in contrast to classical skin testing, which had a sensitivity of only 26.6 percent and specificity of 85.5 percent when compared with a provocative food challenge. In his 1986 presidential address to the

American Academy of Allergy and Immunology, John Salvaggio (1986) referred to " 'fringe element' societies such as the otolaryngologists' allergy society, in which unproven methods of immunodiagnosis and therapy are used." Otolaryngologists who practice provocation-neutralization are not appreciative of the organized write-in campaigns by allergists that have successfully persuaded the Health Care Financing Administration to deny payment for provocation-neutralization for foods. Some merely regard these campaigns by the allergists as a turf battle between the allergists and clinical ecologists.

Sublingual provocation and neutralization is used much less often than injection techniques. Blinding is more difficult for sublingual provocation than for injection. Moreover, many ecologists feel that the clinical result is not as good. Nevertheless, sublingual testing and treatment have high patient acceptance and low risk of adverse reactions. Recent sublingual treatment studies using house dust mite (Scadding and Brostoff 1986) show promise for this approach. Many think the venous network beneath the tongue is responsible for uptake of foreign substances, and research using animals points toward a more direct pathway from the oropharynx to the brain (for example, the hypothalamus) involving very rapid substance transport (Kare 1968; Maller et al. 1967). Such a mechanism might help to explain the rapid alterations in mental status patients report with provocation and neutralization by the oral route, as well as the rapid onset of symptoms they experience when ingesting or inhaling incitants.

The definitive study of provocation-neutralization has not yet been done, and the studies purporting to prove its ineffectiveness have not been free from substantial flaws. Most convincing are individual cases in which symptoms appear dramatically with provocation (Miller 1977; Rapp 1978a, 1978b). The technique may work best in a select subgroup of patients. Indeed, the collective strength of the dozen or so positive studies done to date may be greater than that of any individual study; the statistical technique of meta-analysis may have relevance here as a tool for evaluating them further (Louis et al. 1985; Wachter 1988).

David King, University of California, San Francisco, (1984, 1988) has carefully reviewed studies of provocation-neutralization, and his work is important reading for anyone wishing to understand this subject. He reviews two of the studies upon which the American Academy of Allergy and Immunology relied for its position statement against provocative testing. One was a study by Caplin (1973) sponsored by the American College of Allergists. King (1984) renanalyzes Caplin's data and finds that the reported statistical analysis was incorrect. In fact, the validity coefficient *is* significant, implying that results of provocation *were* related to results of feeding challenges. Similarly, King examines a study by

Lehman (1980) that reported that sublingual testing was not reliable. Lehman did not analyze the data statistically to reach this conclusion. King found that the testing was in fact reliable for two of the four foods tested, even though a very restrictive dependent measure was used for evaluation (nasal mucosal changes for which the reliability of the experimenter's judgment was unknown). King (1984) concludes: "A close examination of other frequently cited evaluation studies reveals similar flaws, making firm conclusions about provocative testing premature." In another paper King (1988) reviews other studies that have found provocative testing unreliable or invalid, including the often cited study by Jewett et al. (1990). In their study three "active" and nine placebo intradermal injections were administered to 18 clinical ecology patients by clinical ecologists. Only patients who consistently had symptoms provoked during open challenges were studied. Small doses ("underdoses") of food injections were administered double-blind, and the subjects guessed which were active and which were placebo. The results of guessing by subjects were no better than chance. Prior to its publication King raised several important concerns about the Jewett study.

1. Patients may have been avoiding the food in question and thus have lost their prior sensitivity (see Chapter 2 regarding adaptation).
2. The use of an either-or, dichotomous measure (guessing active or placebo) coupled with single-subject data analysis will detect only relatively strong effects and may work only for a highly accurate test, not one prone to a certain amount of error.
3. Possibly underdoses are ineffective, and larger doses may provoke reactions in some patients.

King (1981) himself conducted a study of provocation testing and found that allergenic extracts under double-blind conditions could provoke cognitive-emotional symptoms in selected individuals. However, his enthusiasm for provocation-neutralization is carefully tempered. He notes that symptoms are frequently reported by subjects given placebos, "a finding which should concern clinicians employing the test" (D. King 1988). Conceivably, the high rate of placebo reactions may reflect background fluctuation in chronic masked reactions to a less than optimal test environment and suggest the need for trials to be conducted in a controlled, less chemically contaminated environment. King concludes:

> Most studies of provocative food testing contain serious flaws which limit inferences regarding reliability and validity. Thus, whether these tests are sufficiently reliable and valid for clinical use cannot, strictly speaking, be determined from the research available, since the appro-

priately designed studies have not yet been conducted. However, the research that has been reviewed would seem to suggest that both [intradermal and sublingual] methods of testing can provoke symptoms above placebo levels, but that these effects are generally quite small. Such subtle effects, when combined with symptoms that naturally vary over time and with placebo effects, would make it unlikely that their use in the clinical setting is as accurate as some proponents claim. On the other hand, the data do not support the conclusion that these methods cannot provoke genuine symptoms. Rather, the problem is distinguishing the "signal" from the noise. Averaging across many trials, as in group research, aids this process, but this fact is of little use in the clinic, in which every individual test is interpreted.

King's reservations concerning the clinical utility of provocation-neutralization are crucial. Although provocation-neutralization may be in the beginning stages of its evolution, much like traditional allergy immunotherapy was earlier this century, its continued use should depend on objective demonstration of efficacy. Patients should not be held hostage by controversies in this area. With so few available therapies, those that may offer benefits should not be barred but investigated further. Both otolaryngologists and ecologists realize there are limitations to provocation-neutralization's effectiveness. Rea (1989), who probably sees the most severe patients, finds that 30 to 40 percent of patients are *not* helped by provocation-neutralization techniques but feels that its failures should not preclude its use in the 60 to 70 percent who may receive benefit. Most clinical ecologists continue to stress the importance of avoidance as the primary and most efficacious treatment with concomitant use of a rotary diet, but acknowledge compliance may be difficult for many patients.

We asked several ecologists whether, by using neutralizing doses on a frequent basis, one might not simply be masking patients' symptoms, that is, inducing adaptation that might obscure chronic damage caused by administering incitants on a regular schedule. They were concerned over this point, but most discerned a big difference between a patient's taking neutralizing doses on a daily basis and a patient's eating the food on a daily basis. They argue that the former generally would not result in symptoms, whereas actual food ingestion would.

Even if provocation-neutralization were proven valid, extension of this technique from inhalants and foods to chemicals such as formaldehyde, automobile exhaust, phenol, and tobacco smoke is a major leap of faith that needs much further investigation. Exposing patients to levels of chemicals normally encountered in everyday life may be justifiable. However, injection of potentially carcinogenic substances such as formaldehyde or auto exhaust is of concern. Ecologists argue that their patients normally are exposed to these substances at even higher doses, and low doses will not increase risk to any measurable degree.

If provocation-neutralization were to be established as efficacious in clinical trials, evaluation of its long-term efficacy (the longest trials have been on the order of a few weeks) versus avoidance alone would be a next, important step.

A further consideration with regard to provocation-neutralization is that the background level of pollutants in the testing room or those brought in on patients' clothing or skin (such as traffic exhaust or cigarette smoke) might interfere with accurate provocation and/or neutralization. In addition, the time interval since the patient was last exposed to the test substance may affect the provoking and neutralizing doses, just as an individual's response to ozone or other substances may be affected greatly by recency of exposure (see Chapter 2). Thus, adaptation or acclimatization must be considered as a potentially important, if not crucial, variable. If the testing room where provocation-neutralization is being done contains volatile organic compounds, cause-and-effect relationships could be obscured. At present, we have no reason to suppose that skin testing or sublingual testing would be any different in this respect from oral or inhalation challenges conducted in an environmental unit. Background levels and the time interval since the last exposure to a substance must be rigorously addressed in any future studies.

To some, the most convincing bits of evidence in favor of provocation-neutralization come from the many anecdotal cases reported by ecologists. Doris Rapp has filmed several such cases using double-blind procedures. Nevertheless, traditional allergists raise concerns about appropriate control of conditions and her objectivity. Some discount her work by implying that the adults must be acting and the children either hungry or in need of a nap. We wonder whether these anecdotal cases, in which reactions to foods or chemicals seem to be turned on or off by a tiny amount of incitant, might not represent individuals whose sensitivity is very high and thus the reaction is easily observed. If so, these individuals may present a unique opportunity to study and document this phenomenon. Provocations in these individuals must be done with sufficient iterations to satisfy statistical requirements, as was attempted by Jewett et al. (1990). Again, the effect of background noise on the testing must be assessed because inadvertent exposures (food or chemical) could interfere with test results. Perhaps such testing is most sensitive and specific if performed on patients in the deadapted state in an environmental unit where background noise is negligible. A further difficulty, reported by a number of patients, is a tendency for their "endpoints" to shift over time, which may lead to an increase in their symptoms. Although their physicians might recommend retesting, many simply discontinue treatment when this occurs.

Detoxification

The EPA, in its ongoing program to monitor levels of toxic chemicals in human adipose tissue, has found many volatile organic compounds and pesticides in all parts of the body, including the brain and nervous system. Some of these chemicals may persist for decades; for example, beginning in 1973, Michigan residents were exposed to PBB (polybrominated biphenyl) a toxic fire retardant that accidentally had been substituted for a nutritional supplement in farm animals (Wolff et al. 1982). A clinical research team from Mount Sinai School of Medicine found that 97 percent of more than 1,000 state residents had detectable PBB in their fat (0.2 ppb or more). Because serum levels taken 12 to 18 months apart in 1977 and 1978 from the same individuals were not significantly different, the researchers concluded that the PBB in their tissues would remain there indefinitely (Wolff et al. 1979).

A detoxification method employing sauna, exercise, polyunsaturated oils, and various nutrients that has been used in the field of drug rehabilitation for drug accumulations in fatty tissue was offered to seven healthy male volunteers from Michigan. Following the detoxification regimen, fat biopsies from these individuals showed significant reductions in 16 chemicals (averaging 21.3 percent reduction), including PBB. Four months later, after no further treatment, the same subjects nonetheless had additional decreases in chemical fat stores: the average decrease in the 16 chemicals studied was 42.4 percent. Schnare of the EPA et al. (1984) hypothesized that this continued decline might suggest recovery of the body's own ability to eliminate toxic substances. Others have reported use of detoxification therapy for toxicity subsequent to exposure to dioxin (Roehm 1983), PCBs and their by-products (Schnare 1986; Tretjak 1989; Kilburn 1989), and other chemicals (Root 1987). Specifically, the detoxification regimen involves seven components (Schnare et al. 1982):

1. Aerobic exercise for 20 to 30 minutes to increase fat mobilization

2. Low temperature sauna (140–180° F) for 2 or more hours (preferably 5 hours) after exercise to enhance skin excretion of toxic substances

3. Nutritional supplements with gradually increasing amounts of niacin to enhance lipolysis, and proportionate amounts of other vitamins and minerals

4. Water, salt, and potassium replacement

5. Polyunsaturated oil, 2 to 8 tablespoons a day as tolerated, to decrease uptake of toxins in the intestine and facilitate their excretion

6. Calcium and magnesium supplements

7. A daily routine of exercise and sauna for a period of several weeks, balanced meals, adequate rest, and no drugs, alcohol, or medications

This approach is not considered a cure but is claimed to facilitate the recovery of certain patients with multiple chemical sensitivities. Randolph (1980) cites several cases in which significant improvements in patients' food and chemical sensitivities have occurred and states:

> The task of defining the relationship between exogenous and endogenous chemicals in particular patients remains. At this time, we can say that reducing endogenous accumulations of toxic chemicals appears vitally important to the effective treatment of some environmentally ill patients; and the development of a safe and effective method for reducing these burdens gives us a welcome new tool for treatment.

A study of 14 firemen exposed to PCBs in a transformer room fire employed an extensive battery of neurobehavioral tests that was administered 6 months after the fire and again 6 weeks later after the exposed firemen underwent a 2- to 3-week sauna detoxification program (Kilburn 1989b). Controls were firemen from the same department not exposed to the PCB fire. The 14 exposed firefighters showed impaired short-term memory, interpretation of designs, spatial relationship integration, decision-making, and coordination. Following detoxification, cognitive function and memory improved significantly ($p < 0.05$), but other measures did not. Nevertheless, the subjects' own perception of their difficulties did not improve. The effectiveness of detoxification could not be determined conclusively, and the authors urged caution in attributing improvement to detoxification. Serum and body fat PCB content before detoxification did not correlate with results of neurobehavioral tests. The PCB levels were not repeated after detoxification. Thermal decomposition of PCBs, as occurs in a fire, yields polychlorinated dibenzofurans and dioxins that may be 100 to 10,000 times more toxic than PCBs. The authors comment that tissue levels of these by-products, if they could be measured, might correlate better with symptoms and neurobehavioral indices. Clearly, such trials need to be replicated. The mass balance of chemicals must be tracked carefully to be sure that chemicals are not migrating to other parts of the body and that they are being excreted.

Xylene, a solvent that off-gases from paints, varnishes, glues, printing inks, and other sources, is one of the most prevalent indoor air contaminants. Riihimaki and Savolainen (1980) exposed healthy male volunteers to constant (100 or 200 ppm) and varying (200 or 400 ppm hourly peaks) concentrations of xylene, adjusting baseline concentrations in the latter case so that a mean concentration of 100 or 200 ppm was maintained. Exposures occurred over a six-hour period (with a one-hour break at noon) for five days, followed by a two-day weekend and one to three more days of active exposure to xylene. A variety of psychophysiologic parameters were measured, including reaction time, body balance, manual dexterity, and nystagmus. Following cessation of exposure, breath xylene concentrations fell rapidly at first (half-life about 0.5 to 1.0 hours) during the first few hours after which the elimination rate slowed markedly (half-life about 20 to 30 hours). Elimination from fat was estimated to be even slower, around a half-life of 58 hours. Indeed after six days of exposure (5 days + weekend + 1 day), the concentration of xylene in gluteal subcutaneous fat was ten times higher than the blood concentration at the end of the last day of exposure. Thus some accumulation of xylene in fat occurs over several weeks of repeated daily exposure. The data underscore the potential importance of a detoxification method that would accelerate the elimination process. Indeed, elimination of xylene from the body is relatively rapid compared to many xenobiotics whose half-lives are on the order of weeks, months, and years. Most indoor air contaminants are solvents whose tissue levels will be minimal (though from a health standpoint still potentially significant) after several days' avoidance of exposure. In future studies it will be important to document changes in blood levels and tissue levels of target substances, such as xylene, in those who enter an environmental unit or undergo sauna detoxification.

Of particular interest, Riihimaki and Savolainen (1980) observed that most of the adverse effects of xylene upon their normal subjects "tended to disappear after a few succeeding days of exposure." However, "after the weekend away from exposure, the effects were again discernible." They conclude: "This phenomenon suggests that tolerance had developed over a few days with regard to psychophysiological effects by xylene." Parallel to this, the authors observed that fluctuating (as opposed to continuous) concentrations of xylene provoked EEG changes consistent with decreased vigilance. In one subject who had had a normal EEG when he entered the study, fluctuating xylene concentrations provoked bilateral spike and wave complexes suggestive of marked interference with brain electrical activity. Thus exposures occurring minutes or even days before testing may influence the response to a test reexposure. Adaptation, which figures prominently in responses to xylene and ozone, likely affects responses to other xenobiotics. Avoidance of exposure (by having patients change residences or jobs or by utilizing an environmental unit) begins the process of deadaptation. Clearly, finding some way to hasten this process would be helpful, particularly during the late, slow phases of elimination of chemicals from bodystores, for example, from adipose tissue. For this reason, sauna detoxification, if effective and safe, might prove an attractive treatment alternative.

Nutritional Approaches

Extensive data now indicate that vitamins and minerals influence the toxicity of environmental incitants. Amino acids and fat content of the diet also may be important. For example, vitamin E deficiency increases ozone toxicity in rats. An awareness of the role of nutrition in allergy is also developing. Low vitamin B_6 concentrations have been found in adult asthmatics, and supplementation with B_6 produced a dramatic drop in the frequency and severity of wheezing or asthma attacks (Reynolds and Natta 1985). Another recent paper reports significant improvement in atopic dermatitis with vitamin C supplementation (Kline et al. 1989).

Rea and other clinical ecologists (Johnson and Rea 1989; Rogers 1990) routinely measure vitamin and mineral levels in their patients with chemical sensitivities and supplement as indicated. In a sample of 118 patients studied in the environmental control unit in Dallas, mineral levels outside the normal range were found in 53 percent (higher than normal) of the patients for magnesium, 88 percent (lower than normal) for chromium, and 47 percent (higher than normal) for aluminum (Johnson and Rea 1989). A number of abnormal vitamin levels were also found. In some cases. testing for deficiencies involves more than a routine blood test; Rea feels that red blood cell and plasma magnesium levels are poor indicators and prefers an intravenous magnesium challenge to assess magnesium status (Rea et al. 1986b). When ecologists recommend nutritional supplements for their patients, most exercise extreme caution to avoid vitamins derived from food sources that might trigger symptoms; for example, vitamin C from sago palm may be substituted for the usual commercial vitamin C preparations, which often contain corn. McLellan (1987) cautions that before nutritional therapies are embraced too quickly, one must recognize the lack of human data and the fact that most available research pertains to the interaction between single nutrients and single toxins in relatively high doses in contrast to the mixed exposures at lower levels encountered by the patient with multiple chemical sensitivities. However, nutritional status is relatively easy to measure, supplementation can be done fairly safely, and at least a theoretical basis exists for using supplements. Animal and human research have supported use of antioxidants, vitamins A, C, and E, and selenium to protect against certain pollutants (Calabrese 1978; Shakman 1974). Such antioxidants may prevent free radical production that could trigger synthesis of inflammatory prostaglandins (Metz 1981; Cross 1987). Levine (1983) advocates use of antioxidants such as vitamins A, C, and E, zinc, and selenium to prevent free radical formation, which can result in cell membrane damage, release of inflammatory mediators, and perhaps formation of antibodies to altered tissue macromolecules.

Galland (1987) reports several nutritional abnormalities, most notably decreased excretion of essential amino acids in 40 percent of his chemically sensitive patients despite a high-protein diet. Erythrocyte superoxide dismutase activity was decreased in 89 percent (24 patients) versus 79 percent (15) in allergic controls (not significant); erythrocyte glutathione peroxidase activity was decreased in 48 percent (11) versus 36 percent (5) allergic controls (not significant). Some of Galland's controls might have been misclassified because all had either allergies or somatic complaints such as fatigue. Nevertheless, Galland reports that supplementation with antioxidants, including selenium, copper, zinc, and sulfur-containing amino acids, produced major clinical improvement in 25 percent (14) of chemically sensitive patients.

An enormous number of other therapies too numerous to mention have been invoked by clinical ecologists and others. These include dietary, neutralization, and pharmaceutical treatments for candidiasis (indeed, some *traditional* allergists mentioned they found nystatin beneficial in treating certain patients for systemic candidiasis, and they wished for more data to help evaluate this treatment), acupuncture, pancreatic enzymes for food intolerance, oral sodium cromoglycate for food intolerance, and transfer factor. These therapies are outside the scope of this book, but clearly a host of therapies, many of which have been severely criticized for being "unproven," are being offered to patients with multiple chemical sensitivities in an effort to improve their outcome. Some patients report that they obtain small increments of benefit, occasionally more, from each intervention, but none is curative.

For the patient who is having an acute reaction to an environmental incitant or food, ecologists recommend certain "first aid" treatments. In the event of an acute reaction, particularly to a food, some patients take baking soda or a combination of sodium bicarbonate, potassium bicarbonate, and calcium carbonate (so-called tri-salts) with water. They claim these measures relieve their symptoms within moments. Critics caution against possible dangers in using such treatments indiscriminately, for example, as the treatment for IgE-mediated food anaphylaxis. The ecologists' rationale for this therapy, which patients claim can be quite effective acutely, was discussed in Chapter 4. Some use powdered vitamin C mixed in water to mitigate a reaction. In severe cases, for example, to stop a seizure resulting from a food or chemical challenge, some ecologists administer bicarbonate or vitamin C intravenously. Administration of oxygen has also been used in severe reactions. Randolph (1987, pp. 50–51) examined scleral blood vessels of patients before, during, and after reactions to foods and noted increased sludging of red cells; he reasoned that such clumps decreased the red blood cells' ability to carry oxygen. Others have criticized the use of oxygen in these patients without first obtaining a blood oxygen level (Terr 1986). However, oxygen is accepted by many as an effective drug for certain medical conditions, even if the blood oxygen level is normal. The manner in which these treatments are administered to patients may also be important. Plastic or rubber face masks and fresh plastic tubing commonly used for oxygen and intravenous lines may leach small amounts of plasticizers or other substances and provoke symptoms in some chemically sensitive patients.

In summary, clinical ecologists employ a wide variety of treatments, some unproven, in their efforts to help their chemically sensitive patients. When first used, any medical therapy is experimental, but because placebo effects may be significant, ultimately careful, blinded clinical trials are essential for establishing a therapy's value. Increasingly, it is argued that patients with disabling diseases demand and deserve the opportunity to try new, albeit unproven, treatments, provided these do not result in serious harm. Few medical therapies are without some hazard: over the past four decades, at least 46 deaths have occurred following conventional allergy shots or skin testing (Lockey 1987).

Arguments concerning ecological therapies must be kept in perspective. Salvaggio, an allergist, reflected upon the role of unproven therapies in medicine (Salvaggio and Aukrust 1981):

> The practice of medicine will, to be sure, remain primarily an art rather than a science, and physicians will of necessity continue to use clinical judgment and weigh benefit/risk ratios in prescribing a large number of therapeutic procedures that have not been proved to be efficacious by controlled studies. Indeed, one could fill several pages with a list of commonly employed therapeutic procedures in all fields of medicine that have not been proved to be efficacious.

Psychological Interventions

The discussion in Chapter 4 on possible psychogenic mechanisms argues that chemical sensitivity may have physiological causes, psychogenic causes, or both. The search for a cause in a specific patient is most likely to lead a physician to pursue one avenue before investigating the other. Often, however, only one avenue is pursued. The investigator or diagnostician could make either of two kinds of mistakes: in pursuit of an environmental cause, true psychogenic causes could be ignored or, alternatively, in pursuit of a psychogenic cause, true environmental causes could be ignored. The consequences of making those mistakes are different. Pursuing the psychiatric route first may subject the patient to the complexities of establishing a therapeutic relationship and/or the prescribing of psychoactive drugs, and both may generate doubts concerning the patient's mental health. In addition, psychotherapy may be unproductive if environmental causes are at work. Labeling a patient as having a psychiatric illness may be pejorative from the perspective of an employer, co-workers, and family. That psychiatric records are kept separate from the medical records of patients is no accident. In the event that psychoactive drugs are used, unraveling an environmental cause or contribution to the patient's underlying condition may be greatly complicated.

Alternatively, if environmental causes of the illness are investigated first, especially with double-blind, placebo-controlled study in an environmental unit, the patient may discover an environmental cause; even if the patient does not, the confidence or justification with which a psychogenic etiology could be pursued is strengthened. Workup in an environmental unit is unlikely to interfere with or complicate subsequent psychiatric workup, and thus a mistake made in choosing this option (investigating environmental causes first) can be more easily remedied. Black et al. (1990) and Rosenberg et al. (1990) suggest approaches which physicians who are skeptical about the existence of chemical sensitivity may employ to help establish a therapeutic relationship and keep their patients "within the medical fold" (Black 1990). Galland (1990) argues "One cannot empower a patient and at the same time dismiss his or her puzzling symptoms as psychogenic." Black acknowledges that the patients he studied who had seen clinical ecologists were generally satisfied with their ecologist and dissatisfied with the approaches of traditional medicine. Ecologists offered support and understanding of their pain and suffering and a physical explanation for their symptoms.

When adequate controlled studies are done, it may be revealed that some or even the majority of individuals with multiple chemical sensitivities have had episodes of depression or other psychological symptoms years prior to the onset of major disability. For example, Simon et al. (1990) report that plastics workers who developed environmental illness were more likely than controls to have a prior history of anxiety or depression (54 percent versus 4 percent) and a larger number of medically unexplained illnesses (6.2 versus 2.9). It bears repeating that prior psychiatric symptomatology neither proves a psychogenic etiology nor disproves an environmental one. Patients who are disabled by multiple chemical sensitivities may represent a more sensitive subset of the population because of their genetic endowment or even exposures as children. Members of this group may experience depression, asthma, or headaches at low levels of chemical exposure for years without being aware of the cause, yet after a major exposure they might become disabled and exhibit greatly magnified sensitivities to subsequent low level exposures. Thus the persons most likely to develop multiple chemical sensitivities may in fact be those who were more sensitive to begin with, though not recognized as such.

In summary, one can remain agnostic about which route is likely to uncover the truth regarding causation, but the costs of erring are significantly different regarding the two routes of investigation. We think that these facts are sufficiently compelling to justify the investigation of environmental causes first, before committing patients to potentially detrimental psychiatric interventions, such as long-term psychodynamic psychotherapy or long-term medication. Certain short-term or focused cognitive or behavioral therapies may be beneficial but should not be relied on to the exclusion of evaluating the chemical component. Once diagnosed, chemically sensitive patients may find psychotherapy, biofeedback, and other approaches supportive while they make lifestyle changes.

Areas of Agreement and Disagreement between Allergists and Clinical Ecologists

On the basis of interviews with key individuals in allergy, clinical ecology, and occupational medicine, as well as the literature we reviewed, we have discovered both areas of common ground regarding the chemically sensitive patient and areas of disagreement. Although some of the tension between allergists and clinical ecologists may stem from a competition for patients, the differences in their scientific and medical viewpoints are also more fundamental. All physicians agree that chemical exposure can be harmful to any and all systems of the body. Disagreements exist as to what levels of exposure are necessary to cause health effects, what particular symptoms or diseases are associated with specific chemical exposures, and what mechanisms of causation come into play. The range of opinion is wide as to the extent to which the problems of the chemically sensitive patient are psychogenic in origin.

Physicians we interviewed concurred that isolation of the patient in an appropriate environmental unit away from chemical substances in food, air, and water is essential to unraveling the myriad substances that may be causing a variety of effects. More specifically, all of the traditional allergists with whom we spoke acknowledged that in the study or workup of patients with possible environmentally-induced disease, attention must be paid to the potential role of adaptation. Low-level exposure to chemicals must be avoided prior to testing patients for chemical sensitivities in order to avoid adaptation and the loss of a measurable effect. All allergists acknowledged the necessity of controlling for adaptation in any rigorous study of chemical sensitivity, as has been done in the study of ozone (see Chapter 2). Further, all agreed that an environmental unit such as that formerly operated by Dr. Selner in Denver would be an important tool for future investigation and understanding of chemical sensitivities.

Some allergists tend to favor psychiatric referral for patients who do not improve, whereas clinical ecologists are of the opinion that patients' problems, although difficult to solve, are nonetheless likely to be physical in nature. Ecologists feel that environmental factors must be carefully excluded (in an environmental unit if necessary) prior to invoking psychiatric diagnoses.

All physicians agree on the need for studies to clarify unproven therapies, and some physicians in both allergy and clinical ecology think both specialties ought to work together to design the necessary protocols, conduct the studies, and evaluate the results. A few allergists are embracing the fundamentals of clinical ecology such as adaptation and avoidance, but decline to identify their views with those of the clinical ecologists. Allergists have been openly hostile to clinical ecology in the past (AAAI 1980, 1981, 1986). Recently, however, some physicians have become tired of name-calling and legal entanglement, which they recognize as contrary to their patients' best interests, and increasingly want to air and resolve their differences and identify avenues of cooperation. For a fuller understanding of the differences in viewpoints between allergists and clinical ecologists, see Bell's (1987b) article and Terr's position paper on clinical ecology for the American College of Physicians (1989).

For a discussion of avoidance as a treatment modality, see the section on "Avoidance" in Chapter 10.

PART · III

Responding to the Problem

CHAPTER 6

Needs, Concerns, and Recommendations

Research Needs

> The time has come to give to the study of the responses that the living organism makes to its environment the same dignity and support which is being given at present to the study of the component parts of the organism. . . . Exclusive emphasis on the reductionist approach will otherwise lead biology and medicine into blind alleys.
>
> René Dubos

Earlier chapters of this book focused on the magnitude and nature of the chemical sensitivity problem, possible mechanisms, diagnostic approaches, and therapies. These chapters addressed various issues from the perspective of an individual patient or, perhaps more correctly, from the perspective of a physician-scientist looking at individual patients. A problem with such a variety of possible causes and multitude of possible effects might seem to be hopelessly complex to sort out, but in a very real sense the complexity and multifactorial nature of the problem may contribute to its clarification.

Patients with chemical sensitivity differ among themselves and present themselves differently to different physicians. The diagnosis patients receive appears to depend in a very real sense upon which physician's door they enter. Moreover, very different people enter particular

147

physician's doors, manifesting a referral and selection bias. For example, Rea sees individual referrals from other physicians and self-referred patients, 20 percent of whom go on disability, whereas Terr published data on patients referred to him for evaluation, mostly for compensation purposes. Physicians seeing patients with problems stemming from tight buildings or industrial workplaces may see still different groups of affected individuals. The proverbial problem of the blind men and the elephant is the result.

An important research goal is not only the accurate characterization of symptoms and their relationship to specific chemicals but, more fundamentally, characterization of the various populations or groups that appear to be chemically sensitive. We have attempted a preliminary categorization in Table 6-1. Refinement of this categorization is essential and must be the first step in sorting out the myriad chemically caused sensitivities, some of which may represent classical toxicity, some classical allergy, and some what we term multiple chemical sensitivities. Of special importance is the identification of sensitizing events, when they occur.

The four rather distinct groups of patients are: industrial workers; workers and schoolchildren in tight buildings; members of communities exposed to air and water pollution from toxic waste dumps, aerial pesticide spraying, groundwater contamination, or other industrial exposures; and a heterogeneous collection of individuals whose exposure may come from domestic indoor air, consumer products, pesticide use, or other personal contact.

As the description of the patient demographics in Table 6-1 reveals, these patients often may differ greatly in employment or professional characteristics, socioeconomic status, sex, and age. They are also likely to see very different categories of physicians. Industrial workers are much more likely to see occupational physicians or private physicians; the sickest workers may eventually consult clinical ecologists. Individuals suffering from sick building syndrome are not as likely to seek out or be referred to clinical ecologists, even though they are conscious of the fact that their problems stem from tight buildings. People in polluted communities may find themselves going from physician to physician before seeing a clinical ecologist who then may determine that their problems are related to chemicals; by the time these individuals see the clinical ecologist, they may be frustrated, angry, and confused. Finally, the host of other individuals whose exposure to chemicals comes from domestic indoor air, consumer products, pesticide use, and the like may vary greatly in the type and the seriousness of their symptoms and are likely to have seen a series of physicians including allergists and clinical ecologists. The most difficult patients encountered by the clinical ecologists and some allergists may ultimately be referred to Rea, who often sees those most seriously afflicted, persons whose condition therefore may be less reversible. He also probably sees a greater diversity of persons with chemical sensitivities than are represented in the other three groups discussed above. In contrast, those allergists who see chemically sensitive patients referred for worker's compensation evaluation may view only a small segment of the most ill patients.

Multiple chemical sensitivities thus encompass a broad spectrum of people. The allergist sees them either because the patients believe they have an "allergy" or because they are referred by insurance companies or employers for workers' compensation purposes. Which physicians see which patients seems to affect greatly the acceptability of their problem as a bona fide physical illness, or at least not a problem of psychogenic origin. Relatively few physicians today would call hypochondriacal those who are affected by tight buildings. Most workers with chemical sensitivities seek workers' compensation as a matter of last resort. They would prefer to be able to work (Davis 1989).

The exposure-patient profile of people suffering from chemical sensitivities has to be characterized accurately in order to fashion an appropriate response. Such a categorization may also be useful in suggesting areas of research that might be undertaken by federal agencies such as the National Institute for Occupational Safety and Health (NIOSH), the National Institute of Environmental Health Sciences (NIEHS), the Environmental Protection Agency (EPA), and the Agency for Toxic Substances and Disease Registry (ATSDR), which is concerned especially with community-based pollution related to toxic waste facilities and contamination of water supplies and has cooperative agreements with 11 states to undertake surveillance studies in contaminated communities for exposure and disease. In addition, state-based efforts, usually involving state departments of health and possibly state departments of environmental protection, need to be encouraged. In the event that the federal government is not willing to undertake research as to both the nature and the etiologies of chemical sensitivity disorders, a multistate effort may well be advisable.

Most scientific criticisms of clinical ecology have been directed toward the efficacy of provocation-neutralization therapies. To focus initial research efforts in this area would be of limited value because most traditional practitioners question the diagnosis itself. Clinical ecologists must continue to develop objective means of measuring symptoms and relating them to exposures.

A properly constructed environmental health unit could serve as a focal point for studying chemically sensitive patients in the deadapted or unmasked state. Allergists, clinical ecologists, and toxicologists should

TABLE 6-1. Spectrum of Multiple Chemical Sensitivities

Aspects	Industrial Exposure	Tight Buildings
Recognition of problem	Workers themselves; unions; occupational health clinics	Office or school workers themselves; parents of school children; school nurses
Place of diagnosis	Occupational health clinic; private physician's office	Private physician's office; occupational health clinic
Nature of exposure	Industrial chemicals; acute or chronic exposure	Off-gassing from construction materials, office equipment, or supplies; tobacco smoke; inadequate ventilation
Demographics; awareness	Primarily males, blue-collar, 20s to 60s; conscious of classical workplace hazards; often aware of relationship between symptoms and exposure (e.g., better on weekends/vacations, worse at times of peak production or during certain processes, etc.). Awareness via word of mouth, union, occupational physician	Females more than males; children; white-collar office workers and professionals; 20s to 60s; aware of symptoms associated with change in building environment, e.g., new construction, carpeting or seasonal illness. Awareness via word of mouth, media, field study
Manifestations	Multiple symptoms involving multiple systems with marked variability in type and degree of symptoms. CNS symptoms common. Physical exam most often unremarkable	Multiple symptoms involving multiple systems with marked variability in type and degree of symptoms. CNS symptoms common. Physical exam most often unremarkable. Many have symptoms of eye, nose, throat irritation, and malaise
Average severity of illness or disability[a] (varies greatly with individual)	Moderate to severe	Mild

[a] Those individuals whose symptoms are most persistent and disabling often see a series of physicians before they finally see an ecologist. Rea's clinic represents a kind of tertiary referral system for ecologists who send him the most severe cases. Thus the patients Rea sees are markedly different, i.e., more disabled, than the typical patient seen by an allergist.

Community-based Air and Water Pollution	Individual Exposures
Individuals in community; government sponsored field surveys	Individuals themselves
Private physician's office; state-supported clinical study	Private physician's office
Toxic waste dumps, aerial pesticide spraying, ground water contamination, air contamination by nearby industry, and other community exposures	Heterogeneous, personal; indoor air (domestic), consumer products, pesticide use
All ages, males and females; children/infants may be affected first/most; middle to lower class; community awareness via word of mouth, community action groups, media, or field study	70–80% females; 50% in 30-to-50 age bracket (Johnson 1989); white; middle class, upper middle class, and professionals; awareness via word of mouth, patient groups, physicians
Multiple symptoms involving multiple systems with marked variability in type and degree of symptoms. CNS symptoms common. Physical exam most often unremarkable.	Multiple symptoms involving multiple systems with marked variability in type and degree of symptoms. CNS symptoms common. Physical exam most often unremarkable.
Mild to moderate	Mild to moderate if referred to allergists or clinical ecologists. Severe if seen by Rea in Dallas. Disabling if admitted to environmental unit or an applicant for workers' compensation or disability.

contribute to the design of any such study to ensure acceptance of the results. The unit should be constructed and operated following the highest academic standards. For reasons discussed in Chapter 1, epidemiological studies on chemically sensitive populations have to be designed with extreme care, or no evidence will emerge at statistically significant levels. The study group must be appropriately defined, and more than one symptom may need to be counted as a health effect. Biomarkers of both sensitization and sensitivity should be identified, when possible. If immunologic or biochemical mechanisms are involved, this may be a promising area; biomarkers involving the limbic or nervous system in general, however, may be more difficult, if not impossible, to identify, especially since there may not be a blood-borne marker.

Given clinical ecology's low status and reputation in the scientific community at the present time, independent researchers have serious disincentives to examine its tenets. Yet a critical and unbiased airing of the problem of chemical sensitivities is needed. All aspects of the problem —from documentation of the sensitivity itself to diagnostic approaches for its discovery to the range of possible therapies—need attention. The National Academy of Science Panel on the Interrelationships of Toxic Exposures and Immune Response should be encouraged to study the problems of multiple chemical sensitivities as well as problems of immune system damage or dysfunction.

Government and university scientists must be allowed and even encouraged (by grants) to participate in research in this field without being hamstrung by the opinions of traditional medical practitioners. Science is not served by continuing to deny the probable existence of the problem in the face of massive and growing circumstantial evidence, although admittedly subjective in many respects. A better approach would be to acknowledge that something appears to be going on, that low levels of chemicals can affect the body in subtle ways that currently escape our understanding, and that individual susceptibility to environmental agents may vary by several orders of magnitude.

The widely circulated journals representing traditional medicine must also allow unbiased airing of this problem, as they have with other issues, for example, medical versus surgical treatment of coronary artery disease. Numerous university and government scientists who are knowledgeable about chemical sensitivity feel it is worth taking seriously. However, many fear for their own professional careers and are reluctant to write or speak openly on the subject. Fortunately, this appears to be changing. Debate rather than unilateral criticisms of unproven ideas is what is needed to encourage defensible research on these ideas. Certainly, criticisms of therapies should not be used to foster a denial of the existence of chemical sensitivity altogether. Recently, physicians and scientists appear to be more willing to accept the concept of chemical sensitivity in the context of occupational exposures or tight buildings. Perhaps because most patients suffering from these exposures do not seek clinical ecologists, mainstream physicians are probably more accepting of the problems in those contexts. If clinical ecologists are involved, there seems to be more of a desire to shoot the messenger than to take the problem seriously. As similarities are recognized among patients whose exposures arise in different contexts, we hope a more scrutinizing evaluation will be forthcoming.

Patient and Community Concerns

In this section we articulate the concerns and needs of the chemically sensitive patient and the needs of the community in preventing illness triggered by or associated with low-level exposures to chemicals. These needs include information; health care; alternative schooling, employment, and housing; medical insurance; compensation for disability; social and legal services; and the control of chemicals in the office, industrial workplace, home, and consumer products.

Information

Information about Chemical Sensitivity. Chemically sensitive persons need information and guidance concerning the recognition of chemical sensitivity and the availability of diagnostic tools and possibly effective therapies. They need to understand that chemicals can cause both classically recognized environmental and occupational disease (such as lead or solvent poisoning or allergic reactions to organic dust) and less understood, but nonetheless real, problems associated with low-level exposures. Industrial workplaces can give rise to both kinds of environmental illness, and unraveling the vagaries of causation there may be especially difficult. Home and office environments also present a mixture of illnesses, perhaps more often characterized by lower exposure levels. Episodic exposures to chemicals present still different challenges. Chemically sensitive persons need assistance in understanding their condition so that they can make reasoned choices about the health care or preventive actions they might pursue.

Information about Chemicals. Chemically sensitive patients need to know the potential hazards of the chemicals they work with or may be exposed to. Federal and state right-to-know legislation and the Superfund

Amendments and Reauthorization Act (SARA) Title III reporting requirements are important legal avenues for information held by government agencies, employers, and manufacturers and producers. However, patients may need more specific information that perhaps could be provided by industrial hygiene surveys by OSHA, state agencies, or possibly insurance carriers. Access for individuals to state and federal sources for information and services is needed.

Health Care

Access to Appropriate Care from Private Physicians and Clinics.   In one profile of chemically sensitive patients seen at a clinical ecology clinic, 30 percent of the patients had seen six to ten physicians before coming to the unit (Johnson and Rea 1989). Brodsky (1987, p. 695) observes:

> Many of these patients shift from one specialist to another, going from family physicians to allergists to neurologists and other medical specialists, and to chiropractors, acupuncturists, homeopaths, and even faith healers. Both the patients and their physicians feel frustrated and dissatisfied, the patients because they remain convinced that their symptoms signal a physical disorder for which a medical explanation must exist and the physicians because they have been impotent as healers, unable to help these obviously distressed individuals or to reassure them that they do not have a serious disease. Such patients are time-consuming and in clinic settings not infrequently objects of derision.

Our investigations make clear that the chemically sensitive patient finds medical care by trial and error and by word of mouth. A more directed path and identification of helpful physicians and clinics are needed. Patients whose problems stem from occupational, sick building, or home environments may each need different care, as might persons suffering from episodic exposures. No sensible referral system exists, and the segmented nature of medical care and the inability of some physicians to acknowledge a disease they do not understand contribute to the personal suffering of the sometimes bewildered patient. Multi-specialty clinics, health maintenance organizations (HMOs), and preferred provider organizations (PPOs) unfortunately rarely include specialists in diseases caused by chemicals. Occupational medicine clinics have the potential for expanding their concerns and services into illnesses caused by low-level exposures, but specific initiatives are needed to bring about this expansion. Taking work and/or environmental histories is essential to delivering appropriate medical care. Industrial hygiene services in both the home and workplace may also be indispensable.

University-based clinics, such as that at the Robert Wood Johnson Medical School in New Jersey, have the potential for rapidly incorporating new knowledge and perspectives more quickly for the recognition, diagnosis, and treatment of the chemically sensitive patient. The potential needs to be realized, however.

Avoidance. The first line of defense for the chemically sensitive patient is avoidance of offending chemicals and substances found or suspected to cause problems. Some kinds of avoidance are easy, but others are very difficult. Persons suffering from sick building illness may have to abandon employment or residence in those buildings. Some new EPA offices have been planned, which have windows that open, no carpeting, no copier machines, and other features, in an effort to accommodate sensitive employees. In the case of children who are chemically sensitive, an alternative to home tutoring may be the state provision of environmentally safe classrooms where students with documented and suspected chemically related disorders could be educated and observed for follow-up. This practice is currently being pursued in Canada (Rapp and Bamberg 1986). In Maryland's guidelines for indoor air quality in schools, modification of workplace exposure limits is recommended (Maryland 1987, p. 6).

Alternative Employment and Housing

Alternative Employment. Federal and state laws require that employers provide handicapped workers reasonable accommodation. Deciding whether a chemically sensitive person is handicapped is done on a case-by-case basis. In some instances, work at home may be possible; the EPA seems to have reached this accommodation with some of its chemically sensitive employees (Hirzy 1989b). Patients who are not considered handicapped nonetheless need rehabilitation and return to gainful employment.

Alternative Housing. Some chemically sensitive patients cannot live in their prior domiciles. Because of severe economic consequences, some assistance is obviously needed. Options include halfway houses for the severely affected and the establishment of experimental communities in less polluted areas.

Medical Insurance

Receiving reimbursement for medical expenses incurred when the diagnosis and treatment of chemical sensitivity is performed by clinical ecologists is becoming increasingly difficult (Davis 1989). Even though a prominent allergist and critic of clinical ecology, Abba Terr (1989b) has stated that "what we do is as unproven as some of the things we are criticizing," allergists have organized letter-writing campaigns to urge the Health Care Financing Administration (HCFA) to deny reimbursement for "unproven" techniques. Such efforts do not seem to deter the desperate patient from seeking medical care; they just make the patient more desperate. Recognition of diagnosis and treatment of chemical sensitivity for insurance purposes is necessary on grounds of fairness and, in the case of some patients, to enable them to receive adequate care. For traditional medical practitioners to throw up their hands and not be able to help these patients and, at the same time, to lobby vigorously to deny them therapies that sometimes, if not often, relieve their suffering cannot be justified.

Compensation

Chemically sensitive patients are sometimes unable to continue in a specific workplace, often an industrial workplace, or may not be able to work at all. Workers' compensation systems have been painfully slow to provide coverage for occupational diseases. Employers and their insurance carriers have historically denied the work-relatedness of disease associated with chemical exposures; for example, 60 percent of occupational lung disease claims are contested (U.S. Department of Labor 1980). Compensation for the chemically sensitive worker is vigorously resisted, and in some cases patients have to be labeled with a psychiatric disorder such as posttraumatic stress disorder in order to receive compensation for their illness. Earon Davis (1989), an attorney and executive editor of the *Ecological Illness Law Report* believes that only about 1 percent of severely affected chemically sensitive workers will file a workers' compensation claim because they do not want to be labeled as psychiatric cases. In addition, many workers leave jobs because of chemical sensitivity only to find themselves unable to tolerate a new job and unable to file a claim against either the new or old employer.

In most instances chemically sensitive people cannot trace their problems to a specific work-related exposure and may seek disability payments under Social Security. William Rea (1989), who sees the most severely affected patients, reports that about 20 percent of those he sees go on disability. In cases of illness stemming from pesticide exposure or from some other episodic exposure involving industrial or consumer products, the patient may seek compensation through a suit in tort. In 1987, the Consumer and Victims Coalition Committee of the Trial Lawyers of America adopted a resolution supporting "environmental illness" victims. Recovery of damages based on alleged immune system damage is becoming more commonplace, and the manufacturers and insurance industry are reacting vigorously. Dennis Connolly (1988), an insurance executive, writes:

> Courts as well as scientists are routinely grappling with the problems of harmfulness and causation. A disturbing trend from the point of view of those who might be looking toward providing insurance is the increased use of various forms of marginal science to overcome difficulties in proving causation. "Clinical ecology" is a "science" offering broad support for causation in bodily injury cases, but the science has been repudiated by many in the medical establishment and cited as an example of poor science flourishing in the courtroom.

Academics who have joined the criticism include Donald Elliott (1988), who writes that "plantiffs in toxic tort cases are increasingly relying on testimony by a small group of professional witnesses called 'clinical ecologists' (whose views are repudiated by the scientific establishment). Lay juries and the public are vulnerable to being misled by such 'experts.' "

A more thoughtful analysis of the use of clinical ecology in the courts is provided by Sheila Jasanoff. After tracing the successes and failures of patients in their attempts to establish harm to their immune system by using the testimony of clinical ecologists, Jasanoff (1989, pp. 86–87) notes: "That medical societies [the American Academy of Allergy and Immunology and the California Medical Association] might not be wholly disinterested in their efforts to discredit a competing, and apparently successful, specialty does not seem to have been a worry [to the court]." She concludes that these societies "proved effective because these organizations defined the boundaries of valid medical science in a way that courts could not readily ignore." Nonetheless, the principles established by clinical ecology are enjoying some success in the courts (Cornfeld 1989).

Powerful economic and industrial forces have joined to deny the chemically sensitive patient compensation, just as they did earlier in this century for occupational injury and later for occupational disease, by accusing the worker of malingering and bad faith. The issue of compensation may seem peripheral to the scientific-medical debate over chemical sensitivity, but it is actually central to the resolution of public policy in this area. Economic issues heighten the conflicts between allergists and clinical ecologists and need to be resolved if an optimum scientific consensus is to be achieved. The existence or origins of the patients' disease are contested in conflicts over who should pay while the patient continues to suffer. Brodsky (1987, p. 696) observes: "Private and public agencies that provide disability benefits argue that these patients are not truly disabled, although those dealing with them recognize that they are in great distress."

Social and Legal Services

As with others debilitated by disease, chemically sensitive people need psychological, financial, and legal counseling to enable them to manage their affairs, seek help from appropriate government agencies, and cope with stress. Some of these services can be provided by state government and private patient-support groups. Earon Davis (1989) reports that many chemically sensitive persons suffer neuropsychological defects often resulting from continuing unavoidable exposures and have extreme difficulty interfacing with the legal or social service systems. He argues that these persons need social workers, not lawyers, who can guide them into avenues that improve their situation. Such guidance could be provided at halfway houses or special communities.

The Regulation of Chemical Exposures and Other Preventive Initiatives

The community has an interest in preventing and limiting the problems of chemical sensitivity. For chemical sensitivity that has its origin in exposures to chemicals in the workplace, pesticides, chemical spills, and the like, adherence to and enforcement of *existing* environmental regulations is necessary to prevent sensitization of more individuals. The existing standards of OSHA, EPA, and state agencies do not, however, protect those individuals already sensitized. New regulations governing inadequately regulated substances or unregulated applications of chemicals, such as pesticides or other chemicals applied in office buildings, schools, or apartment complexes, are also needed. (Following health complaints associated with new carpeting at EPA headquarters in Washington, D.C., the EPA union petitioned EPA to regulate 4-phenylcyclohexene, the chemical suspected of causing health problems [National Federation of Federal Employees 1989]. However, the EPA denied the petition and instead embarked on a program involving voluntary testing of all carpet-related volatile chemicals by carpet companies [EPA 1990].) At a minimum, regulators should require that application of chemicals, such as pesticides, be accompanied by adequate notice so that people can avoid the exposure. Currently most OSHA and EPA regulations control exposures at the parts per million (ppm) level. More stringent regulations may be needed to protect both sensitized (and hence chemically sensitive) individuals and those who may become sensitized. The mandates behind environmental regulation do indeed require the protection of sensitive populations (Friedman 1981). If regulations are imposed to *prevent sensitization,* it may be less necessary to control exposure more stringently in the future in order to *control sensitivity,* because people will not be sensitized in the first place.

The appearance of similar kinds of health problems in widely divergent populations exposed to chemicals (see Table 1-1) illustrates that the failure to regulate adequately or prevent exposures to chemicals in the environment, workplace, and consumer products has resulted in the present chemically sensitive population. In order to protect this population from further or continuing damage, some chemicals, such as formaldehyde, will need to be controlled at the part per billion (ppb) range or banned outright for some uses (Massachusetts 1989, pp. 74–102). Although the regulation of chemicals traditionally has been viewed as a federal government initiative with states as secondary partners, states may have to take vigorous regulatory action in order to protect the chemically sensitive. Massachusetts, for example, banned the use of urea formaldehyde foam insulation. California regulated vinyl chloride levels in ambient air, even though the federal government issued only emission limitations. Other states may need to examine the adequacy of their regulations.

The Role of Medical Practitioners and Their Societies

The roles that primary care physicians, occupational and environmental health physicians, allergists, and clinical ecologists can play in addressing the needs of the chemically sensitive patient differ, depending upon the group of patients in need. Table 6-2 depicts the strategies that might be followed for each group of chemically sensitive patients. At this time, patients typically consult clinical ecologists and allergists out of desperation, rather than as a result of referrals. Our considered opinion is that a structured, sensible referral strategy needs to be developed.

Primary care physicians are in the best position to provide knowledgeable referrals for chemically sensitive patients by referring them to the health professional most likely to be of help to the patient. (For a general discussion of the role of the primary care physician in occupational and

TABLE 6-2. Strategies for Primary Care Providers

Group	Primary Referral[a]	Subsequent Referral[a]
Workers	Referral to occupational health physicians or clinics Work histories Industrial hygiene surveys	Clinical ecologists Allergists Detoxification programs where appropriate
Occupants of tight buildings	Adults: For office workers, as above Children: Clinical ecologists or allergists	For office workers, as above
Contaminated communities	With help of State health department, EPA, and/or ATSDR[b] Referral to environmental/ occupational health physicians to take an environmental exposure history	Clinical ecologists Allergists Detoxification progams where appropriate
Individuals Pesticides and other toxic substances	As for contaminated communities	As for contaminated communities
Indoor air (domestic)	Clinical ecologists Allergists	

[a] Selected with great care. In our view, this means selecting physicians who take the problem seriously, who do not dismiss these patients' problems as psychiatric without adequate work-up, and who rely upon careful avoidance and judicious reexposure to help determine their patients' sensitivities.

[b] Agency for Toxic Substances and Disease Registry.

environmental medicine, see Institute of Medicine 1988.) Workers exposed to industrial chemicals should be referred to occupational health clinics or occupational physicians. The coupling of industrial hygiene services and a detailed work history help occupational physicians decide what can be done for the chemically exposed patient. In the absence of or in cooperation with an occupational physician, the industrial hygienist may aid primary care physicians in identifying possible illness and relevant exposures. If the problems the worker is experiencing are those of classical toxicity, such as chronic lead poisoning, the occupational physician can help the worker directly. In special cases, such as polybrominated biphenyl (PBB) exposure, some occupational physicians might refer the patient for detoxification therapy to remove the bioaccumulated toxins (Schnare 1986). When the worker is seen to exhibit chemical sensitivity of a nontraditional nature, the occupational physician, if knowledgeable about multiple chemical sensitivity problems, may be able to help the patient. Indeed, many occupational physicians are developing their knowledge in this emerging area. Alternatively, the occupational physician may refer the patient to either a clinical ecologist in whom he has confidence or to an allergist who accepts (recognizes) the problem of multiple chemical sensitivities as real.

Occupants of tight buildings who could be suffering from either classical sensitivity, for example, to molds, or from multiple chemical sensitivities, can also be referred to an occupational health clinic or an occupational physician. The occupational physician may then manage the patient personally or provide the appropriate referral.

For patients that comprise part of a contaminated community, the primary care physician should, ideally, involve the state health department and the EPA or the Agency for Toxic Substances and Disease Registry (ATSDR), which could document exposures and watch for a pattern of illness in that community. (In 1988 the ATSDR [1989, p. 8] awarded a cooperative agreement to support the Association of Occupational and Environmental Health Clinics to expand chemically based information and physicians' educational opportunities relating to toxic substances.) With the assistance of these agencies, the primary care physician can make appropriate referrals to physicians expert in occupational and/or academic environmental medicine. These physicians take an environmental history in much the same manner an occupational physician takes a work history, and this history needs to be coupled both with disease patterns recognized by and with exposure measurements made by the state health and environmental protection departments, EPA, or ATSDR. At that point the occupational or environmental medicine physician can make appropriate referrals to clinical ecologists or allergists in whom he has confidence. Relatively few physicians specialize in environmental medicine. Because environmental medicine and occupational medicine have similar knowledge bases and require many of the same skills, efforts should be directed at developing professionals who span *both* fields in order to serve the chemically sensitive patient.

Finally, for the divergent group of individuals whose illness results from indoor air in the home, pesticide applications, or other chemical exposures, the primary care physician may need to find ways to identify those clinical ecologists and allergists who are able to help the chemically sensitive patient. This group of patients is most challenging because they are diverse and may not fit a particular pattern of illness like the patterns often seen in the workplace, in tight buildings, or as part of a contaminated community. Indeed, some of these patients may not recall a sensitizing event, although they recognize chemical triggers of their symptoms and are polysymptomatic.

The strategies we have outlined for dealing with these four groups of patients need to be carefully developed and refined. The weakest link in affording the patient proper medical care involves raising the consciousness of primary care physicians or those specialists whom the patient might see in a random manner, such as ear, nose, and throat specialists, neurologists, and rheumatologists. However, engaging the primary care physician is the first essential step in sending the patient down a directed pathway of proper referrals. The primary care physician's level of knowledge and concern regarding this problem must be given immediate attention.

The role of the medical specialty societies is central in facilitating the success of these referral strategies. For primary care physicians, including those in family practice, internal medicine, and pediatrics, a clear understanding of the problems of the chemically sensitive patient is requisite. For other specialists, their societies need to address the particular problems of chemical sensitivity that relate to their specialty.

The allergists need to adopt broader perspectives (Kniker 1985), which several allergists seem to be doing. Selner and Staudenmayer (1985b, p. 666) observe:

> It is time for allergy to claim its interest in [the chemical environment] and assume a more active role in the field of toxicology. Allergy is in a position to bring the same disciplined commitment to the principles of scientific investigations to the area of chemical intolerance that has resulted in the remarkable contributions to the field of immunology over the past two decades.

The allergy societies need to commit themselves to a critical but fair appraisal of those techniques and approaches of clinical ecology that may be useful in expanding the practice of allergy beyond its present boundaries. Selner and Staudenmayer (1985a), for example, have stressed the importance of an environmental care unit. Allergists need to be able to take comprehensive work and environmental histories, learn about toxicity and chemical sensitivity, and familiarize themselves with appropriate diagnostic and therapeutic approaches and techniques. Bardana and Montanaro (1989), for example, have suggested that industrial hygiene evaluations of both the workplace and the home "may prove invaluable in identifying chemical sensitivity." The allergy societies should promote the practice of allergy with a broader vision through continuing education efforts and by trying to build on common ground shared with clinical ecology and occupational medicine. Doris Rapp (1985), a board-certified practicing allergist for 18 years and also a clinical ecologist, in response to the position papers of the American Academy of Allergy and Immunology on clinical ecology, cautions:

Try not to wedge the academy in a corner with statements that will haunt allergy in the years to come. If the thinking, leading allergists do not listen, [then] soon, the whole specialty of allergy will be lost. Use your mighty caches of money, and brains to help elucidate our impressive observations and successes. Help us refine what we are doing. Not only will you gain, but the patients, who should be the bottom line of whatever we discuss, will be helped.

The clinical ecologists also need to learn to take better work and environmental histories; to be thorough and not overlook other concomitant medical conditions, for example, hypomagnesemia resulting from a prior partial gastrectomy (Bardana and Montanaro 1989); to engage in continuing educational activities in this rapidly developing area; and to put their work and techniques into a form that would serve as a useful primer for others. The environmental unit is an essential tool for both allergists and clinical ecologists, and their knowledge should be combined in developing new units. The societies whose members practice clinical ecology need to develop rigorous standards for its practices and shun mystical approaches. Bell (1987a) concludes:

Clinical ecology thus needs well-designed, systems-oriented, rigorous interdisciplinary studies. The work must focus on specific diagnostic subgroupings and syndromes as well as on specific immunological and physiological concomitants of adverse reactions. Clinical ecology needs the input of scientists and clinicians from many fields such as public health, occupational medicine, and behavioral medicine, to refine its concepts, treatments, and goals. It otherwise runs the risk of extinction as a fad with several good ideas mingled with too many pernicious and unsubstantiated beliefs.

Finally, clinical ecologists should not simply invoke traditional toxicity as a way of legitimizing the case for avoiding chemical exposures. Mechanisms for multiple chemical sensitivities may be different. Although clinical ecologists may feel pressured to develop a theory of causation, reliance on classical toxicity or allergy, as they are currently understood, may be misplaced.

Recommendations

Having identified the needs of the chemically sensitive person and the community concerned with preventing an increase in the number of chemically affected individuals, we turn to specific recommendations.

Research

States with ATSDR cooperative agreements should establish a registry of chemically sensitive persons with the help of physicians, industrial hygienists, labor organizations, patient-support groups, and others that would be based on physician reports and be as broad as possible to collect data on persons and exposures to be refined and stratified in later analysis. The purpose of the registry is to characterize the nature of the problem and trends over time and to provide a basis for linkage to geographical information system analysis at some time in the future in order to discover sources of exposure.

The federal government should provide funding for a statistically useful questionnaire survey of these persons that stratifies respondents by group, for example, occupationally exposed, occupants of tight buildings, members of contaminated communities, schoolchildren, and the like, and, if possible, by the kind of exposure thought to be responsible for the person's condition, for example, new carpeting, pesticides, and so on. Additionally, states should solicit the financial support of health insurance companies doing business in the state for this effort. State departments of health and ATSDR should analyze the results of the survey in order to identify problem chemicals and affected groups that might serve as the focus for specific field studies.

The federal government should undertake controlled studies of the economic and social effects of indoor air quality on the workforce.

With the assistance of ATSDR, states should undertake field studies of various subgroups of chemically sensitized persons to document their illness. The groups should include occupational groups, contaminated communities, office workers, and children. Studies should involve incidents in which exposures have led to recognized problems, such as certain workplace exposures, toxic waste dumps, contaminated communities, and tight buildings.

The federal agencies (NIH, NIEHS, EPA, and ATSDR) should construct a patient profile of those with chemical sensitivity by evaluating the Environmental Health Center in Dallas. William Rea has agreed in principle to such a study.

The NIH, NIEHS, EPA, NIOSH, and ATSDR should plan a national conference to identify key areas for research into chemical sensitivity that would include allergists, immunologists, clinical ecologists, occupational and environmental physicians, and key governmental researchers.

States should create interagency working groups of state agency professionals to guide the development of state initiatives relevant to the problems of chemical sensitivity.

Information

States should designate one or more professionals to staff a 3-year effort addressing low-level exposures to chemicals. The designated professional (and necessary support staff) should be responsible for preparing written guidelines for the chemically sensitive person designed to provide the affected individual with a clear understanding of the condition and the options for diagnosis, treatment, and compensation. State departments of health should provide a telephone hot line for chemically sensitive individuals in order to guide their inquiries to the appropriate state agencies and offices.

States should request their medical centers to identify, compile, and maintain a list of physicians and clinics interested in handling chemically sensitive patients with consideration, understanding, and relevant medical or other interventions.

State departments of health should prepare educational materials and hold short courses in conjunction with local medical associations to give guidance to primary care physicians in the recognition, diagnosis, treatment, and referral options relevant to chemical sensitivity. Details of possible referral strategies were discussed earlier in this chapter.

The federal agencies should compile and make available emissions data from building materials and consumer products.

States should convene a meeting of those concerned with the design and construction of public and private office buildings and homes to inform them of the problems of indoor air pollution.

Health Care

States should seek funds to enhance the capabilities of existing occupational health clinics to address problems of chemical sensitivity through financial and professional support.

State departments of health should encourage insurance carriers to provide industrial hygiene services for homes and workplaces where multiple chemical sensitivities are suspected. Schools, where problems are indicated, should be investigated by the state.

The federal government, assisted by those experienced in establishing and operating a successful unit, should establish a pilot or demonstration environmental health unit.

The U.S. Department of Education, in consultation with other federal agencies, should evaluate the extent of the problem among schoolchildren and assist states in establishing special classrooms for chemically sensitive children. These special classrooms should be used to study and document the impact of avoidance measures for this subpopulation.

Alternative Employment and Housing

States should educate employers about the chemically sensitive and encourage them to make reasonable accommodations and/or provide alternative worksites within their places of employment and, in some cases, to allow employees to work at home while they improve. States should identify employment options for the chemically sensitive. States should also inform employers and employees of their obligations and rights under federal and state legislation for the handicapped.

Vocational rehabilitation programs, coordinated with programs and activities of state departments of labor and workers' compensation boards, should be established for the chemically sensitive worker.

The state interagency working groups recommended in the Research section ought to be convened to coordinate efforts related to alternative employment and to study housing needs. One option to be studied should be the establishment of halfway houses where newly diagnosed persons or less severely affected persons can recover and receive guidance. Options for the establishment of experimental communities in less polluted environments should also be seriously investigated.

Medical Insurance

The federal government should undertake a study of economic savings that might result from timely and effective medical intervention for chemically sensitive persons.

State departments of health and insurance should use their good offices to express their disapproval of attempts to curb reimbursement for health care for chemically sensitive patients. This effort should be directed towards HCFA, Blue Cross/Blue Shield, and other health insurance carriers. As the problems of the chemically sensitive become better understood, states should do all within their power to facilitate recognition of chemical sensitivity for both health insurance and disability purposes.

Compensation

State departments of health should convene a meeting with the departments of insurance and workers' compensation boards to explain the work-relatedness of chemical sensitivity.

Social and Legal Services

The state interagency working groups recommended in the Research section should study state options for providing access to medically-related social and legal services to those whose illness stems from chemical sensitivity.

Regulation of Chemicals

Both state and federal government should consider revising or adding standards to deal with both chemicals that cause initial sensitization and chemicals that trigger sensitivity, that is, low-level exposure to chemicals in the environment, industrial workplace, office, home, and consumer products. Just as no-smoking areas are provided in public and private facilities, environmentally acceptable areas could be required. States should work closely with the EPA's Office of Indoor Air Pollution to establish federal policy for chemical sensitivity.

Resolution of Conflicts among Medical Practitioners and Their Societies

Federal and state governments should facilitate dialogue and an easing of antagonisms among allergists, clinical ecologists, and occupational and environmental physicians through educational efforts and through co-sponsorship of conferences on chemical sensitivity.

For further discussion of research needs and recommendations, see sections on "Research Needs" and "Research Recommendations" as well as "Help from Governmental Agencies and Other Organizations" in Chapter 10.

PART • IV

Update since the First Edition

CHAPTER 7

Recent Developments

Introduction

More than five years have passed since the publication of the first edition of this book. Interest on the part of the scientific and medical communities in chemical sensitivity has increased markedly. Four major national meetings sponsored by U.S. government agencies, meetings in Canada, and one international scientific meeting on MCS have been held. Numerous professional meetings in the United States and abroad have featured panels, workshops, and individual presentations on the topic. The terms "multiple chemical sensitivity" and "environmental illness" are now listed on Medline, the bibliographic database of the National Library of Medicine, and as of April 1997 this database contained 120 MCS references dating back to 1994. In all the scientific literature, more than half of the papers addressing issues relevant to multiple chemical sensitivity have been published since 1992 (Donnay 1996). Quantity, of course, does not guarantee quality, but it is clear that advances in understanding are occurring.

We therefore thought it timely to review and reflect upon the most important developments and literature. Part IV of this edition selectively updates the original material by adding four new chapters.

A Note on Terminology

As we discussed earlier (see Table 2–1), in North America a number of names have been used for heightened reactivity to one or more chemicals, which in this volume we have called chemical sensitivity. "Multiple chemical sensitivity" or "MCS" is currently the name most commonly used to describe patients with multiple intolerances. An advantage of this term (read as "sensitivity to many chemicals") is that it describes the most distinctive feature of the illness without presuming a particular etiology or mechanism. Nevertheless, some observers think that this name implies a chemical cause or attempts to "legitimize" an illness for which there is insufficient proof of chemical origins. Many patients still use the term "environmental illness" or "EI." Critics of MCS who think the illness is psychogenic have proposed other names (in addition to those in Table 2–1), including "Multiple Symptom Complex" (Gots et al. 1993), "Environmental Somatization Syndrome" (ESS) (Göthe et al. 1995), "Toxic Agoraphobia" (Kurt and Sullivan 1990), "Idiopathic Environmental Intolerance" (IEI) (Anonymous 1996), "Multi-organ Dysesthesia" (Cohn 1994; Kavanaugh 1996), and "Odor Aversion" (Amundsen et al. 1996).

The term "chemical sensitivity" admittedly lacks precision (Miller 1996a). It has been used loosely by various authors to describe the *symptom* of feeling ill from chemical odors, a clinical *syndrome,* and a hypothetical disease *mechanism.* We propose reserving the term "chemical sensitivity" or "chemical intolerance" to describe the symptom of feeling ill when exposed to chemicals. Because a "syndrome" is a "group of symptoms or signs typical of a [single] disease" (*Webster's* 1986), and these patients report incredibly diverse symptoms, it may be technically inaccurate to label the phenomenon a "syndrome." Implicit in discussions of MCS is an unstated supposition that we may be dealing with an emerging new mechanism or *theory of disease.* According to this theory, a two-step process occurs: (1) an initial, salient exposure event(s) interacts with a susceptible individual, leading to loss of that person's prior, natural tolerance for everyday, low-level chemical inhalants, as well as for specific foods, drugs, alcohol, and caffeine; (2) thereafter, such common, formerly well-tolerated substances trigger symptoms, thus perpetuating illness.

To describe this two-step mechanism, and to distinguish it from chemical sensitivity or MCS possibly caused by psychological or physical trauma, Miller (1996a) recently proposed the term "toxicant-induced loss of tolerance" or "TILT" (see the later discussion in Chapter 10). She draws parallels between TILT as theory of disease and the germ and immune theories of disease, and suggests that doctors may be at an early observational stage with respect to their understanding of TILT, much as they were with respect to infectious diseases during the Civil War, the last American war fought without knowledge of the germ theory of disease. Research on the specific mechanism(s) underlying TILT is just beginning. Indeed, it may be years before the precise mechanism(s) are clearly understood, just as decades elapsed between articulation of the germ theory and the fulfillment of Koch's postulates proving that particular organisms were the cause of particular diseases. In the interim, until the specific mechanism(s) underlying chemical sensitivity are elucidated, it seems wise to construe the theory in the broadest possible sense; hence the selection of "toxicant-induced loss of tolerance" as a name for this postulated general mechanism. In this context, we could regard chemical sensitivity as a sentinel symptom for the TILT family of illnesses, just as fever is a sentinel symptom for most infectious diseases. Unlike "chemical sensitivity" or "multiple chemical sensitivity," "TILT" both offers a descriptive or phenomenologic label ("loss of tolerance") and postulates a generic cause or origin ("toxicant-induced"). The term, however, avoids the label "syndrome," in keeping with the notion that the condition may be a class of diseases, rather than a single, well-defined clinical entity.

Closely related to the issue of terminology is the continuing problem of the lack of a symptom-based case definition for MCS. We continue to believe that a rigid case definition is premature at this time. We are not persuaded that multiple symptoms involving several organ systems are the only manifestation of toxicant-induced loss of tolerance. Single organ systems may be involved. Further, subsets of conditions with other labels, such as intrinsic asthma, migraines, depression, or chronic fatigue syndrome, may well be due to a toxicant-induced loss of tolerance. Thus, the possible chemical origins of conditions with other labels should not be excluded from investigation at this stage of our knowledge. Finally, we do not reject the possibility that psychological or physical trauma may induce chemical sensitivity. As we discuss in subsequent chapters, however, the evidence for such a pathway has remained meager in the five years since our original writing. In contrast, the evidence increasingly points toward a physical pathway that we describe here as toxicant-induced loss of tolerance to chemicals.

North American Workshops and Increased Governmental Interest in MCS

Developments in the United States

In March 1991, in response to growing public and professional interest in chemical sensitivity—in no small part spurred by two state-sponsored reports on MCS (Ashford and Miller 1989, 1991; Bascom 1989)—the National Research Council (NRC) convened a workshop to develop research recommendations for multiple chemical sensitivity. Clinicians, toxicologists, immunologists, epidemiologists, psychiatrists, psychologists, and others with relevant skills or interests were invited to attend. The Environmental Protection Agency (EPA) and National Institute of Environmental Health Sciences (NIEHS) shared sponsorship, and the Agency for Toxic Substances and Disease Registry (ATSDR) provided ancillary support for the meeting. Participants offered diverse perspectives and achieved consensus as to future research directions, despite general concern that "definition of the phenomenon was elusive and its existence as a distinct clinical entity had not been confirmed" (NRC 1992a). Recommendations from the meeting are summarized in Chapter 10 of this book.

In September 1991, the Agency for Toxic Substances and Disease Registry (ATSDR) in conjunction with the Association of Occupational and Environmental Clinics (AOEC) convened the second federally sponsored meeting in the United States devoted exclusively to MCS, inviting occupational medicine physicians from across the country (AOEC 1992). Participants agreed that there were many unanswered questions about the illness, and that further research was needed (see Chapter 10).

With growing public concern and litigation over MCS, several federal agencies now find themselves facing important policy questions related to the condition. One of these agencies is the Environmental Protection Agency (EPA), whose mission includes preventing adverse human health effects from pesticides and indoor air pollution. The EPA is tasked with enforcing more than a dozen major environmental laws, including the Clean Air Act, the Toxic Substances Control Act (TSCA), the Federal Insecticide, Fungicide and Rodenticide Act (FIFRA), the Clean Water Act (CWA), the Safe Drinking Water Act (SDWA), the Resource Conservation and Recovery Act (RCRA), and the Comprehensive Environmental Response, Compensation and Liability Act (CERCLA or Superfund). Many MCS patients point to pesticide exposures or air contaminants in sick buildings as the initiating causes of their illness (Miller and Mitzel 1995). In 1987–88, when the EPA installed 27,000 square yards of new carpeting and painted and remodeled office space in its Waterside Mall headquarters in Washington, D.C., some 200 agency employees developed symptoms associated with sick building syndrome. Several dozen EPA workers subsequently reported developing MCS (see Chapter 3). These individuals complained of being unable to tolerate tobacco smoke, perfume, engine exhaust, and other low-level exposures that had not been a problem for them before the remodeling took place. Some left the agency claiming that they could no longer work. Others went to new jobs or obtained permission to work at home. Some moved into specially furnished offices that the agency provided, which had no carpeting, disinfectants, copiers, perfume, or similar exposures nearby, and where occupants could open windows. Litigation ensued (see later discussion in this chapter). Nationwide, indoor air pollution costs tens of billions of dollars annually (EPA 1989). MCS cases are part of this costly burden, exacting a significant financial toll on patients, building owners, and employers. In addition to helping sponsor the 1991 NRC meeting on MCS, the EPA has initiated its own in-house study to characterize the condition (Kehrl 1996) and plans to conduct chemical challenge studies in the future (Koren 1996).

In 1992, the U.S. Congress asked the Agency for Toxic Substances and Disease Registry (ATSDR) to direct $250,000 from its fiscal 1993 budget toward chemical sensitivity and low-level environmental exposure workshops. In the spring of 1993, ATSDR convened a panel of physicians, scientists, and MCS patients who recommended that a conference be held to explore the extent to which the central nervous system might be involved in the disorder. In response to this recommendation, ATSDR convened a conference entitled "Low Level Exposure to Chemicals and Neurobiologic Sensitivity" in Baltimore in April 1994. The purpose of the conference was to initiate a two-way dialogue between clinicians who work with patients with MCS and investigators with innovative laboratory tools and approaches that might be applied to this area. Two white papers provided participants with background information concerning (1) the history and phenomenology of MCS (Miller 1994a) and (2) an overview of the neural sensitization model for chemical sensitivity (Bell 1994), originally proposed by Bell, Miller, and Schwartz (1992). Patients and representatives of patient support groups participated actively in the dialogue. Since that meeting, perhaps a dozen or so researchers have been funded by various government agencies and private foundations to explore possible neurobiological underpinnings for multiple chemical sensitivity. In addition, the EPA sponsored a workshop to explore possible animal models of nervous system susceptibility to indoor air contaminants (EPA 1995), and several animal models for MCS are in the early stages of development (see Chapter 8).

Over the past five years, ATSDR has taken a leadership role in fostering sound scientific research in this area. Superfund monies fund ATSDR to investigate and provide information regarding health effects related to toxic wastes in communities. Many citizens who live near Superfund hazardous waste sites report being ill, yet their exposures frequently are "low-level," that is, well below generally accepted safe limits (i.e., those traditionally thought necessary to avoid adverse health effects). Consequently, ATSDR is interested in a wide range of possible health effects from low-level chemical exposures, including MCS. Thus far, more than 1,300 toxic waste dump sites have been placed on a national priority list for remediation out of an estimated 400,000 sites throughout the United States. Nevertheless, there is a paucity of data concerning health effects associated with most of the exposures involved. Billions of dollars have been spent for cleanup of Superfund hazardous waste sites over the past decade, and results of research on MCS could affect future policy and expenditures in this area in important ways.

Since Operations Desert Storm and Desert Shield in 1990–91, the

Department of Veterans Affairs (DVA) and the Department of Defense (DOD) also have been drawn into the MCS debate. Many Persian Gulf War veterans returned from the war complaining of multisystem symptoms, including fatigue, depression, irritability, memory and concentration difficulties, muscle aches, shortness of breath, skin rashes, and diarrhea. Some attributed their conditions to exposures they had while in the Gulf, including combustion products from oil well fires, paints, fuels, pesticides, solvents, contaminated water, immunizations, insect repellents, and an anti–nerve agent pill, pyridostigmine bromide, that contains a carbamate drug that inhibits acetylcholinesterase and is chemically related to organophosphate pesticides and nerve agents.

In 1996, the DOD acknowledged that thousands of U.S. troops may have been exposed to chemical nerve agents when they blew up Iraqi bunkers at Kamisiyah, in which unmarked missiles containing the organophosphate sarin later were found. Some question remains as to whether U.S. bombing of vast Iraqi weapons production and storage facilities during the air war in January 1991 (Tuite 1996) or destruction of thousands of other Iraqi bunkers after the war also may have disseminated low levels of chemical agents over areas occupied by troops. Of the approximately 700,000 U.S. soldiers who were deployed to the Gulf, some 10 percent have sought special government health screening examinations to determine whether they suffer from illnesses related to the war. Growing numbers of veterans have expressed dissatisfaction with the DOD's and the DVA's inability to link their illnesses with their wartime exposures and have sought help from private practitioners, including but not exclusively clinical ecologists (who now prefer to be called "doctors of environmental medicine"), who have diagnosed them as having MCS. Some of these veterans subsequently have tried to obtain medical benefits and compensation for war-related injuries, only to be told that MCS is not a recognized medical condition. Angry, frustrated, and sick, more and more veterans have turned to legislators and lawyers for assistance.

In 1994, the DVA established Environmental Hazards Research Centers in three VA medical centers—in Boston, Massachusetts; East Orange, New Jersey; and Portland, Oregon. Research projects in these centers focus on the role of environmental exposures in the Gulf veterans' health problems, including, to a limited degree, chemical sensitivity (Fiedler et al. 1996a). More funding is devoted to chronic fatigue syndrome, post-traumatic stress disorder, and other diagnoses that do not implicate chemicals as possible causes than is given to study of chemical sensitivity (for further discussion concerning the role of chemical sensitivity in the Gulf veterans' illnesses see the section on "Origins of Chemical Sensitivity" in Chapter 8).

In September 1995, the Rutgers University Environmental and

Occupational Health Sciences Institute (EOHSI) and the National Institute of Environmental Health Sciences (NIEHS) Superfund Hazardous Substances Basic Research and Training Program co-sponsored a workshop in Princeton, New Jersey, entitled "Experimental Approaches to Chemical Sensitivity." The meeting was attended by academic and governmental scientists working on MCS or in related areas. Five working groups were charged with developing research strategies to explore various mechanistic hypotheses that have been advanced to explain MCS: toxicant-induced loss of tolerance; conditioning and learning; psychoneuroimmunology; neurogenic inflammation; and kindling and time-dependent sensitization. Presentations from the meeting, as well as consensus research strategies and recommendations formulated by the working groups, recently appeared in a 1997 supplement of Environmental Health Perspectives, published by NIEHS (NIEHS 1997).

Thus, a growing number of U.S. agencies involved in occupational and environmental health issues, including principally the EPA, ATSDR, DOD, and DVA, but also NIEHS, OSHA (Occupational Safety and Health Administration), and NIOSH (National Institute for Occupational Safety and Health), have encountered MCS, and several of these agencies have taken steps to advance scientific knowledge concerning this condition. The Interagency Work Group on Chemical Sensitivity was formed in 1994 to look at MCS from a federal standpoint, with the mission of examining published reports, findings, and agency recommendations in order to evaluate where the federal government stands on the issue and to recommend areas for future agency activity and cooperation. Its members include representatives from the ATSDR, CDC National Center for Environmental Health, NIEHS, National Institute for Occupational Safety and Health, EPA, Department of Defense, Department of Energy, and Department of Veterans Affairs. The group plans to publish a policy paper in the Federal Register in the near future. The workgroup is co-chaired by Richard Jackson, M.D., Director of the National Center for Environmental Health, and Barry Johnson, Ph.D., Assistant Administrator of the ATSDR.

While the U.S. federal government has been grappling with MCS, it has also addressed other issues of possible relevance to MCS. In 1992, the National Research Council (NRC) Subcommittee on Immunotoxicity made the following recommendation (NRC 1992b):

> Because sick-building syndrome appears to be a real phenomenon caused by contamination of VOCs that cause discomfort to a substantial number of persons, programs should be developed to establish indoor air pollution standards for homes, schools, and workplaces. These standards should restrict VOCs or other chemicals involved in indoor air pollution

below those [levels] at which significant numbers of occupants suffer headaches, mucous-membrane irritation, eye and nose irritation, lethargy and difficulty with concentration [page 138 of the report].

We know, particularly from the TEAM (Total Exposure Assessment Methodology) studies conducted by the EPA (see pages 12–15) and from Scandinavian work already cited, that mixtures of VOCs, in which the concentration of each component is well below the established occupational exposure limit, cause some building occupants to suffer. However, no progress has been made in establishing indoor air quality standards. If we were to establish standards to prevent sick building syndrome, we would in all likelihood create an improved indoor air environment—reducing the concentration of substances that trigger chemical sensitivity—for persons with MCS as well.

After a nationally televised news program aired in October 1992 on indoor air problems, the National Institute for Occupational Safety and Health (NIOSH) received 814 requests for investigations of indoor air quality. As a result, NIOSH undertook a study of 105 "problem" office buildings (Crandall and Sieber 1996; Malkin et al. 1996; Sieber et al. 1996). Although NIOSH did not evaluate the direct cause of "non-specific symptoms" in the buildings, many were documented (Malkin et al. 1996), and their associations with various building environmental factors were noted (Sieber et al. 1996). Unlike the EPA TEAM studies discussed above, no industrial hygiene measurements of chemicals and volatile organics were made in this study.

In 1990, the U.S. Congress Office of Technology Assessment (OTA), relying on input from experts in neuroscience, issued a report on neurotoxicity drawing attention to the fact that neurotoxic disorders are one of the nation's ten leading causes of work-related disease and injury, as reported by NIOSH, and that 17 of the top 25 substances then listed on EPA's Toxic Release Inventory have neurotoxic potential (OTA 1990). Yet it was not until October 1995 that EPA proposed guidelines for evaluating neurotoxic risk from chemicals (Federal Register 1995, 60 (192) 52031-56). If, as some evidence suggests, MCS (or at least TILT) is initiated via a neurotoxic mechanism, adequate control of neurotoxins is needed. Current pesticide regulations, in particular, may be insufficient to prevent the neurotoxic effects of pesticides. Various pesticides, particularly some organophosphates and carbamates, have been implicated anecdotally as initiators of MCS. For years it has been recognized that "[t]here is also considerable evidence that toxic encephalopathy may be caused by high-level, prolonged, and repeated exposure to some organic solvents" (OTA, p. 299, 1990). Kilburn (1993a) noted that about 20 percent of his patients with chemical encephalopathy reported chemical intolerances consistent with MCS, and that cacosmia (illness from odors) has been identified among solvent-exposed workers (Ryan et al. 1988). He suggests that the next question to be addressed is: "What proportion of all patients with chemical encephalopathy have MCS and how many MCS patients have chemical encephalopathy?" In essence, there may be considerable overlap between the conditions, or toxicant-induced MCS may be a subset of all chemical encephalopathies.

In the vacuum created by the lack of scientific data and consequent paralysis of federal policy in this area, many state health departments now are grappling with the complex policy questions posed by MCS and are struggling to find ways to meet the needs of these patients. In 1994, the Environmental Health Investigations Branch of the California Department of Health Services, with funding from ATSDR, convened a national, multi-disciplinary panel consisting of 19 members with divergent views on the subject to assist the department in developing a research protocol for evaluating individuals reporting sensitivities to multiple chemicals. Based upon the panel's recommendations, four questionnaires were developed: a brief screening instrument, a household survey for use in a community following a chemical spill, a detailed questionnaire for population-based research, and a follow-up questionnaire (Kruetzer and Neutra 1996). In addition, two protocols were designed, one for a population-based epidemiologic study of chemical sensitivity and the other for a post–chemical spill investigation. Some of the screening questions for chemical sensitivity were included in the 1995 California Behavioral Risk Factors Surveillance (BRFS) telephone survey of 4,000 Californians. This study provided the first statewide, population-based data on perceived chemical sensitivities (see Chapter 8, section on "Magnitude of the Problem").

In 1994–95, the legislature of the State of Washington authorized the Washington Department of Labor and Industries to invest $1.5 million in research projects on chemically related illnesses. The department awarded $300,000 for a three-year (1995–98) contract with the University of Washington's Occupational and Environmental Medicine Program at Harborview Hospital for a center that combines research, education, and clinical care for chemically exposed people. In January 1996, the Department of Labor and Industries awarded the remaining $1.2 million for competitively-reviewed research proposals on chemical illnesses. Three of the six contracts awarded focus specifically on MCS: (1) reliability and reproducibility of immune and lymphocyte tests in MCS patients and controls ($296,220 to Johns Hopkins School of Hygiene and Public Health); (2) time-dependent sensitization and cognitive dysfunction in chemical sensitivity ($218,370 to University of Arizona College of Medicine); and (3) SPECT imaging of the brain in patients with MCS and controls ($272,607 to Brigham and Women's Hospital in Boston). Investigators planned to present results of these studies at a special conference held in the State of Washington in 1997. Other states are also launching a variety of initiatives (*MCS Referral & Resources* 1997).

Developments in Canada

In 1992, Health and Welfare Canada (now Health Canada) sponsored an MCS workshop to which were invited medical experts, health researchers, representatives of professional and lay organizations and of provincial and federal governments, and individuals with a special interest in the issue. The purpose of the meeting was to stimulate further research on MCS; to advise on research priorities; to consider the operational definition proposed by Ashford and Miller, and, if it were thought fitting, to promote its use in future research; and to assist practitioners in managing these patients by providing an appropriate background document (for recommendations from this workshop, see Chapter 10).

In 1994, the Ontario Ministry of Health earmarked $2.5 million over a five-year period to establish a clinical and research program on environmental hypersensitivity. The program consists of two components, a research effort within the Department of Preventive Medicine and Biostatistics at the University of Toronto and a collaborating Environmental Health Clinic located at Women's College Hospital, a teaching hospital of the University of Toronto.

This research effort, called the "Environmental Hypersensitivity Research Unit" (EHRU) (although it does not include the environmental medical or control unit that its name might suggest), is focusing on the etiology, diagnosis, and treatment of environmental hypersensitivity using epidemiological approaches. Priority is being given to the establishment of diagnostic criteria and the development and assessment of diagnostic tests. As a first step, a questionnaire exploring relevant clinical features was developed and administered to over 2,500 individuals in order to test its discriminant validity. Results from this study will be used to refine research criteria for identifying patients most likely to have environmental hypersensitivity. Reproducibility of the questionnaire is also being assessed by retesting a random sample of 200 of the original participants. Selected diagnostic tests will later be evaluated in a blinded manner for their ability to distinguish between patients with high and low likelihoods of being environmentally hypersensitive. The Environmental Health Clinic, which opened in March 1996, is mandated to evaluate patients who attribute their symptoms to environmental exposures, to collaborate with the Environmental Hypersensitivity Research Unit, and to develop a network of physicians throughout Ontario who are knowledgeable about environmental hypersensitivity. Currently Canada, like the United States, lacks an environmentally controlled, hospital-based research facility (environmental medical unit) for blinded challenge testing.

Government Recognition and Accommodation

Recognition of multiple chemical sensitivities by government agencies and the courts has slowly been increasing. A number of government agencies have supported research on MCS and have either recognized MCS or adopted policies accommodating people with MCS. Claims involving MCS have had a mixed reception in the courts.

U.S. Federal Agencies

Department of Justice   The Americans with Disabilities Act (ADA), passed by Congress in 1990, protects from discrimination individuals who have a physical or mental impairment that substantially limits one or more major life activities, or who are perceived (by employers, landlords, and others) to have such an impairment. Employers must provide "reasonable accommodation" under the ADA, which may include "job restriction, part-time or modified work schedules, reassignment to a vacant position, acquisition or modification of equipment or devices . . . ." The Equal Employment Opportunity Commission (EEOC) further defines reasonable accommodation as:

> (1) Any modification or adjustment to a job application process that enables a qualified individual with a disability to be considered for the position such qualified individual with a disability desires, and which will not impose an undue hardship on the . . . business; or
> (2) Any modification or adjustment to the work environment, or to the manner or circumstances which the position held or desired is customarily performed, that enables the qualified individual with a disability to perform the essential functions of that position and which will not impose an undue hardship on the . . . business; or
> (3) Any modification or adjustment that enables the qualified individual with a disability to enjoy the same benefits and privileges of employment that other employees enjoy and does not impose an undue hardship on the . . . business. (EEOC Interpretative Guidelines)

In proposed regulations related to enforcement of the law, those providing commentary asked that both MCS and allergy to cigarette smoke be recognized as disabilities. In final regulations, the Department of Justice declined "to state categorically that these types of allergies or sensitivities are disabilities," as this determination must be made on a case-by-case analysis. It noted that "[s]ometimes respiratory or neurological functioning is so severely affected that an individual will satisfy the requirements to be considered disabled under the regulation. Such an individual would be entitled to all of the protections afforded by the Act . . . . In other cases, individuals may be sensitive to environmental elements or to smoke but their sensitivity will not rise to the level needed to constitute a disability" (*Code of Federal Regulations* 1996).

The department has issued two letters of finding stating that a fragrance-free policy is not a required accommodation under the Act.

Housing and Urban Development   The Department of Housing and Urban Development (HUD) has been in the forefront of federal agencies in acknowledging and accommodating people with MCS. Since 1990, HUD has recognized MCS as a disability under the Rehabilitation Act of 1973 and under the Fair Housing Amendments Act of 1988 (HUD 1990, 1992). Housing providers are therefore required to provide reasonable accommodation to chemically sensitive individuals. HUD does not have a written policy defining reasonable accommodation for MCS, but makes determinations on a case-by-case basis, taking into account feasibility and practicality.

A 1991 HUD technical guidance memorandum (HUD 1991) from the General Deputy Assistant Secretary for Fair Housing and Equal Opportunity (FHEO) to all Regional FHEO Directors provides a description of Multiple Chemical Sensitivity Disorder (MCSD), noting that the "most common substances that are believed to cause adverse reactions in people with MCSD are solvents and other volatile compounds, pesticides, formaldehyde, natural gas, disinfectants, detergents, plastics, tobacco smoke, and perfumes." For housing providers, the memorandum continues, "acts which are necessary and accepted business practices, such as cleaning, painting, exterminating the building or fertilizing the lawn, may be threatening events to people with MCSD since exposure to the various chemicals involved can cause severe symptoms."

The memorandum goes on to provide several examples of accommodations that are considered reasonable:

- A tenant with MCS who is sensitive to chemical pesticides requests that the housing provider notify him before fumigating the apartment building and substitute boric acid for the chemicals normally used to spray the tenant's apartment.
- An applicant with MCS has a sensitivity to chemicals found in certain types of carpeting. She inquires about an available apartment in a building in which all the apartments have wall-to-wall carpet-

ing. She asks that the housing provider inform her as to the type of carpeting used in the building so that she can determine whether the apartment would suit her needs before renting it.

In practice, the agency has followed the contours of this memo, treating requests of advance warning of pesticide application as reasonable, and usually treating as reasonable requests that an individual's housing unit not be sprayed. It would probably treat as unreasonable, for example, a request by a person sensitive to exhaust fumes that no traffic be allowed outside his or her apartment.

In one of a number of representative cases dealing with housing, a conciliation agreement approved by HUD was reached in which a Chicago realty company agreed to make several accommodations for chemically sensitive residents, including using integrated pest management, providing notification of maintenance or repair work that might create chemical exposures, and using less toxic cleaning products in common areas of the building (National Center for Environmental Health Strategies 1992). In another case, an action initiated by the Department of Justice, originating in a complaint filed with HUD, resulted in a consent order in which an association of apartment owners agreed to use only cleaning products that a tenant with MCS could tolerate, as well as to give the tenant written notice of repairs and unusual cleaning and to refrain from using pesticides (United States District Court for the District of Hawaii 1993).

In the fall of 1994, the first U.S. government-subsidized housing for people with chemical sensitivities opened its doors in San Rafael, California. The $1.2 million, two-story, 11-unit complex for low income persons, named "Ecology House," was the culmination of six years of work spearheaded by Susan Molloy, an activist for better access for people with MCS, and housing consultant Kate Crecelius. The pair began with a $5,000 Community Development Block Grant. Ultimately, Section 811 HUD (Housing and Urban Development) funding provided more than $800,000 in construction costs, as well as rental assistance for tenants (Molloy 1996). The architect and the contractor on the project consulted extensively with people with MCS to choose building materials and construction techniques that would keep the building clean and nontoxic. For example, concrete was poured without a curing agent, and the tile grout used was pure Portland cement without additives (Ecology House, Inc. 1995).

Ecology House was envisioned by many as the prototype for future apartment complexes needed to house people with MCS throughout the country. Eleven tenants were selected by lottery from nearly 100 applicants. Those chosen were encouraged to spend time in a unit before signing the lease and moving in, in order to determine whether they could tolerate the interior finishes and materials. However, few did so.

According to Molloy, there were several problems from the outset. The original plan had been to choose a hillside site in Marin County with ocean breezes that would help minimize outdoor air pollution. But in order to comply with HUD requirements under the Fair Housing Amendments Act (FHAA), the site had to have a grade of less than 1:12 to permit wheelchair access and had to be located close to a busline. No available land in Marin County qualified. Eventually, a lot in a densely populated suburban area of San Rafael was chosen, even though there is a fair amount of traffic there, and it is on top of Bay landfill, barely above sea level.

Another problem was that many of the people with MCS who qualified for the federal preference list were extremely disabled and had few resources. Molloy was of the opinion that these individuals simply did not have the stamina or the support systems needed to go into new housing with new materials that were still outgassing. Many complained that these exposures were making their illness worse but, because they had nowhere else to go, they felt constrained to stay despite their worsening health. Currently, the housing is fully occupied, and there is a waiting list for it, with tenant health improved (see below).

Design features and materials used in Ecology House include: tile floors, cementitious roof, metal siding, formaldehyde-free fiberglass insulation, hydronic baseboard heating, unpainted interior plaster, less allergenic landscaping (versus commonly used plantings), concrete (versus asphalt) parking lot, and solid board floor sheathing (versus plywood). In addition to the 11 one-bedroom units, Ecology House has a community room, an office, a laundry room, and an airing room (for outgassing new items). Tenant rules restrict smoking, burning, use of fragrances, keeping pets, idling of engines, and use of paint, glues, and odorous laundry and cleaning products. Only baking soda can be used in one of the three washers.

Molloy identifies two principal exposures that she thinks may have made many of the initial occupants ill. First, the wall plaster did not adequately seal the drywall core. Second, although metal kitchen cabinets with a baked-on finish had been specified, metal cabinets with a painted finish that outgassed were installed.

According to Molloy, Ecology House is more tolerable for MCS patients now that the materials in it have aged and outgassed. However, the adverse national publicity that occurred when it was first occupied almost certainly affected HUD's willingness to undertake similar projects. Molloy remains hopeful that in the future HUD will be able to assist people with MCS in purchasing homes or living outside existing federal housing projects in settings conducive to restoring their health. Although she never lived in the building, Molloy points out that certain aspects of the project were successful, such as the shared laundry room, lawn-free landscaping (no herbicides needed), management practices (including nontoxic maintenance), and scent-free and smoke-free requirements.

Social Security Administration   Although the Social Security Administration has not recognized MCS as a disability per se, its operations manual discusses MCS and notes that claims based on MCS "should be made on an individual case-by-case basis to determine if the impairment prevents substantial gainful activity" (Custer 1996). An increasing number of administrative law judges in the Social Security Administration are becoming aware of MCS; paradoxically, the Secretary of Health and Human Services, of which the Social Security Administration is part, seems reluctant to allow disability benefits in MCS cases (HUD 1992).

The Court of Appeals for the Ninth Circuit has ruled that MCS is a disability under the Social Security Disability Act (HUD 1992, citing *Kornock v. Harris* 1980). The Eighth Circuit Court of Appeals also ruled this way in a 1990 decision (HUD 1992, citing *Kouril v. Bowen* 1990), but reversed itself in a 1994 case (Custer 1996, citing *Brown v. Shalala* 1994). The district court for the Northern District of Illinois denied benefits to a claimant alleging MCS, rejecting the testimony of her doctor, a clinical ecologist, and siding instead with members of the medical establishment who view the diagnostic and treatment approaches used in clinical ecology as scientifically invalid (HUD 1992, citing *Lawson v. Sullivan* 1990, 1991).

Department of Education   The Department of Education has issued two letters of finding under the Rehabilitation Act concluding that MCS can be a handicap (HUD 1992). In one (*San Diego (Cal.) Unified School District* 1990), the DOE concluded that a school district violated the Act by refusing to reasonably accommodate a school bus driver who was chemically sensitive to petrochemical fumes. The school district had refused to allow the driver to wear a respirator while driving. However, the DOE concluded that the bus driver was handicapped, and that the accommodation he requested was reasonable.

In the other letter of finding (*Montville (Conn.) Board of Education* 1990), the DOE concluded that a guidance counselor with MCS was handicapped under the Act, but that the school district had provided reasonable accommodations to her (NDLR 1990). (The HUD memo provides no details on the nature of the accommodation.)

Equal Employment Opportunity Commission   In the *Americans with Disabilities Handbook* (1991), jointly published by the EEOC and the Department of Justice, a discussion of what constitutes a disability in the context of employment discrimination included this statement: "[S]uppose an individual has an allergy to a substance found in most high rise office buildings, but seldom found elsewhere, that makes breathing extremely difficult. Since this individual would be substantially limited in the ability to perform the broad range of jobs in various classes that are conducted in high rise office buildings within the geographical area to which he or she has reasonable access, he or she would be substantially limited in working."

Between late 1993 and mid-1995, approximately 170 MCS employment discrimination complaints were filed with the EEOC. The majority of them alleged reasonable accommodation violations, with the second-largest group alleging unlawful discharge. Forty-four of the complaints have been administratively discharged without a hearing; in 14 other cases, the EEOC found no discrimination; and six of the complaints have been settled (National Center for Environmental Health Strategies 1996).

Environmental Protection Agency   Although the Environmental Protection Agency's Indoor Environments Division has referred to MCS as an issue in some of its publications, it has not taken action or developed any policy on MCS.

However, scientists at the EPA's National Health and Environmental Effects Research Laboratory in Research Triangle Park, North Carolina, stated in 1996 that they planned to launch within the next year a clinical study measuring the responses of chemically sensitive subjects to various chemical exposures (Kehrl 1996; Koren 1996). In a 1996 paper entitled "What Can Research Contribute to Regulatory Decisions about the Health Risks of Multiple Chemical Sensitivity?" (Dyer and Sexton 1996), a scientist with that EPA laboratory and a former EPA researcher wrote that "our present lack of scientific understanding about MCS is so acute that it is not possible to ascertain" the cause of MCS, and that "unless steps are taken to improve the quantity and quality of the existing scientific data base, we cannot, with any acceptable degree of certainty, evaluate the extent to which regulatory decisions about MCS are either protective of public health or cost-effective." Dyer and Sexton called for further research into MCS, following the public health model in which the highest goal is prevention.

Other Federal Agencies   Numerous other federal agencies have taken action on MCS, as mentioned earlier. In 1992, the National Center for Environmental Health Strategies (NCEHS) in Voorhees, New Jersey, announced that Congress had appropriated $250,000 for workshops on MCS, to be coordinated with NCEHS (American Council on Science and Health 1994). With funds from the 1993 budget for the Agency for Toxic Substances and Disease Registry (ATSDR), research protocols were also to be developed, along with a national registry of people with MCS. Although no registry has been created, ATSDR has taken a leading role in furthering understanding of MCS, by helping to organize a number of important conferences devoted to the illness. The National Institute of Environmental Health Sciences (NIEHS), part of the National Institutes of Health in the Department of Health and Human Services, has also supported research relevant to MCS.

In a report issued in 1996, the National Council on Disability, an independent federal agency, frequently made reference to MCS (National Council on Disability 1996). The report notes in its introduction: "Some disabilities are less acknowledged and less understood than others. For example, people with multiple chemical sensitivities have a particularly difficult time securing recognition for their disability. Most people do not understand the chemical and environmental barriers that preclude such persons' access to the most basic and essential areas of life, such as housing and education."

In its recommendations for national policy, the report notes that the "promotion of clean air and use of nontoxic substances (such as industrial cleaners) in public places is of particular concern to people with multiple chemical sensitivities and should be addressed within the public health agenda." The report also recommends increased research and data collection on ways to expand the effective design of housing for people with MCS.

U.S. State Agencies and Courts

Several state legislatures and agencies have taken actions related to MCS. For example, in 1991 the Attorney General of the State of New York, backed by 25 other attorneys general, requested that the U.S. Consumer Product Safety Commission issue safety standards and warning labels governing the sale of carpets, carpet adhesives, and padding suspected of causing MCS and other illness. Agencies of several states—California (see below), Washington, and New Mexico among them—have issued reports noting the increasing incidence of MCS and calling for research and action (*MCS Referral & Resources* 1996).

In California, the Senate Subcommittee on the Rights of the Disabled of the California Legislature issued a report entitled "Access for People with Environmental Illness/Multiple Chemical Sensitivity and Other Related Conditions" in September 1996. Spearheaded by State Senator Milton Marks, chairman of the committee, the report followed several years of study by an advisory panel that included people with MCS, representatives of state regulatory agencies, industry representatives, building owners and managers, architects, and physicians and other health professionals. Interestingly, the report notes that since the Subcommittee began work on this issue, "it has become increasingly apparent that people with allergies, asthma, emphysema, other respiratory diseases, immunological and neurological conditions," in addition to those with MCS, "also can be seriously affected by chemicals in the home, work and outdoor environment and in the foods we ingest" (California Legislature 1996).

After enumerating common barriers to access for people with MCS in public buildings, institutions, transportation, and employment, the report makes recommendations for improvement of the situation in 31 categories, such as: public meetings; building standards; research; heating, ventilating, and air conditioning systems; education; pesticides; changes in cleaning methods; transportation; health care access; regulation of consumer products; and airlines. For example, in the area of research, the report recommends promoting research intended to reexamine "safe" exposure levels for chemicals in various forms, promoting research that takes into account synergistic and cumulative effects of chemicals in the environment, and performing surveys to assess EI/MCS and multiple-disabled populations. In the personnel area, the report recommends that metal detector operators and security and direct service personnel—in public buildings, airline terminals, and other public places—who come into close contact with clients or travelers with MCS be smoke- and scent-free.

At least a dozen states—including Florida, Colorado, Maryland, West Virginia, Pennsylvania, Louisiana, California, Massachusetts, Michigan, Wisconsin, Connecticut, and New York—have established pesticide registration laws that document what pesticides are being sprayed and by whom and, in some cases, require posting of intended spraying or notification of spraying to those who request it (Fletcher 1996). Florida's law specifically provides for notification of individuals whose names are on a list of persons with chemical sensitivities, requiring them to pay a small annual fee for the information. The pesticide registries of most other states do not refer to MCS by name but can be helpful to those MCS patients who experience adverse reactions to pesticides by allowing them to take precautions to protect themselves from exposure. California's stringent law requires commercial applicators to file detailed monthly reports. However, there are complaints that the information provided by some of the registries may be vague or confusing, and that only commercial applicators—not homeowners—are bound by the law (Fletcher 1996).

Integrated Pest Management (IPM), an approach to controlling insects that keeps pesticide use to a minimum, has become increasingly widespread, with substantial benefits to persons adversely affected by low-level exposures. Although not generally legally required, IPM has become a part of the policies of numerous states and localities. For example, an understanding of the principles of IPM is required by most states in the certification and training examinations given to pest control applicators (Bravo 1996). The General Services Administration uses IPM in federal buildings, and the EPA has encouraged the use of IPM, especially in schools (EPA 1993).

In October 1996, the San Francisco Board of Supervisors passed one of the toughest pesticide ordinances in the country (Wilson 1996a). The legislation applies to pesticide use on city property and bans certain pesticides immediately; it bans all pesticide use by 2000; it provides for an Integrated Pest Management specialist to work with city employees in making the transition away from pesticides; it requires four-day notification of pesticide applications by the city both prior to and after a spraying; and it improves the reporting of pesticide use.

Information about pesticide ingredients will be more readily available as the result of a lawsuit brought in 1994 in the U.S. District Court for the District of Columbia by the Western Environmental Law Center, representing the Northwest Coalition for Alternatives to Pesticides and the National Coalition Against the Misuse of Pesticides (Wilson 1996b). Ruling that the EPA must provide information about the identity of "inert" ingredients in four pesticides, the court found that the EPA had improperly relied on unsubstantiated claims by manufacturers that the identity of these ingredients were trade secrets or confidential business information. In addition to solvents, EPA is aware of over 2,300 substances that are added to pesticide products and labeled as inert, but not otherwise identified on product labels. They often comprise most, up to 99 percent, of a pesticide product.

Pennsylvania, California, and Ohio state courts have interpreted their state civil rights statutes that prohibit discrimination against the handicapped to apply to people with MCS (HUD 1992). In a notable decision involving housing discrimination, a Pennsylvania trial court found that a tenant unable to tolerate various chemical compounds was handicapped under the Pennsylvania Human Relations Act (HUD 1992, citing *Lincoln Realty Management Co. v. Pennsylvania Human Relations Commission* 1991). The court affirmed the Pennsylvania Human Relations Commission's order insofar as it required the defendant to give notice to the plaintiff of pesticide application and painting and to permit the plaintiff to modify her apartment at her own expense by installing a kitchen ceiling fan and a washer and dryer. The court vacated the rest of the order's required accommodations, some of which the complainant had not requested.

New York state has provided funding for several initiatives. The N.Y. Department of Labor directly supported the New York Labor Institute's production of an MCS training manual and video, and through a general grant supported the N.Y. Coalition for Alternatives to Pesticides, which published a 16-page booklet on MCS. The N.Y. Department of Health provided $100,000 to Mt. Sinai for research on MCS.

Local Governments

Some local authorities also have addressed MCS. For example, the Fairfax County, Virginia public school district made accommodations for a teacher with MCS, including eliminating or controlling aggravating exposures to such substances as art materials, cleaning products, and air fresheners; the Santa Cruz City Council in California made several provisions in accordance with the Americans with Disabilities Act (ADA) to accommodate people with MCS, including purchasing less troublesome materials where possible and posting notices at entrances to public buildings at which construction, remodeling, or cleaning activities are taking place; the Northwest Air Pollution Authority, in Washington state, wrote to residents apprising them of a neighbor who was very sensitive to wood smoke and asking them to try to reduce their outdoor burning and heating with wood (*MCS Referral & Resources* 1996).

U.S. Court Cases

Workers' Compensation   Decisions on MCS in workers' compensation cases are mixed although claimants with MCS have generally been more likely to prevail under workers' compensation statutes than in other actions for damages (see below) (Custer 1996). Several state courts and workers' compensation boards—including those of California, Oregon, Arizona, and Louisiana—have made decisions recognizing MCS as a work-related injury or illness (*MCS Referral & Resources* 1996). The Supreme Court of New Hampshire has stated that MCS due to workplace exposure to chemicals is an occupational disease compensable under that state's workers' compensation law (Custer 1996, citing *Appeal of Denise Kehoe* 1994). The Nevada Supreme Court upheld a workers' compensation award to 23 employees of a casino who developed chemical sensitivities following pesticide application in their workplace (Brazil 1993; Cone and Sult 1992; *MCS Referral & Resources* 1996, citing *Harvey's Wagon Wheel, Inc. DBA Harvey's Resort Hotel v. Joan Amann, et al.* 1995).

Tort and Other Lawsuits   People with MCS have filed lawsuits against such parties as building owners, contractors, architects, chemical manufacturers, and employers. The plaintiffs in these lawsuits seek damages on the basis of a variety of legal theories, including negligence, product liability, and intentional tort. Defendants in these cases have prevailed more often than plaintiffs *in those cases that were decided in the courts.* However, many cases are settled out of court and do not have the fact of settlement or their terms of settlement disclosed in the law reporters. Often court cases that are settled have the details of their outcomes "sealed." Obviously, defendants, especially those likely to be sued by many plaintiffs, prefer to settle rather than have cases decided adversely in the courts, thus establishing legal precedent and encouraging others to sue. Therefore, the cases described below do not present the whole picture. They do, however, illustrate different approaches taken by the courts.

One recent lawsuit came out of the highly publicized indoor air quality problems at the EPA headquarters building in Washington, D.C. (see additional discussion in Chapters 3 and 6). In that case, 19 EPA employees claiming MCS filed suit against the owners of the building, alleging that the owners were negligent in failing to provide adequate ventilation and in scheduling renovation work near occupied offices. The court ordered the parties to choose five of the employees to be plaintiffs in a summary jury trial before addressing the other cases. The jury found that one of the plaintiffs had been physically injured by the defendants' negligence and that four suffered only psychosomatic disorders, but it nonetheless rendered a verdict for all five for an amount totalling about $900,000. However, the court granted a post-trial motion to set aside the verdict for the plaintiffs found to have experienced only psychosomatic disorders; and it upheld a $232,000 award to the fifth plaintiff (Toxics Law Reporter 1995 and Custer 1996, citing *Bahura v. S.E.W. Investors L.P.* 1995). All five cases are under appeal, but must await the disposition of the claims of the remaining fourteen plaintiffs (*Toxics Law Reporter* 1995).

In a jury trial in Florida (*Melanie Marie Zanini v. Orkin Exterminating Company Inc. and Kenneth Johnston* 1995), the plaintiff was awarded $632,500 in damages for negligent application of the pesticide Dursban resulting in the plaintiff's illness. The following year, a public lease-back corporation in California was held responsible for 14 awards of partial to permanent disability based on MCS and various other health complaints that started after "offgassed" toxic emissions from extensive renovations were inadequately ventilated due to the malfunctioning of half of the air conditioning units in the facility (*Ruth Elliot et al. v. San Joaquin County Public Facilities Financing Corp. et al.*). Individual awards ranged from $15,000 to $900,000.

In another case, *Bradley v. Brown* (1994), the U.S. Court of Appeals for the Seventh Circuit affirmed a trial court finding that the plaintiffs had failed to establish that the etiology of MCS is "known or tested." Attempting to show that they were suffering from MCS as a result of exposure to a pesticide applied in their workplace, the plaintiffs had sought to introduce the testimony of two clinical ecologists. The trial court had excluded the doctors' testimony on the basis that their opinions regarding the causes of the plaintiffs' MCS lacked sufficient scientific basis. The Court of Appeals agreed. The case is important because it is the first appellate decision applying the new standard for admissibility of scientific evidence in the federal courts set forth by the U.S. Supreme Court in its 1993 decision in *Daubert v. Merrill Dow Pharmaceuticals, Inc.* (1993) to determine whether a diagnosis of MCS may be admitted into evidence (Custer 1996). The Court in *Daubert* held that for such evidence to be admissible, it must constitute "scientific knowledge" that is supported by appropriate validation, and listed factors to be considered in evaluating a scientific theory or technique, including whether it can be and has been tested, whether it has been subjected to peer review, and whether it has gained acceptance in the scientific community. The Supreme Court in *Daubert* agreed that trial courts could and should act as gatekeepers for scientific evidence by barring fringe science, but the Court also held that science that departed from mainstream scientific views was not necessarily inadmissible, thereby broadening rather than narrowing the rules of evidence in the courts. Nonetheless, the trial court in *Bradley* concluded that the diagnoses of MCS came nowhere close to meeting the criteria for "scientific knowledge," and the appeals court found that the trial court had adhered to the approach set out in *Daubert*.

Similar reasoning was followed in *Cavallo v. Star Enterprise,* a Virginia U.S. district court case (*Cavallo v. Star Enterprise* 1995), which was upheld by a panel of the Fourth Circuit Court of Appeals. In this case, a woman who was exposed to jet fuel fumes as she crossed a parking lot sued the petroleum distribution facility that had spilled the fuel near the lot. She claimed that the exposure had caused her to suffer from chronic conjunctivitis, sinusitis, pulmonary disease (diagnosed as Reactive Airways Dysfunction Syndrome), and increased sensitivity to petroleum hydrocarbons. In finding for the defendant, the court ruled that, under *Daubert,* Cavallo's expert witnesses had failed to ground their opinions on a scientifically valid methodology. One witness, a toxicologist, relied on studies on chemicals that had only some overlap with the components of the jet fuel and that failed to demonstrate dose–response relationships, the court said. The court also noted the controversial nature of MCS, as well as the literature upon which the toxicologist relied in attributing Cavallo's sensitivity to exposure to petroleum hydrocarbons and various volatile organic compounds. The court found that the testimony of the second expert witness, an immunologist, was also based largely on speculation. The plaintiffs have appealed the panel decision to the Court of Appeals for the Fourth Circuit, requesting an en banc (full court) hearing, arguing that the panel's decision both was a misapplication of *Daubert* and conflicted with other federal appeals court precedents applying *Daubert*, namely, *Benedi v. McNeil-P.P.C. Inc.* (1995), *City of Greenville v. W. R. Grace & Co.* (1987) and *Ferebee v. Chevron Chemical Co.* (1984) (*Toxics Law Reporter* 1996a). The Supreme Court has since agreed to clarify the general legal standard for appellate review of the admissibility of expert testimony (*General Electric Co. v. Joina; Toxics Law Reporter* 1997a).

One plaintiff and her son, alleging MCS manifesting as toxic brain encephalopathy, won a $6.6 million settlement in 1995 in a consolidation of three suits filed in Louisiana District Court, Orleans Parish, against her university where she was employed as a research associate and exposed to chemicals (*Mealey's Litigation Reports*).

Litigation related to silicone (mostly) breast implants has resulted in 18 verdicts for plaintiffs since 1984, totaling up to $25 million and an undisclosed number of out-of-court settlements (Begley 1996; *Mealey's Litigation Reports* 1995). Facing more than 19,000 lawsuits, Dow Corning Corporation filed for bankruptcy-court protection after a multi-plaintiff action settlement proposal failed. It also asked for a "science trial" using a court-appointed panel of scientific experts and a jury to determine whether implants can in fact cause disease (*Toxics Law Reporter* 1996b). (An Alabama federal judge is also using a panel of experts to evaluate conflicting scientific evidence.) This move was no doubt stimulated by recent court victories for Dow Corning. In October 1996, two district court judges ruled that "silicone implants . . . do not cause classical recognized diseases" (Begley 1996). In December 1996, a U.S. district court judge in Oregon, on advice from a scientific panel he appointed, and applying *Daubert*, ruled (*Hall v. Baxter Healthcare Corp.* 1996) that expert witnesses who claim a link between silicone breast implants and a wide spectrum of serious immune system diseases, such as lupus, rheumatoid arthritis, and scleroderma, should be barred from testifying in lawsuits in his court (*Toxics Law Reporter* 1997b; *Washington Post* 1996). These cases will be no doubt be appealed. As we go to press, the outcome of the bankruptcy request has not been decided for Dow Corning, although a Michigan court has dismissed claims against the parent company Dow Chemical (*Toxics Law Reporter* 1997c).

In the United States, legal action and liability concerns unfortunately drive much of the acerbic rhetoric and distortions of science that retard the advancement of understanding about chemical sensitivity. The same is also true in parts of Europe, notably Germany and the United Kingdom.

Developments in Canada

A number of Canadian government agencies have implemented policies that strongly support persons with MCS. The Canadian Human Rights

Commission has stated that it regards environmental illness as a disability (*MCS Referral & Resources* 1996). The Canadian government has declined to set up detoxification clinics to treat its ill Gulf War veterans (Spence 1996). Canadian medical boards have in recent years made numerous attempts at—and have succeeded in—taking away the licenses of some doctors who practice environmental medicine (Taylor 1995). Reportedly, they have also been successful in making "alternative" remedies such as nutritional supplements and homeopathic remedies less available to the public. Thus, generally speaking, people in Canada who suffer from MCS may have less access to alternative therapies than patients in the United States. On the other hand, the provincial government in Nova Scotia has funded a clinic dedicated to environmental illness (see the discussion below).

Since 1984, the Canadian Mortgage and Housing Corporation (CMHC) has been researching ways to improve indoor air quality for all Canadians via improved construction and informed selection of building materials. As part of this process, it has designed and built several different kinds of housing units for environmentally hypersensitive persons (Rafuse 1995). CMHC hopes to interest designers and manufacturers in the results of its studies. The agency's flagship project is a prototype one-bedroom, 850-square-foot research house designed for the chemically sensitive, sited at its national office in Ottawa. Low- or no-cost solutions to maximize indoor air quality for persons with chemical sensitivities or allergic or respiratory disease were sought. The spartan home features: no carpets, curtains, or wall hangings; unpainted hard plaster veneer over dry wall; sealed wood trim and cabinets; tile floors; wood furniture with cotton cushions; mattresses without flame- or stain-retardant finishes; radiant underfloor heating; vented closets and cupboards; a thermal insulation envelope that prevents condensation; a work/entertainment cabinet that is exhausted for housing television, computers, and other electronic equipment; a drying room for clothing; a refrigerator that drains to a pipe instead of an open tray; and a toilet tank set away from the wall to minimize condensation.

Other Canadian housing projects targeting environmentally hypersensitive persons include a seven-unit, two-story building that is a part of the 41-unit Barrhaven Multi-unit Housing Project in Nepean, Ontario, and a single, specially constructed unit in an apartment building in Victoria, British Columbia. The Barrhaven Project was sponsored by the Barrhaven United Church with financial assistance from the Ministry of Housing for Ontario as a nonprofit housing program (Sharp 1994). The one- to three-bedroom units, which are geared toward low- and modest-income families, were completed in the spring of 1993. The church's interest in the project evolved because one of its members had environmental hypersensitivities. The project architect, who specializes in designing housing for the disabled, expressed interest in researching and designing some units for persons with environmental hypersensitivities.

A testing panel made up of eight environmentally hypersensitive persons from the area helped with the selection of building materials and finishing products for the Barrhaven Project. Features of the complex, which won the American Institute of Architects and the International Union of Architects "Call for Sustainable Solutions" design competition, include: siting near open space, away from vehicular traffic; use of inert or low-emitting materials; slab construction so as to avoid basement or crawl spaces that can harbor dust or mold; an entry mudroom/airlock for changing footwear and airing outer clothing; a large, mechanically ventilated cabinet with sliding glass doors on both sides that serves as a wall between the living room and the kitchen and can house a computer workstation, television, and other items of electronic equipment that outgas; access to the back of the refrigerator and other appliances for cleaning; a mechanically ventilated clothes closet separate from bedroom areas; concrete slab without plasticizers or other additives and second-story solid-core precast concrete slabs for which organic soap was used as a form release agent; hardwood partitions, doors, stairs, cabinets, and counters; ventilation ductwork that can be disassembled for cleaning; and cured silicone caulking.

Other CMHC projects that are being implemented under Canada's National Strategy for the Integration of People with Disabilities involve: training indoor air quality specialists to help homeowners identify and solve problems; doing research on building materials; providing information for builders; publication of *The Clean Air Guide*, which contains checklists to help homeowners identify common indoor air quality problems (CMHC 1993); and release of a video, *This Clean House*, that complements *The Clean Air Guide* and illustrates indoor air problems and how to correct them (CMHC 1995).

In 1989, the Environmental Health Center at Dallas established a part-time clinic in Nova Scotia for patients with environmental illness, led by Dr. Gerald Ross, a former Canadian family practitioner who works with Dr. William Rea of the Dallas Environmental Health Center and for years commuted to Halifax every other month to treat these individuals. By 1994, the clinic had grown to a full-time operation, and in that year became affiliated with Dalhousie University. Governance is provided by a steering committee with representatives from the Dalhousie University Faculty of Medicine, the Nova Scotia Department of Health, the Nova Scotia Medical Society, other health professionals, and representatives of the community. A new 8,500-square-foot center specially designed, constructed, and furnished for patients with environmental illness opened in May 1997. Approximately 600 patients have been treated at the clinic, and over 1,000 are on a waiting list. The number of referring physicians has increased to around 450 (Fox 1996).

Dr. Roy Fox currently serves as director of the center. Fox, a geriatrician, himself developed MCS while working at Halifax's Camp Hill Medical

Center in Nova Scotia. Between 1988 and 1993, up to several hundred Camp Hill employees became sick. Contemporaneous with onset of their symptoms, a corrosion inhibitor (a blend of amines including cyclohexylamine and morpholine) was added to the boiler to prevent scale buildup at ten times the recommended concentration. The steam was piped to all buildings for heat and to two buildings for humidification (Fox 1996). Fox was treated at Rea's clinic in Dallas and later was funded by the province to spend a year there training in environmental medicine. Subsequently, he returned to Nova Scotia to practice. With initial funding of $1 million from the Nova Scotia government and with expected annual operating costs of $1 million, the new Nova Scotia Environmental Health Center (NSEHC) is intended "to become a national resource providing leadership in the prevention and treatment of environmental illness." Fact sheets about the clinic state: "Controversy about these issues and the lack of expertise in Canada has led to confusion and distrust. Many affected by EI (environmental illness) have little faith in the usual providers of health care and even less in the agencies responsible for compensation. This situation is unacceptable and will be greatly improved as the NSEHC fulfills its mandate and pursues the outlined goals."

Fall River, Nova Scotia will be the site for the new clinic. Features will include: three occupancy zones based on relative degree of cleanliness and/or the potential for activities within a space to adversely impact air quality; use of nonoutgassing materials, including tile and glass blocks; a challenge testing area; pre-offgassing of furnishings and equipment prior to installation; and separate access to mechanical spaces in order to facilitate maintenance of equipment and air filters without adversely affecting air in occupied spaces. Clinical research will focus on refining a definition of the condition, validating diagnostic approaches, and identifying effective treatments. Treatments to be examined include enzyme-potentiated desensitization and detoxification therapies such as sauna, nutrient replacement, and exercise (Robb 1995).

European Perspectives

Three teams of investigators recently completed an exploratory study of chemical sensitivity in Europe for the European Commission (Ashford et al. 1995). The purpose of their investigation was to investigate the existence and the nature of chemical sensitivity in nine selected countries. No prior systematic study of the occurrence or the magnitude of chemical sensitivity had been undertaken in any European country, and there were no case definitions or agreement on the criteria for diagnosis of the condition. However, it was thought that cross-country studies might yield fresh insights into the problem, which appears to be influenced by a number of social and cultural factors. In the United States, where chemical sensitivity has received the most attention, some of these social and cultural factors have, to varying degrees, hindered study and understanding of this condition. These include partisan biases among physicians concerning the etiology and the relevance of chemical sensitivity; lawsuits; disagreements with respect to who should pay for diagnosis and treatment; chemical manufacturers' concerns about liability; the presence of well-informed, networked, and activated patient groups; and a citizenry with an acute awareness of and concern for environmental exposures. Not all of these factors are present to the same degree in Europe although conditions in the United Kingdom and Germany most resemble the situation in the United States. It was felt that a cross-country investigation in Europe might provide a new perspective on the subject, as well as afford an opportunity to examine differences between countries in terms of their pattern and use of various chemicals, building construction and ventilation practices, and differing traditions of occupational and environmental medicine.

The study was not designed to test any specific hypothesis, but to collect and compare information from several countries that might suggest hypotheses for future research. Definitive conclusions about the nature and the etiology of chemical sensitivity were not sought. Following similar protocols, three teams collected data and reported findings from: Denmark, Finland, Norway, Sweden, and the United Kingdom (Team A); Belgium, Germany, and the Netherlands (Team B); and Greece (Team C). A computerized literature search was undertaken, and persons thought likely to have some knowledge or experience with chemical sensitivity, including ministries of environmental or public health, environmental groups, labor unions, and professional medical associations, were contacted and interviewed according to general guidelines. Anecdotal clinical observations and non-peer-reviewed "gray" literature reports were included in the analysis for the additional insights and opportunities they might provide for future study. [The literature review should be considered complete through 1994 although selected additions were made to the report through November 1995. See also two books, which have been published in Dutch (Peereboom 1994) and German (Maschewsky 1996).]

In an attempt to describe the population of interest in the European study, the investigators formulated the following taxonomy to guide data collection activities and analysis, in which chemical sensitivity was defined to encompass three relatively distinct categories:

1. The response of normal subjects to known exposures in a traditional dose–response fashion. This category includes classical allergy or other immunologically mediated sensitivity.

2. The response of normal subjects to known or unknown exposures, unexplained by classical or known mechanisms. This category includes:

   (a) Sick building syndrome, in which individuals respond to known or unknown exposures but their symptoms resolve when they are not exposed to the building.

   (b) Sensitivity, such as that induced by toluene diisocyanate (TDI), which begins as specific hypersensitivity to a single agent (or class of substances) but may evolve into nonspecific hyperresponsiveness, described in category 3 below.

3. The heightened, extraordinary, or unusual response of individuals to known or unknown exposures, whose symptoms do not completely resolve upon removal from the exposures and/or whose "sensitivities" seem to spread to other agents. These individuals may experience:

   (a) A heightened response to agents at a given exposure level compared to other individuals;

   (b) A response at lower levels than those that affect other individuals; and/or

   (c) A response at an earlier time than that experienced by other individuals.

The European investigation focused primarily on categories 2b and 3 above. This focus essentially excluded traditional sick building syndrome (category 2a) although hypersensitive individuals who became ill in tight buildings (i.e., those individuals who did not recover, but who experienced subsequent sensitivities) were thought to constitute a potentially useful group that might provide important information on low-level chemical sensitivity (Chester and Levine 1994).

Despite the potential usefulness of exposure or event-driven information (see Chapter 8), the research teams were unable to discover many situations or incidents that could provide useful data relevant to chemical sensitivity as defined above. There is no paucity of events or exposures; there is simply little information available about the *outcomes* in terms of the development of chemical sensitivity. Information on the temporal features of the development and disappearance/waning of the problem would be important, but was difficult to obtain. A variety of factors may explain this relative lack of information. For example, the research tended to focus on physicians and the medical literature as sources of data. In general, physicians interact with individual patients and have little reason to recognize that their patients may be part of a larger group of individuals who have experienced a common exposure or event (and perhaps little interest in doing so). Second, physicians, researchers, and health authorities who are involved in events or exposure situations (e.g., a "sick building" or exposures of a particular workplace/occupation) are unlikely to focus on chemical sensitivity and thus have little reason to (1) follow the affected individuals for extended periods of time, (2) identify subsequent sensitivities, or (3) distinguish between initiating and subsequent triggering exposures. Despite this, the research teams did identify some exposure- or event-driven information that may be suggestive of low-level chemical sensitivity.

The predominant loci of the alleged initiating exposures/events in this investigation were industrial, office, and domestic environments. Agricultural exposures resulting in chemical sensitivity were mentioned in several countries, and hairdressers appeared to be affected in several countries.

A relatively small number of substances were specifically associated with the onset of chemical sensitivity (Table 7–1). The substances most often mentioned as initiators included pesticides, solvents, paints and lacquers, and formaldehyde. Repeated or continuous low-level exposure, rather than a single event, characterized most of the experience. Psychosocial stressors were also mentioned as initiating chemical sensitivity.

A unique situation was reported in Germany, where exposure to emissions from treated wood has been associated with its own clinical entity— "wood preservative syndrome" (or "pentachlorophenol syndrome") (Schimmelpfennig 1994). Some individuals exposed to wood (or rooms with wood) treated with pentachlorophenol (PCP) and lindane (contaminated with dioxins and furans, and dissolved in solvents at a concentration of about 5 percent) have experienced a multitude of symptoms commonly associated with chemical sensitivity. These include immunologic, dermatologic, neurologic, psychiatric, endocrinologic, and ophthalmologic symptoms (Huber et al. 1992). Many of the physicians surveyed in Germany reported that pentachlorophenol and wood preservatives initiated illness, and they described subsequent sensitivities (e.g., to odors, solvents, and, sometimes, foods) in their patients.

Although these investigations were neither exhaustive nor comprehensive, some interesting observations can be made. Pesticides, organic solvents, formaldehyde, and stress were mentioned as causes of chemical sensitivity in many countries, whereas anesthetic agents were mentioned repeatedly only in Greece. Problems with hairdressing chemicals were cited in Denmark, Sweden, and Greece.

Of course, the categories "organic solvents" and "pesticides" are overly broad. Identification of specific substances in these categories would be more informative, but in many cases more definitive information simply was not available. With the possible exception of pentachlorophenol (but see McConnachie and Zahalsky 1991), these are the same sources associated with the onset of chemical sensitivity in North America.

TABLE 7-1 Some Exposures Reported as Associated with the Onset of Chemical Sensitivity in Europe

Exposure	Denmark	Sweden	Norway	Finland	Germany	Holland	Belgium	U.K.	Greece
Amalgam/mercury		√	√		√	√			
Anesthetic agents									√
Carpets and glue		√				√			
Diesel exhaust	√								
Formaldehyde	√		√		√				√
Hairdressing chemicals	√	√							√
Indoor climate	√	√		√					
Industrial degreasers	√								
Methyl methacrylate		√	√		√				
New/renovated buildings		√	√		√				
Organic solvents	√	√	√	√	√	√	√	√	√
Paints/lacquers	√	√			√				√
Pentachlorophenol/ wood preservative					√	√	√		
Pesticides	√		√		√		√	√	√
Pharmaceuticals			√						√
Printed material	√	√							
Stress/psychosocial factors	√	√			√			√	

A much larger number of chemically diverse substances were reported to trigger symptoms in persons who were already alleged to be chemically sensitive than were reported to *initiate* the condition (Table 7–2). These substances parallel the "triggers" frequently reported in the United States and include perfumes, detergents and cleaners, smoke, cooking odors, car exhaust, new clothing, nail polish, and newspaper print. Reactions to these substances were reported in each country. Symptoms frequently included: mucous membrane irritation; gastrointestinal problems; joint pain; respiratory difficulties such as chest tightness and rhinitis; fatigue; and central nervous system problems such as headache, dizziness, memory loss, and difficulty with concentration. Physicians reported a higher than average prevalence of symptoms associated with chemical sensitivity among women 30 to 50 years old than among men in Scandinavia, Germany, and Greece.

The European research group recommended that serious research efforts be undertaken to clarify the nature of chemical sensitivity (see Chapter 10). However, they believed that:

> Until the nature of the condition is better understood, reasonable preventive and accommodative action should be taken. These may include: 1) serious public health intervention efforts to reduce exposures to possible "initiators" of chemical sensitivity, suggested in part by the experience collected to date, and 2) avoidance, as much as possible, in public places of substances known to trigger symptoms in persons who already report chemical sensitivity. Reasonable accommodation should be made in housing and employment, such as limiting and warning occupants about pesticide application in buildings and providing less-contaminated places to work.

TABLE 7–2. Some Sources Reported to Trigger Symptoms in Patients with Purported Chemical Sensitivity in Europe

Air fresheners	Nail polish
Alcohol	New car interiors
Automobile exhaust	New clothing
Carpets	Newly painted rooms
Cleaners/detergents	Newspapers/printed material
Cooking odors	Perfumes/fragrances
Cosmetics	Solvents
Diesel	Stress
Drugs/pharmaceuticals	Tobacco smoke
Foods	White spirits
Gasoline	

Comparison of European and North American Experiences with Low-Level Chemical Sensitivity

The limited data available at this time from North America and Europe suggest that low-level chemical sensitivity is not a single, distinct clinical entity. Clinical presentations are extraordinarily diverse, a major reason why consensus on a case definition for the illness has been difficult to achieve despite numerous attempts (Miller 1996a). Symptoms appear to involve any and every organ system or several systems simultaneously although central nervous system symptoms such as fatigue, mood changes (irritability, depression), and memory and concentration difficulties predominate. Even among persons who have shared the same initiating exposure, symptoms and severity differ markedly. Ultimately, chemical sensitivity may be more accurately characterized as a general class of disorders, like infectious diseases, which share a common general mechanism; yet within the class, particular members may involve different symptoms, agents, and specific mechanisms.

From European and North American observations, a wide range of environmental exposures appear able to *initiate* the problem. Implicated chemicals are structurally diverse, but certain ones appear again and again on both continents:

1. Pesticides are frequently cited in North America and Europe, with the exception of Sweden, Finland, and the Netherlands, where indoor use of pesticides may be less frequent as a consequence of cooler temperatures and reduced insect populations. Organophosphate and carbamate pesticides are those most often reported as causing illness in the United States, but this may simply reflect the fact that these are among the agents most commonly applied. The greater symptom severity reported by chemical sensitivity patients exposed to organophosphates versus indoor air contaminants associated with remodeling (summarized in Chapter 8) suggests that some compounds in this class (organophosphates and carbamates) might be especially potent sensitizers, at least for a subset of the population.
2. Organic solvent exposure was cited in every European country surveyed and is commonly cited in North America. Such exposures frequently occur in the workplace and are more often chronic than acute in nature.

Although there are consistent observations regarding the causes of chemical sensitivity between continents, there are also notable differences, for example, the so-called wood preservative syndrome associated with pentachlorophenol use in Germany (Schimmelpfennig 1994). (But see McConnachie and Zahalsky 1991.)

Although Sick Building Syndrome (SBS) is widely recognized in the Scandinavian countries, where a number of internationally known researchers are engaged in its study, instances of SBS per se did not generally reveal chemically sensitive subgroups. Conceivably, preoccupation with immediate effects may have obscured their discovery. Certainly, there was no indication of a large problem in those instances. Initiating experiences with carpets were noted, however (Table 7–1). If future inquiry were to reveal that chemical sensitivity does not occur in even a subset of individuals in European SBS episodes, this finding might suggest the importance of other factors, for example, the use of wall-to-wall carpeting (common in the United States and relatively infrequent in Europe), or the use of certain fragrances, air fresheners, cleaners, and/or extermination practices.

In both Europe and North America, patients report the spreading of their sensitivities to an array of common exposures, including fragrances, cleaning agents, engine exhaust, alcoholic beverages, foods, and medications they formerly tolerated without difficulty. The fact that many of these individuals voluntarily forgo pizza, chocolate, beer, or other favorite foods because they make them feel so ill warrants consideration, as there is little secondary gain to be garnered from such forbearance. Many participants in one North American study (Miller and Mitzel 1995) reported that drugs, ingestants containing chemical additives (monosodium glutamate, chlorinated tap water), and food–drug combinations (alcoholic beverages or caffeine/xanthine-containing foods) made them ill, a finding consistent with a hypothesis that these individuals exhibit amplified responses to pharmacologic doses of a variety of substances (Bell et al. 1992; Bell et al. 1993a).

Generally speaking, awareness of chemical sensitivity may be enhanced in countries with greater environmental activism, but illnesses resembling chemical sensitivity were described in every European country that was studied. Clinical ecology's origins in the United States and its spread to other English-speaking nations, including Canada and the United Kingdom, no doubt have influenced the numbers of patients receiving a diagnosis of chemical sensitivity in those countries. Discord among physicians as to what constitutes appropriate diagnostic and therapeutic approaches in these countries permeates professional meetings, medical journals, and court proceedings. Where patients must "prove" that a particular exposure caused their illness in order to receive workers' compensation or reimbursement for medical expenses (as in the United States, where there is no national health care system), disputes between medical practitioners (who may testify on opposing sides) are most contentious.

Cultural practices may affect the prevalence of chemical sensitivity. In some European countries, people typically spend several hours each day outdoors, for example, walking to work or shopping, and windows in homes and offices may be left open part or most of the day. In contrast, on average, Americans spend 90 percent or more of the day indoors, often in tightly sealed structures, where levels of certain volatile organic air contaminants can be orders of magnitude higher than they are outdoors (see Chapters 1 and 3).

Choices of building construction materials and furnishings also vary greatly between countries, including the use of wall-to-wall carpeting versus washable throw rugs or no floor coverings at all; solid hardwood furnishings versus particle board or pressed wood; paint, wallpaper, and adhesive constituents; and the amount of office equipment, including photocopiers and computers.

Ventilation practices may be similarly diverse. The building of tightly constructed buildings with little fresh makeup air in North America since the oil embargo of the mid-1970s could be a factor in the apparent increase in chemical sensitivity cases over the past two decades in the United States and Canada. The experience with SBS but not chemical sensitivity in Scandinavia merits closer examination to determine whether the latter condition has thus far escaped attention, or whether environmental or perhaps genetic or cultural differences may prevent its development.

The use of chemicals also varies from country to country, in particular, pesticides, cleaners, and personal care products, including fragrances. Comparing differing rates of consumption of these products, as well as pharmaceuticals, and the incidence of chemical sensitivity among countries could provide further clues.

Conclusion

Complex questions concerning the origins and mechanisms of chemical sensitivity will not be resolved by retrospective survey studies—indeed, probably not by retrospective studies of any kind. Perhaps more informative would be prospective observations on the natural history of chemical sensitivity associated with particular *incidents* or *exposure* events rather than isolated case reports. Nevertheless, enlightening similarities and instructive differences can be gleaned from future, better-directed cross-country comparisons of experiences with chemical sensitivity.

In the past five years in the United States, controversies surrounding chemical sensitivity have exploded far beyond the narrow confines of a medical debate into a national debate with far-reaching policy and regulatory implications. Most recently, a number of U.S. Persian Gulf War veterans have reported multisystem health problems and new-onset intolerances to chemicals, foods, and other substances since returning from the war (Miller 1994b). (See the discussion in Chapter 8.) Some have received a diagnosis of chemical sensitivity from private physicians and now seek medical care and compensation for the condition. Such trends in North

America could be mirrored in European countries over the next few decades. Indeed, this condition, popularly called "Gulf War Syndrome," has surfaced among Persian Gulf veterans in Canada and several European countries, including the United Kingdom and Czechoslovakia.

Understanding chemical sensitivity is pivotal to establishing sound environmental policy. If there is a subset of the population that is (or can become) especially sensitive to low-level chemical exposures, a strategy for protecting this subset must be found. If it were to be determined that certain chemical exposures could lead to sensitization, then perhaps these exposures could be avoided. Possibly by preventing chemical accidents, prohibiting occupancy of buildings prior to finish-out or completion, avoiding the use of certain cholinesterase-inhibiting pesticides indoors, and adopting other measures, society could protect especially vulnerable individuals from becoming sensitized in the first place. It would make little sense to regulate chemicals at the parts per billion level or lower if what is required is to keep people from becoming sensitized in the first place. Indeed, by understanding the true nature of chemical sensitivity and who is at risk, we may prevent unnecessary and costly regulation of environmental exposures in the years to come.

Chemical sensitivity could be a new paradigm for disease (see Chapter 8) that has the potential to explain many chronic and costly illnesses, including fatigue, depression, headaches, and asthma, or it could continue to elude definition. Not understanding the causes of chemical sensitivity, we take an immense gamble—but knowledge will not come cheaply. Future studies on chemical sensitivity that involve blinded challenges in a controlled environment, that utilize brain imaging, state-of-the-art immunological testing, or other sophisticated tests, and that compare adequate numbers of patients and controls, will be costly. Although small sums, on the order of a few million dollars, have been invested in research on MCS, funding agencies will need to make a much greater financial commitment if progress is to be made in this area, as it has been for other diseases such as breast cancer and AIDS. Until sufficient research funds become available, chemical sensitivity no doubt will continue to pit physician against physician, perplex policy makers, impoverish patients, and plague industry.

CHAPTER 8

Key Research Findings since the First Edition

Introduction

Clinical Data: Inherent Limitations and Unwarranted Extrapolations

In researching low-level chemical sensitivity, it is useful to distinguish contrasting ways in which observations might be recorded. First, physician reports of individual patients can be examined. Because chemical sensitivity was "discovered" by observant physicians, this might seem like a useful place to start, but there are difficulties with this approach. Although physician reports contain much information about patients' symptoms and complaints, they usually contain little information about possible *initiating exposures or events, triggering exposures,* and *outcomes* of various interventions—both clinical and nonclinical. Moreover, information differentiating initiating events/exposures from subsequent "triggers" of symptoms is often lacking or conceptually muddled. Because the precise nature of and mechanisms for chemical sensitivity remain ill-defined, information on possible initiating factors and effective interventions (e.g., avoidance) is crucial to improving our understanding of this bewildering condition. Also, each of the more prevalent MCS symptoms, such as headache or shortness of breath, could be caused by any of several different biological mechanisms and a variety of environmental exposures.

Most physicians do not obtain occupational or environmental histories from their patients, and the patients themselves may not be aware of possible precipitating events or exposures. Moreover, physicians approach patients with their own disciplinary orientations and biases, making it difficult to compare reports on individual patients from different physicians. (Of course, different patients with their own convictions about the cause of their condition may also influence their physicians' diagnoses.) For example, pulmonary physicians will tend to focus on respiratory symptoms and airborne contaminants, perhaps overlooking or discounting somewhat more subjective (but perhaps equally bothersome) central nervous system (CNS) symptoms. Indeed, chemically sensitive patients often go from physician to physician, acquiring different diagnoses and labels—from organic brain syndrome to chronic fatigue syndrome to psychosomatic disease. Because most physicians see very few MCS cases and there seem to be few proven, effective medical interventions for these patients, the eventual outcome of the condition and the possible success of various interventions (such as avoidance, food rotation, or simply tincture of time) may not be known to the diagnosing physician or clinic. Physicians may erroneously surmise that patients who do not return to see them must have improved when, in fact, many MCS patients have consulted dozens of physicians with no discernible improvement in their condition.

Finally, isolated case reports suffer from being symptom- or syndrome-focused, overlooking the possibility that patients' health problems may be induced by a wide variety of initiating exposures or events. This has compounded the difficulty in understanding the origins of chemical sensitivity. We suggest that low-level chemical sensitivity might be more correctly described as a general class of disorders, like infectious diseases, the members of which may share similar symptomatology, but whose different causes and pathways may need to be particularized for physicians to successfully understand and treat them (see the discussion in Chapter 10). The different forms of chemical sensitivity may be precipitated by different physical or chemical exposures or by psychosocial events. The presenting symptoms and signs, if present, are most often nonspecific and not indicative of etiology.

In making diagnoses, physicians frequently invoke, without being conscious of what they are doing, the "representativeness heuristic," a judgmental shortcut or rule of thumb that is used to render complex problems manageable (Gilovich and Savitsky 1996). For example, for decades physicians attributed ulcers to stress because they "knew" that when people are under stress, their stomachs hurt and feel acidic. Only recently has it been shown that a bacterium, *Helicobacter pylori*, causes ulcers that can be cured by antibiotics, not stress reduction. Now some physicians are applying the representativeness heuristic to MCS: because depressed (or otherwise psychologically disturbed) people commonly report fatigue, memory and concentration difficulties, mood changes, and the like, and because MCS patients report these symptoms, these physicians conclude that MCS patients must be depressed (see Davidoff and Fogarty 1994 for a critique of common logical errors committed in studies claiming a psychologic basis for MCS). Investigators seeking simple answers to complex problems are sometimes led down the wrong path.

Dimensions of an Illness

The causes, symptoms, and interventions for illnesses can each be characterized as physiological (P) or psychological (Ψ). Physiological and psychological events can precipitate either physiological or psychological symptoms, or both. Psychological interventions such as biofeedback and social support can alleviate some aspects of physical disease. Neither the nature of symptoms nor the successes of interventions are dispositive of the origins of a condition. Schematically, the three factors—causes, symptoms, and interventions—can be represented as separate "dimensions" of illness (Fig. 8-1). Different physicians and researchers may operate in different "quadrants." For example, a physician may believe that the cause of a particular patient's chemical sensitivity is physiological, observe CNS (psychological) symptoms, and treat with biofeedback or other coping (i.e., psychological/behavioral) strategies. In contrast, a researcher may assume stress as the cause, observe fatigue as a consequence, and investigate the use of new drugs to alleviate the symptoms.

What is disappointing in much of the literature is the continuing failure to distinguish between causes and symptoms of MCS, and unjustified conclusions drawn from successes or failures of particular interventions (Davidoff and Fogarty 1994). Although lip service is given to making these distinctions, both the failure thus far to identify consistent objective markers of disease (Simon et al. 1993) (despite the fact little research on some of the most plausible hypotheses has yet been undertaken) and the finding of a history of childhood abuse in some patients (Staudenmayer et al. 1993b) have led some authors to lean heavily in the direction of psychogenic causes and the recommendation of psychological interventions, rather than physiological causes and the avoidance of further exposure as a treatment modality. Even a recent review of the literature on low-level chemical sensitivity (Sparks et al. 1994a and b), although acknowledging the multifactorial origins of this condition, ends up recommending psychological interventions as the only acceptable treatment modality. Inasmuch as great uncertainty continues to characterize this condition, these views are premature and perhaps even harmful to patients (Miller 1995).

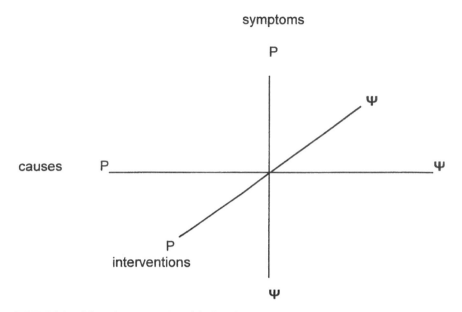

FIGURE 8-1.   *Schematic representation of the three dimensions of illness. (UTHSCSA © 1996)*

Empirical Approaches to Understanding Chemical Sensitivity

Given the need to distinguish between causes, presentations, and the success of interventions that has been discussed above, physicians' observations may be more helpful when: (1) the physicians see a large number of chemically sensitive patients, take a complete exposure history, and recognize subgroups that give clues to different origins and successful interventions in each; (2) the physicians happen to see several patients who have experienced the same or similar events or exposures, such as living in the same neighborhood or apartment building or using the same type of product, such as new carpets; (3) the physicians specialize in occupational or environmental medicine and see groups of patients with similar exposures, occupations, or environmental histories; or (4) the physicians are specialists (e.g., pulmonary or ear, nose, and throat physicians) who concentrate on specific organ systems and are thus more likely to recognize subsets of patients with features uncharacteristic of the majority of patients with the same illness. For example, patients whose asthma is precipitated by perfumes, detergents, and clothing stores may constitute a chemically sensitive subgroup of special interest to pulmonologists or allergists. Studies focusing on (1) patients with particular symptoms, (2) patients seen by particular physicians or clinics, or (3) patients meeting a particular case definition for MCS may suffer from referral biases, are likely to overlook the full range of illnesses associated with an exposure, and may be diluted by the inclusion of large numbers of patients with unrelated conditions.

Perhaps more informative than studies on patients as described above would be observations on the natural history of chemical sensitivity associated with particular *incidents* or *exposure events* rather than isolated case reports. Event-driven information includes both (1) disease or symptom outbreaks in particular communities, buildings, workplaces, or occupational groups and (2) chemical exposure events/scenarios involving certain occupations or particular building materials, pesticides, consumer products, or medications. Studies of multiple case reports linked to specific incidents or exposure events might be particularly useful. Events or exposures whose impact could be followed prospectively may be more readily identified by public health or environmental/occupational health authorities, compensation or disability agencies, affected individuals, trade unions, and patient associations than by physicians. Although retrospective investigations may be helpful, prospective studies (e.g., of greenhouse workers, exterminators, or occupants of newly renovated office buildings) might yield useful perspectives, especially if the cohort is followed for a sufficient period of time. Unfortunately, ongoing litigation (in the case of retrospective studies) or the potential for it (in the cases of retrospective and prospective studies) has in the past interfered with access to groups of research subjects who shared a well-defined exposure event, and is likely to continue to do so unless in the future the parties in such cases agree to studies of this kind.

We have previously cautioned about the necessity of taking into account adaptation or masking in observing the symptoms of patients with alleged low-level chemically sensitivity, including the effects of therapeutic drugs and food intolerances. Researchers and clinicians who ignore these concerns and then find no consistent markers, symptoms, or success in recommending chemical avoidance cannot rightfully claim to have tested or investigated the key explanations offered for this condition (Datta 1993).

Characterizing the Patient Population

Epidemiological Studies

As discussed in the preceding section, event-driven studies on MCS, that is, epidemiological studies targeting groups of individuals sharing a common (initiating) exposure event, offer the greatest potential for understanding the origins of this illness, its progression, and the spectrum of health problems associated with it. Below are summarized the several epidemiological studies conducted on MCS since the first edition.

In the largest exposure-driven study of MCS patients to date, Miller and Mitzel (1995) surveyed 75 individuals who reported onset of their illness following remodeling in a building (home or workplace) and 37 who reported onset following exposure to a cholinesterase-inhibiting pesticide (organophosphate or carbamate). They hypothesized that if MCS were predominantly due to a neurotoxic exposure rather than psychogenic in origin, then MCS patients who became ill after an organophosphate or carbamate pesticide exposure should report more severe symptoms than those who became ill after exposure to low levels of mixed solvents in a remodeled building. Although solvents are also neurotoxic, they would generally be considered less so than pesticides in this class. Conversely, if the illness were mainly psychogenic in origin, for example, due to depression, there should be no differences in symptoms or symptom severity between the two groups.

The most frequently reported symptoms in each group were remarkably similar, with a predominance of central nervous system symptoms (Table 8-1). The most common gastrointestinal complaint was "problems digesting food," and the most common respiratory complaint was "shortness of breath or being unable to get enough air." Further, the groups shared similar ordering of symptoms, from most severe to least severe, and identified similar inhalant and ingestant triggers. Notably, however, the pesticide-exposed group reported statistically significantly greater symptom severity than did the remodeling-exposed group, particularly for neuromuscular, mood-related, airway, gastrointestinal, and cardiac symptoms. The authors interpreted their data as suggesting (1) a biological basis for MCS and (2) a distinct pathophysiology or final common pathway for the condition that, while as yet undefined, appeared to be shared by these different exposure groups.

The authors also noted that although subjective, multisystem health complaints characterize both MCS and somatoform disorder (a psychiatric diagnosis), features of the patients they studied were inconsistent with somatoform disorder: Severe cognitive symptoms predominated in their MCS sample, but tend to be far down on the list of complaints, if they appear at all, among somatoform disorder patients. In addition, somatoform disorder almost always begins before 30 years of age, yet 83 percent of their MCS patients reported onset after the age of 30.

Most of Miller and Mitzel's subjects had experienced major disruption of their careers since the exposure event. Eighty percent indicated that they had worked full-time prior to their exposure. Yet at the time of the survey (nearly eight years post-exposure on average), more than 80 percent reported that they were no longer able to work full-time. Consistent with findings in other studies, 80 percent of their subjects were women with an average educational level of nearly four years of college. Forty percent said they had consulted ten or more medical practitioners since their illness began.

TABLE 8-1. Top 20 Symptoms (out of 119 Symptoms) Reported by MCS Patients Attributing Their Illness to Exposure to Pesticides (N = 37) versus Remodeling (N = 75)

	Ranking		Mean Symptom Severity**	
Symptom	Pesticide	Remodel	Pesticide	Remodel
*Tired or lethargic	1	1	2.49	2.44
*Fatigue > 6 months	2	3	2.43	2.10
*Memory difficulties	3	4	2.32	2.09
*Difficulty concentrating	4	2	2.32	2.17
*Dizziness, lightheadedness	5	6	2.19	1.85
*Depressed feelings	6	8	2.19	1.83
*Spacey	7	12	2.19	1.74
*Groggy	8	5	2.14	1.96
*Loss of motivation	9	7	2.11	1.84
*Tense, nervous	10	15	2.11	1.64
*Short of breath	11	18	2.11	1.61
*Irritable	12	10	2.03	1.79
Problem focusing eyes	13	43	2.03	1.27
Chest pain	14	52	2.00	1.19
*Muscle aches	15	11	2.00	1.79
Problems digesting food	16	33	1.97	1.35
*Joint pain	17	9	1.95	1.83
Tingling fingers/toes	18	59	1.95	1.12
*Headache	19	14	1.92	1.67
*Head fullness or pressure	20	19	1.92	1.60
Difficulty making decision	21	13	1.89	1.69
Eye irritation	22	16	1.89	1.64
Slowed responses	34	17	1.72	1.63
Nausea	36	20	1.65	1.56

*Among top 20 symptoms in both pesticide and remodeling patients.
**Symptoms scored on 0 to 3 scale: 0 = not a problem; 1 = mild; 2 = moderate; 3 = severe.
Source: Miller 1994a.

In the first controlled psychiatric study on MCS, Black et al. (1990) contrasted the current and the past psychiatric status of 26 MCS patients and 33 controls. The MCS patients, predominantly middle-aged women with some college education, attributed onset of their condition to "fumes" at work or home (50 percent), pesticides/insecticides (27 percent), oral contraceptives/pregnancy/hysterectomy (19 percent), psychological stress (15 percent), and/or antibiotics (12 percent). Compared with controls, the MCS group exhibited higher rates of major depression (30 percent versus 7 percent), anxiety disorders including simple phobia and panic (43 percent versus 17 percent), and somatization disorder (17 percent versus 0 percent). Sixty-five percent of the MCS patients but only 28 percent of controls qualified for a lifetime psychiatric diagnosis. Consistent with the finding of Miller and Mitzel (see above), although many of the MCS patients in

Black's study had multiple somatic symptoms, they did not meet criteria for somatization disorder because their age at illness onset was over 30 years. The authors ascribed most of the MCS symptoms to "commonly recognized psychiatric disorders," but offered no causal explanation for the 35 percent of their MCS patients who had no history of psychopathology.

In two separate studies, Simon et al. (1990, 1993) compared psychiatric features of MCS patients with those of controls. In a 1990 study of 13 aerospace workers who developed MCS and 23 control subjects from the same plant who did not following a change in manufacturing materials, these researchers found no differences in current psychiatric diagnoses based upon a standardized interview (the Diagnostic Interview Schedule). They identified no cases of somatization disorder, but did find an increased frequency of past major depression and panic disorder in the MCS group. Results of several psychological instruments, however, suggested subclinical, somatization-like illness in this group. The MCS patients reported an average of 6.2 unexplained physical symptoms prior to their workplace exposure versus 2.9 for controls. Likewise, 54 percent of the same MCS patient group reported anxiety or depression prior to their workplace exposure versus only 4 percent of controls. The authors assumed that the low-level chemical exposures identified by the patients were not toxic and that therefore neurotoxicity could not have played a role in these illnesses. Others have pointed out that such an assumption is unwarranted at this early stage in our understanding of MCS (Bell 1994; Davidoff and Fogarty 1994; Miller 1996a, 1997).

For their second study, Simon et al. (1993) recruited from a local allergist 41 patients who had been ill for more than three months, had symptoms in three or more organ systems (including the central nervous system), and reported sensitivity to 4 or more of 14 common chemicals. Controls were 34 patients with musculoskeletal injuries seen at a university clinic who were not assessed for chemical intolerances. The chemically sensitive patients, primarily well-educated women, exhibited higher rates of current panic disorder (24 percent versus 3 percent), with trends toward more major depression (29 percent versus 12 percent) and generalized anxiety disorder (10 percent versus 0 percent). These patients also had higher scores on the Symptom Checklist 90 (revised) subscales for depression, anxiety, and somatization. There were no differences between cases and controls for past psychiatric diagnoses, in contrast with the authors' earlier study (see above). The chemically sensitive group again reported significantly more premorbid, unexplained symptoms than did controls. Chemically sensitive patients were less likely to be using alcohol currently (24 percent versus 56 percent). There were no significant group differences in performance on several neuropsychological tests. Fifty-six percent of the MCS patients had no current psychiatric diagnosis. In contrast with their earlier paper (see above), the authors acknowledged that psychiatric findings in these patients could be either the cause or the result of chemical sensitivity. The fact that 61 percent of the chemically sensitive group used caffeine and that such low inclusion criteria were used (intolerances for 4 of 14 chemicals) suggests that these patients may not have been typical of most MCS patients (Bell 1994), or that they were very masked.

Other studies suggest that many, but by no means all, MCS patients have a lifelong history of medical problems (Davidoff and Keye 1996; Bell et al. 1995a). Fiedler et al. (1992) did not find that premorbid psychiatric conditions accounted for MCS in a group of 11 patients they studied. MCS proponents argue that even if some MCS patients were depressed prior to florid onset of their illness, the question remains of whether MCS is caused by depression, whether depressed people are more susceptible to MCS, or whether the preceding depression was in fact the result of prior, undiagnosed chemical or food sensitivities. Indeed, a cholinergic theory of depression posits that depressed persons are hypersensitive to acetylcholine (Dilsaver 1986), the same neurotransmitter whose breakdown is impaired by organophosphate and carbamate pesticides. Rosenthal and Cameron (1991) suggest that "vulnerability to the effects of acetylcholine might account for both environmental hypersensitivity and the endogenous tendency to become depressed."

Davidoff and Keye (1996) conducted a standardized telephone interview with 60 MCS patients, 20 of whom attributed onset of their illness to organic solvents, 20 to an organophosphate pesticide, and 20 to a sick building. In addition, 10 workers exposed to chlorine dioxide and chloroform who subsequently reported chemical intolerances were studied. Sixty randomly selected controls matched for gender, age, and socioeconomic status to the MCS patients were also interviewed. The four exposure groups were similar to one another with respect to all general health and illness status variables, but collectively they differed significantly from controls. The three MCS patient groups were more likely than controls to report changes in tolerance for odors (85–100 percent versus 27 percent), allergens (70–85 percent versus 32 percent), foods (60–90 percent versus 20 percent), alcohol (40–55 percent versus 8 percent), and medications (30–75 percent versus 13 percent). Changes in tolerance in two or more of these categories were more frequently reported by MCS patients (90–95 percent) than by controls (29 percent). Seventy to 80 percent of MCS subjects attributed daily illness to chemical exposures versus less than 2 percent of controls. On average, MCS subjects attributed 15 symptoms weekly to chemical exposures, whereas controls attributed only one symptom to such exposures. Chronic sickliness in childhood, defined as three or more chronic health conditions before age 18, were reported by 15 to 55 percent of the MCS groups but by less than 2 percent of the general population sample. Whether they had or had not been treated by a clinical ecologist, MCS patients shared similar illness characteristics although those who had seen an ecologist reported greater loss of tolerance than those who had not. In contrast with the MCS groups, no one among the ten chlorine dioxide–exposed individuals had been seen by a clinical ecologist, belonged to a patient support group, or claimed to have "MCS." Nevertheless, these patients reported symptoms and intolerances that were congruent with those of the three MCS groups, leading the authors to suggest that MCS "is not a figment of the clinical ecologists' collective imagination."

MCS patients often are engaged in litigation or compensation cases, and many have been in a sick role for extended periods. Bell et al. (1993a, 1993b, 1994a, 1994b, 1995a, 1996a) have sidestepped these confounding factors by studying instead several thousand college-age students and active, retired elderly persons in Arizona, a subset of whom reported chemical odor intolerances. Over half of the college students and retirees reported that one or more of the following chemical odors made them feel ill: pesticides, drying paint, perfume, car exhaust, new carpet (Bell 1994). Some 15 percent identified three or more of these odors as frequently causing illness. In contrast, approximately 30 percent of both groups answered affirmatively the question, "Do you consider yourself especially sensitive to certain chemicals?" (Compare with results of other population surveys on MCS discussed later in this chapter under "Magnitude of the Problem.") In four out of her five surveys on nonpatient populations, Bell found that women outnumbered men among the subsets reporting the greatest sensitivity to odors. Less than 1 percent of the college students and about 4 percent of the older adults reported a physician diagnosis of MCS (Bell et al. 1994b, 1996a). Psychological variables did not fully explain the presence of odor intolerance in the most sensitive subsets although odor intolerance clearly was associated with psychological distress. In a subset of active retired women, Bell found slowed reaction times on a divided attention task among those who rated themselves chemically intolerant versus those who did not (Bell et al. 1996b).

Among both college students and geriatric adults who were odor-intolerant, there was an increased prevalence of, or trend toward, more lifetime diagnoses of nasal allergies, breast cysts in women, sinusitis, food intolerances, irritable bowel, and migraine headaches (Bell 1994). Among her odor-intolerant college students, there were more diagnoses of hives, ulcers, chronic pain, childhood hyperactivity (males), chronic fatigue syndrome, anxiety disorders, and premenstrual syndrome (females) than in subjects who were not odor-intolerant. A more recent survey revealed more hypertension and juvenile arthritis in chemically intolerant college students than in controls (Bell et al. 1996a). As in these chemically intolerant elderly persons and college students, nasal allergies, irritable bowel, chronic fatigue syndrome, and migraine headaches also were reported more frequently by a group of MCS patients recruited via patient newsletters and newspaper advertisements versus controls (Bell et al. 1995a). In addition, these MCS patients reported more: chronic bronchitis; arthritis; low adrenocortical function; vaginal candidiasis; ovarian cysts; irregular, heavy, or painful menses; chronic pelvic pain; depression, anxiety, and panic disorder—with trends toward more hypothyroidism, osteoporosis, and fibrocystic breast disease. Sixty percent of these patients considered themselves occupationally disabled. In a community-based sample of county government employees, those sensitive to "chemical stressors" were significantly ($p < 0.05$) more likely [expressed as a higher relative risk (RR)] to have seen a physician for sinus trouble ($RR = 5.35$), acute bronchitis ($RR = 5.87$), pneumonia ($RR = 4.52$), heart problems ($RR = 7.05$), or other serious health problems ($RR = 2.73$) (Baldwin et al. 1995).

Family histories of odor-intolerant (cacosmic) individuals and MCS patients may provide other clues regarding the interrelatedness of these conditions. MCS patients reported that their blood relatives had more nasal allergies and diabetes mellitus, with trends toward more candidiasis, nasal polyps, and epilepsy, but not depression, anxiety, panic disorder, or substance abuse (Bell et al. 1995a). Notably, 27.3 percent of chemically sensitive government workers (Baldwin et al. 1995) reported family histories of heart disease ($p = 0.04$).

In most of her surveys, Bell (1994) found that chemically intolerant groups (but not MCS patients) registered more shyness on standardized scales than did comparison groups. The trait of shyness is thought to reflect inherited, neurobiologically based hyperreactivity to novel stimuli or environments. Notably, there is an animal model for shyness that involves partial limbic kindling (Adamec 1990). Additionally, one variable that predicts sensitizability to drugs (time-dependent sensitization) is hyperreactivity to novel physical environments (Hooks et al. 1992). Available data do not permit differentiating between shyness as a cause, an effect, or a correlate of chemical intolerance. It is also unclear why MCS patients studied thus far have not manifested shyness, but perhaps only those who are not shy respond to questionnaire surveys and newspaper ads. Another possibility is that the shy people with chemical intolerance have lifelong problems, not necessarily initiated by an identifiable chemical exposure (see Fiedler et al. 1996b).

In summary, Bell's surveys of large numbers of chemically intolerant college students and retired persons, without the biases of litigation or a sick role, reveal striking similarities to MCS patients, suggesting shared neurobehavioral dysfunction.

Cullen et al. (1992) at the Yale Occupational and Environmental Clinic in New Haven, Connecticut, and Lax and Henneberger (1995) at the Central New York Occupational Health Clinical Center in Syracuse compared their MCS patients with other patients from their occupational health practices. In New Haven, MCS patients represented approximately 1.8 percent (49/2,759) of the clinic population, whereas in Syracuse, 5.8 percent (35/605) fit an MCS case definition, proportionately over three times as many. Patient selection and referral biases are a possible explanation for this difference: the Yale group applies the Cullen criteria for MCS and discourages referrals for MCS patients who live outside the Connecticut area. Both MCS samples were predominantly female and in the middle age ranges. Solvents and indoor air pollutants were the most frequent workplace exposures implicated by MCS patients in both clinics. A relatively small percentage of the MCS patients seen in New Haven (22 percent) and Syracuse (26 percent) worked in traditional high-exposure industries (i.e., mining, construction, and manufacturing). Approximately twice as many MCS patients in the Syracuse clinic (40 percent) were employed in service industries compared with other patients visiting that clinic. Likewise, 46 percent of the New Haven MCS patients came from the service industries—almost all from education and health care—compared with only 5 percent of the other patients.

Lax and Henneberger suggest that the predominance of women in the MCS group (80 percent versus 25 percent of their other patients) could reflect differences in cultural permission to express illness, segregation of women into unique exposure environments (e.g., secretaries in a sick building confined to their workstations), or biological differences in responses to exposures. Cullen et al. also found that 68 percent of Yale MCS patients were women versus 18 percent of their general clinic population. The findings by both clinics of a relatively low percentage of MCS cases coming from traditionally hazardous occupations and a high percentage from service industries warrants further exploration. Lax and Henneberger muse that possible explanations might include: the "healthy worker" selection effect, whereby workers who felt ill from some exposures would tend to leave an industry (or never enter it in the first place), preferring jobs involving less contact with chemicals; gender segregation resulting from the traditional exclusion of women from mining, construction, and other high-exposure jobs; and/or effects of unique exposures such as low-level, repeated exposure to complex mixtures of indoor air pollutants, a finding that would suggest reconsideration of current notions of safe and hazardous work.

Exposure Challenge Studies

Little scientific data is available concerning the responses of MCS patients to chemical challenges. Few controlled studies and no adequately designed studies involving provocative challenges have been conducted despite the fact that clinicians and researchers who have participated in federally sponsored conferences on MCS have deemed such studies both essential and urgently needed. The few human challenge studies that have been conducted and their limitations are discussed in this section.

Leznoff (1993, 1997) performed single, unblinded challenges in 20 patients with self-reported MCS using test agents tailored to each patient, for example, perfume, cigarette smoke, hairspray, or detergent. Ten of 15 MCS patients who had reported breathlessness and lightheadedness with their exposures experienced symptoms with challenges. No changes in pulmonary function measures were observed. Because these patients' pCO_2 (blood level of CO_2) fell and pO_2 (blood level of O_2) rose post-challenge, the authors thought that these individuals were most likely suffering from hyperventilation and chemophobia. Five other patients with similar complaints had neither symptoms nor changes in pCO_2 or pO_2. Chemical challenges were also performed in five MCS patients with throat-related symptoms, but none experienced symptoms, and no changes were seen with laryngoscopy. Because challenges were not blinded, it is difficult to know whether the breathlessness, lightheadedness, and hyperventilation in 10 of the 20 patients studied were caused by chemical exposure or anxiety.

Staudenmayer et al. (1993a) performed 145 double-blind, placebo-controlled chemical challenges on 20 MCS patients evaluated in their private practice in Colorado between 1985 and 1988. Challenges were conducted in an exposure chamber equipped with HEPA filters for particle removal and activated charcoal and other chemical sorbents for VOC removal. Active agents included formaldehyde, natural gas, cleaners, combusted kerosene, fuel oil, trichloroethane, trichloroethylene, Freon, denatured alcohol, printer's ink, oil paint, and an unspecified insecticide. The exposure duration (15 minutes to 2 hours) and challenge chemicals were selected according to each patient's history. Prior to entry into the study, patients had to demonstrate that they did not react during a single-blind control challenge using an olfactory masking agent such as peppermint, cinnamon, or anise.

A chemical challenge was considered positive if (1) any objective sign was observed such as decreased pulmonary function tests or hives, (2) the patient reported a response to the active agent, or (3) a symptom-rating increased from none or mild to moderate or severe. The authors found that "[t]hese 20 patients did not demonstrate response patterns consistent with their presenting symptoms of chemical intolerance nor did they show signs of toxicity, other than those associated with the irritant mucous membrane responses seen in normal individuals."

The authors did not indicate how many of the patients were referred for evaluation for purposes of workers' compensation or litigation, how many had histories of a well-defined initiating chemical exposure event, or how many were referred because of psychiatric problems. Referral biases may greatly affect the makeup of an MCS study population. For example, one investigation found that only one-fourth of MCS patients whose illness began after an identifiable exposure fulfilled criteria for a major psychiatric disorder, whereas two-thirds of MCS patients who recalled no specific initiating exposure met such criteria (Fiedler et al. 1996b). Given that Dr. Staudenmayer is a psychologist and frequent defense expert witness whose views that MCS is psychogenic have been widely published, it would not be surprising if his study population, drawn from his practice, reflected a selection bias.

Patients in this study were not unmasked prior to challenge; that is, no effort was made to control exposures in the hours or days preceding a challenge. Although it admittedly requires more effort than a standard exposure challenge, unmasking prior to challenge is, in our view, the single most crucial variable that has been overlooked by most investigators in this area. Administering chemical challenges to patients without unmasking them first is the equivalent of giving habitual coffee drinkers a cup of coffee to find out whether caffeine is causing their headaches *without stopping their use of all caffeine first.* (MCS patients who have sorted out their sensitivities often report caffeine stimulatory and withdrawal symptoms, making this analogy particularly apt.)

Other experimental design considerations not addressed in this paper include the possibility that exposure to VOCs revolatilized from charcoal or other sorbents used to clean the air in the chamber could have caused false positive responses to sham exposures. Further, many MCS patients report symptoms due to various types of charcoal (wood, coconut) used in respirators and air purifiers they have purchased. Another concern is this study's reliance upon masking agents rather than below-odor-threshold challenges to accomplish blinding. Kay (1996) found that rats given olfactory challenges with mint, a trigeminal stimulant and one of the masking odorants used in this study by Staudenmayer et al., exhibited narrow band, high amplitude oscillations in the limbic region. This disruption of electrophysiological activity in the rats was not as severe or persistent as for toluene exposure, but if MCS were in fact due to limbic dysfunction, the use of a mint masking agent (and perhaps other odorant maskers) in challenge studies could result in false positive responses. Further, if responses were due to limbic sensitization, repeated exposures to a masking agent conceivably might lead to sensitization to that agent (see later section in this chapter on "Mechanisms: Olfactory-Limbic Sensitization"). Thus, an initial challenge to a masking agent could be negative, yet subsequent challenges might result in symptomatic responses. Another problem with this study is that open, unblinded challenges using the active agents of interest do not appear to have been conducted prior to blinded challenges. The purpose of such open challenges would be to ensure that when a sufficient amount of the active agent is administered, the patient does in fact report symptoms. If the concentration is too low or if the patient has been avoiding that agent for a long time (e.g., months), the active agent might not precipitate symptoms. In the latter case, a waning of sensitivity might have occurred. Hence the importance of an initial, unblinded active challenge in which the patient agrees that the investigator has adequately replicated the exposure of concern. Failure to do so is likely to result in false negative responses (no response to active agent). It is telling that 20 percent of the patients in the study by Staudenmayer et al. had exclusively true negative and false negative responses (i.e., no true positives or false positives). These occurred in cases 2, 7, 8, and 19, who underwent ten, eight, eight, and five individual chemical challenges, respectively. If these patients' responses were entirely random and unrelated to exposure, as these authors suggest, then the chances of so many patients showing no true positive responses and no false positive responses would be slim. Yet this is precisely what these authors report: these four patients never experienced symptoms with sham exposures (which is the correct response), but the active agent never provoked symptoms either (which is not the correct response). A likely explanation for so many false negative findings is that the concentrations of the active agents were too low to evoke symptoms. This is a major flaw the investigators could have overcome by conducting open challenges with active agents first. Unfortunately, the reader cannot tell from the paper which active substances and which placebos were used to challenge which patients, nor the concentrations used and how they were chosen. Likewise, it is unclear how the investigators generated certain exposures quantitatively and reproducibly, for example, combusted kerosene and "chemically contaminated dust." There are numerous pitfalls in attempting to generate reproducible concentrations of complex mixtures. This paper lacks requisite experimental details as to how such exposures were achieved or measured.

Another paper by Staudenmayer et al. (1993b), entitled "Adult Sequelae of Childhood Abuse Presenting as Environmental Illness," provides a detailed description of a 45-year-old woman whom they challenged in their chamber using this same protocol. With exposure to peppermint masking agent, she reported reproduction of her presenting symptoms, particularly tremors, weakness, and speech problems, "which were quite remarkable to the blinded observer." They then explained to the patient that anticipation may result in such symptoms and signs. Next, a double-blind sham challenge was performed at an unspecified interval following the first challenge (raising questions about possible masking), and the patient had no symptoms this time. However, when the same sham challenge was repeated (again after an unspecified interval), the patient's symptoms recurred. The authors deemed that further testing, using active substances, was not appropriate. The authors reveal that the patient underwent two years of psychotherapy during which repressed memories of physical and sexual abuse emerged. Notably, on one occasion, her father had forced her to ingest chemicals used in photographic processing, resulting in nausea, vomiting, and severe gastrointestinal irritation, resembling her presenting symptoms. After she spent several months in therapy, a diagnosis of multiple personality disorder was made. The authors note that although she now can tolerate workplace chemicals, "the psychologic sequelae of childhood abuse have not resolved so quickly. Long-term psychotherapy will be required." The paper fails to state what symptoms remain (depression? fatigue? headaches? irritability?). Many MCS patients in a masked state say that they lose their awareness of chemical triggers and experience symptoms such as depression, anxiety, fatigue, confusion, and other psychological symptoms. Thus, at the conclusion of this paper, it remains unclear to the reader whether this patient was in fact helped medically and to what extent she may have been chemically sensitive.

Even if this particular patient's responses to chemicals were wholly psychological, such an observation in one patient cannot be extrapolated to all MCS patients. What if chemicals, other stressors, and genetics each contributed to limbic sensitization, but in varying proportions depending upon the patients and their exposure histories? Then physicians would need to view each patient individually, just as they view patients complaining of back pain or headaches individually. Etiologies for these conditions can be wholly physical, wholly psychological, or varying combinations of the two.

To date, few exposure studies involving MCS patients have been conducted. Flaws in these studies (as described above) include: failure to ensure that patients are at baseline (unmasked) prior to challenge; failure to demonstrate that relevant active challenge substances at relevant concentrations are used; failure to demonstrate that masking agents, filter media, and/or other incidental exposures do not provoke symptoms; failure to consider effects of spacing challenges too closely together (causing acclimatization or habituation); referral biases affecting the makeup of the study population; and failure to provide essential methodologic details in papers.

Using topographic electroencephalography (19 channels of EEG), Schwartz et al. (1994) studied the responses of college students and the elderly to various olfactory stimuli. In one experiment, college students were assigned to four categories—high cacosmia (chemical intolerance) and high depression; low cacosmia and low depression; low cacosmia and high depression; or high cacosmia and low depression—based on scores on scales measuring cacosmia and depression. Cacosmic subjects, independent of depression, had greater decreases in low frequency alpha (8–10 Hz) and greater increases in low frequency beta (12–16 Hz) to the odorless solvent propylene glycol, compared to an empty control bottle. Decreased alpha is a nonspecific indicator of nervous system arousal. The increases observed in the cacosmic students' EEG beta may be parallel to those seen in rodent inhalant challenges to toluene (Kay 1996) and to the increases in EEG beta observed in organophosphate pesticide-poisoned individuals a year or more following their exposure (Duffy et al. 1979) and in organophosphate-exposed primates (Burchfiel and Duffy 1982).

Biomarkers

In the past decade, much attention has been given to the use of biomarkers as indicators of chemical exposure, host susceptibility, and effects of occupational and environmental disease (Ashford et al. 1990). Biomarkers are seen as a potentially useful adjunct to epidemiology and toxicology for MCS and other conditions characterized by idiosyncratic response (Cullen and Redlich 1995). Biomarkers may also provide a means by which disease mechanisms might be better understood.

Acceptance of chemical sensitivity as a bona fide medical illness has been hampered by, among other things, the lack of a diagnostic laboratory marker for the condition (Ashford et al. 1995; Cullen and Redlich 1995). Other illnesses such as fibromyalgia and chronic fatigue syndrome share the same difficulty. In the United States, up to now most clinical studies of MCS patients have focused on markers of immunological, neurological, inflammatory, and psychological responses tested in the absence of a chemical challenge.

To date, neuropsychological testing of MCS patients done in the absence of chemical provocation has not objectively confirmed their cognitive complaints. Fiedler et al. (1992) initially found performance decrements on the California Verbal Learning Test (CVLT) in 6 of 11 MCS patients they tested who had no premorbid psychiatric diagnoses. After repetition of the CVLT world list (five learning trials), recall and the slope of the learning curve for MCS patients did not differ from those of controls. Thus, only initial learning appears to have been hampered. There were no differences between groups with retesting after a 30-minute delay (Fiedler et al. 1994). On the other hand, a comparison of MCS patients and normal controls on the Continuous Visual Memory Task, a complex signal detection task, suggested that MCS patients may be less able to recognize nontarget designs (Fiedler et al. 1994, 1996b). Bell et al. (1996b) found decreased performance on a computerized visual divided attention task in active retired adults who rated themselves as chemically intolerant. Although these findings involved a community sample rather than MCS patients, similar tests may have utility in future studies of MCS patients.

Clinical ecologists, a few other physicians in the private sector, and some commercial laboratories have reported alterations in a number of parameters in MCS patients, including T- and B-lymphocyte counts; helper/suppressor T-cell ratios; immunoglobulin levels; autoimmune antibodies (including anti-nuclear, anti-smooth muscle, anti-thyroid, anti-parietal cell, and other auto-antibodies); activated T-lymphocytes (Ta1 or CD-26); quantitative EEGs; evoked potentials; SPECT and other brain scans; levels of various vitamins, minerals, amino acids, and detoxification enzymes; and blood or tissue levels of pesticides, solvents, and other "pollutants" (Miller 1994a). Flaws in these studies are varied, including: failure to define the study population (no case definition used); failure to compare cases with age- and gender-matched controls; failure to blind specimens; and failure to assess the accuracy and the reproducibility of the test method. For these reasons, results of such studies have been viewed with considerable skepticism by regulatory agencies and academic researchers (Miller 1994a; Kreutzer and Neutra 1996).

Some MCS investigators claim that different immunological abnormalities occur in different patients [for a review of the evidence, see the earlier discussion in this book and Miller (1994a)]. However, if enough tests are done, statistically a certain number will be expected to be abnormal (one in 20 in the case of a 95 percent confidence interval). This is not always appreciated. With regard to claims of immunological dysfunction, to date no consistently abnormal immunological parameter has been demonstrated in these patients.

There are a number of reasons for a biomarker for chemical sensitivity to be elusive:

1. If chemical sensitivity in fact were to involve alterations in brain or limbic function, then salient markers might not be accessible with current technology. For example, biochemical alterations in the central nervous system may not be reflected in blood chemistry determinations. Conceivably, advances in functional brain imaging (including SPECT and PET) someday may provide insight into blood flow or metabolic changes that correlate with symptoms (see discussion below).

2. Biomarkers of interest may be in normal ranges while patients are at baseline, under nonexposure conditions. Provocative chemical challenges with pre- and post-exposure measurement of markers may be necessary to distinguish between patients and normal controls. Just as methacholine challenges are needed to diagnose certain patients with reactive airways disease, it may be necessary to perform low-level chemical challenges with chemically sensitive patients in order to elicit their symptoms and observe a change in a biomarker.

3. Patients may need to be deadapted or unmasked prior to challenge in order for investigators to see the most robust symptoms and changes in biomarkers.

Also, the fact that no consistently abnormal immunological marker has been found in these individuals does not necessarily mean that the immune system is unaffected. It is conceivable that chemically induced limbic/hypothalamic disturbances could alter immune function secondarily but in unpredictable directions, or in ways that vary from person to person. The importance of undertaking longitudinal studies of biomarkers, rather than a single "snapshot," has been emphasized (Heuser 1992). Specific immune cell subsets or cytokines not yet explored in these patients may prove significant in the future.

In the United States, provocative challenge tests performed on chemically sensitive patients (Doty et al. 1988) revealed that patients manifested decreased nasal patency relative to controls, both before and after challenge. In Scandinavia, researchers have also studied nasal mucosal swelling and reactivity among hyperreactive patients and found positive results (Hallén and Juto 1992; Ohm and Juto 1993; Falk 1994). Similar low-level chemical challenge studies that examine other parameters of interest are needed, for examples, immunological, neurological, and endocrinological markers.

Supported by a grant from ATSDR, the Environmental Health Investigations Branch (EHIB) of the California Department of Health Services convened an advisory group to develop empirical approaches for the study of MCS patients in various settings (Kreutzer and Neutra 1996). EHIB cataloged laboratory diagnostic tests previously used in MCS studies in North America (see Appendix B). Agreeing that there is no currently recognized laboratory biomarker for MCS, the advisory group discussed possible laboratory and clinical diagnostic tests (including immunological tests), psychological tests, and neurobehavioral tests for studying chemically exposed populations for possible development of MCS. Little enthusiasm was expressed for the inclusion of PET scans, SPECT scans, and MRI brain studies in a *community* research protocol. Notwithstanding, some practitioners who see individual *patients* are enthusiastic about the techniques (Heuser et al. 1992, 1994). Others are concerned that there are insufficient data to support the use of brainscans of any type for diagnosing MCS (Mayberg 1994).

In other sections we discuss various diagnostic tests investigators have used in an attempt to measure objectively symptoms reported by MCS patients, including changes in vision and pupillary responses to light, nasal resistance, pulmonary function, nasal cytology, porphyrins, and responses on neuropsychological tests. Appendix B is a compilation of laboratory and clinical tests that have been used in studies of chemically sensitive patients. Although such markers help quantify certain aspects of the condition, they should be distinguished from a biomarker that is much more specific for the condition, for example, a finding of HIV antibodies in AIDS patients, increased immunoglobulin E for ragweed in patients allergic to ragweed, or specific autoantibodies in a connective tissue disease. It is worth noting that if MCS in fact represents a general class of chemically triggered illnesses, analogous to the general class of infectious diseases, there may be no single specific biomarker for it, but rather a host of specific biomarkers (analogous to individual types of bacteria, viruses, and other infectious agents) whose identities currently elude us.

Further, if chemical sensitivity is a class of diseases rather than a single disease, applying diagnostic tests—such as those focusing on the immune system—on a population-wide basis should not be expected to yield particularly useful results unless appropriate stratification of the study population into subsets is achieved. Still better would be the application of diagnostic tests in event- or exposure-driven studies, such as following groups exposed to a chemical spill, a recently remodeled building, or a new chemical introduced into a particular workplace (see the discussion of epidemiologic approaches in Chapter 10).

An example of an exposure-driven investigation is provided in a study of 38 individuals in 10 families exposed to pentachlorophenol (PCP) in manufacturer-treated log houses (McConnachie and Zahalsky 1991). The authors observed:

> Comparison of subjects with controls revealed that the exposed individuals had activated T-cells, autoimmunity, functional immunosuppression, and B-cell dysregulation . . . . Even though this study was designed to characterize alterations in lymphocyte phenotypes and functions, extensive interviews with family members revealed that all subjects had experienced an excessive incidence and persistence of cold and flu-like illnesses. Two individuals became asthmatic during the exposure. Also, there were numerous complaints of nausea, vertigo, allergies (in children), skin rashes, and headache . . . . Collectively, these findings support a clinical basis for the immunological tests that were performed.

As mentioned earlier, PCP is the chemical that has been the major focus of "wood preservative syndrome" in Germany (Schimmelpfennig 1994) and was the subject of a lawsuit there. Patients with wood preservative syndrome often report the same intolerances and multisystem symptoms that MCS patients report (Ashford et al. 1995).

Another study (Dayal et al. 1995) demonstrates the value of community-based evaluations for uncovering differences between exposed and unexposed populations. Two sites on the Superfund National Priority List in Texas with different chemical exposures, including heavy metals, aromatic hydrocarbons, polychlorinated biphenyls, halogenated ethanes and ethylenes, and heptachlor, were investigated:

> The prevalence of 29 symptoms reported by 321 individuals who had been highly exposed was compared with symptoms reported by a group of 351 persons from the same community who had limited exposure. A meaningful difference between the two groups emerged for some of the symptoms, the most notable of which symptoms were neurologic. Almost twice as many subjects in the high-exposure group reported five or more neurological symptoms, compared with the low-exposure group. This excess of neurological symptoms is consistent with the known toxic properties of the chemicals at the sites.

A similar approach would be useful for investigating MCS among members of a community following a chemical spill or release. In the study just described, if neurophysiological tests had also been conducted on the subject and control populations as some have recommended (Kilburn 1996), it might have been possible to correlate "subjective" symptoms with more objective findings.

The use of brain imaging in characterizing patients following chemical exposures is receiving increasing attention. Heuser et al. 1994 described 41 patients exposed to neurotoxic chemicals, concluding that compared to controls:

> ... patients exposed to chemicals present with diminished [cerebral blood flow], worse in the right hemisphere [in right-handed subjects], with random presentation of areas of hypoperfusion, more prevalent in the dorsal frontal and parietal lobes. These findings are significantly different from findings in patients with chronic fatigue and depression, suggesting primary cortical effect, possibly due to a vasculitis process.

Heuser observed that significant impairment of brain function may persist for years after exposure to neurotoxic chemicals has ceased. Callender reported abnormal SPECT scans in 33 patients (without controls) following occupational exposure to neurotoxins. Lesions in exposed patients noted on SPECT/PET scans were reported to correlate well with clinical presentations and neuropsychological testing (Callender et al. 1993).

Simon et al. (1994) report brain imaging abnormalities in a group of six ill Gulf War veterans. A group of 40 "neurotoxic subjects" and 3 "clinically toxic subjects with silicone breast implants" also was investigated using SPECT (Simon et al. 1992). Diffuse cortical defects were not seen in normal subjects or depressed control subjects, but were reported in all breast implant subjects and 35 of the 40 neurotoxic subjects.

A physician in Germany (Fabig 1988) performed SPECT scans on 74 wood-preservative–exposed persons and 41 unexposed persons, all complaining of CNS symptoms such as headaches, dizziness, inability to concentrate, and depression. He reported decreased cerebral blood circulation in the forebrains of persons exposed to wood preservatives, corresponding in severity to the duration of their exposure. The results are considered controversial because of the nonspecificity of SPECT scans (Düsseldorf 1990). Since the original investigation, additional persons have been scanned, bringing the totals to 139 wood-preservative–exposed persons and 214 unexposed persons, with similar findings reported (Fabig 1994).

Mayberg (1994) has written a thoughtful critique of the role of SPECT scans in multiple chemical sensitivity. She argues that:

> [w]hile [research studies using SPECT and PET scans] remain extremely important for identifying previously unrecognized brain abnormalities and potential disease mechanisms . . . , their utility in the management of individual patients is still far from clear.
>
> An enthusiastic but cautious attitude . . . seems appropriate in evaluating preliminary SPECT scan findings in patients with toxic exposures, chronic fatigue syndrome, and presumed multiple chemical sensitivity. . . .
>
> Controlled studies of well characterized patients selected using standardized clinical criteria are clearly needed. Appropriate comparison groups are also required. Subjects with similar exposure histories but without subjective complaints may be a more credible control population than age-, sex-, educationally, or socioeconomically matched subjects . . . .
>
> While the standard approach has been to measure patients in a basal resting state, it can reasonably be argued that optimal results will be obtained if patients are "challenged" to reproduce their clinical symptoms.

Although brain imaging may never be used in community-based epidemiological studies, it may have value in correlating subjective reports of symptoms and performance on neurobehavioral tests with brain scan findings before and after chemical challenges in studies of individuals, for example, in an environmental medical unit.

Overlaps with Other Illnesses

Fatigue is consistently one of the most prominent complaints of MCS patients, who frequently acquire a diagnosis of chronic fatigue syndrome during their medical odyssey. Miller (1996a) found that 68 percent of MCS patients who became ill following exposure to a cholinesterase-inhibiting pesticide, 52 percent of MCS patients who became ill during remodeling of a building, and 78 percent of sick Persian Gulf War veterans seen at a tertiary referral center, but only 3 percent of controls reported severe fatigue (scored as "3" on a 0–3 scale).

Buchwald and Garrity (1994) explored similarities and differences among 30 patients diagnosed with chronic fatigue syndrome, 30 with fibromyalgia, and 30 with MCS. Patients with either chronic fatigue syndrome or fibromyalgia frequently reported symptoms consistent with MCS. All three groups were remarkably similar in demographic characteristics and the presence of specific symptoms: Some 60 to 90 percent were female, mean ages ranged from 41 to 44 years, and the mean years of education were 14.7 to 14.9. Not surprisingly, 87 to 97 percent of the MCS patients reported sensitivities to each of four exposure types: air pollution/exhaust, cigarette smoke, gas/paint/solvent fumes, and/or perfumes. Likewise, 53 to 67 percent of patients with chronic fatigue syndrome and 47 to 67 percent of patients with fibromyalgia also reported adverse effects when exposed to these substances. Over 80 percent of the fibromyalgia and MCS patients met the major criteria for chronic fatigue syndrome (Centers for Disease Control and Prevention), whereas 70 percent and 30 percent, respectively, fulfilled the full case definition. More than 75 percent of chronic fatigue syndrome and MCS patients reported musculoskeletal symptoms characteristic of fibromyalgia such as weakness, arthralgias, and myalgias. Nearly two-thirds of the patients in each group reported cognitive difficulties as one of the most frustrating aspects of their illness. The investigators mused: "Despite their different diagnostic labels, existing data, though limited, suggest that these illnesses may be similar, if not identical, conditions . . . . In fact, the diagnosis assigned to patients with one of these illnesses may depend more on their chief complaint and the type of physician making the diagnosis than on the actual illness process."

In 1994, a group of chronic fatigue investigators met in order to examine and update the chronic fatigue syndrome working case definition that had been published in 1988. They modified the definition so as to make it less restrictive (Fukuda 1994). According to the new definition, fatigue of six-months duration or longer continues to be the central criterion, but the requisite number of chronic symptoms was decreased from eight (out of a list of eleven symptoms) to four out of the following eight symptoms: impaired short-term memory or concentration, sore throat, tender cervical or axillary lymph nodes, muscle pain, multijoint pain, new headaches, unrefreshing sleep, and post-exertional malaise. The group's findings specifically mention MCS, stating that MCS patients are not to be excluded from a diagnosis of chronic fatigue syndrome because, like fibromyalgia, anxiety disorders, somatoform disorders, depression, and neurasthenia, MCS is "defined primarily by symptoms that cannot be confirmed by diagnostic laboratory tests."

Fiedler et al. (1996b) compared 23 chemically sensitive patients whose illness reportedly began following a defined exposure (MCS), 13 chemically sensitive patients with no specific time of onset (CS), 18 chronic fatigue syndrome (CFS) patients, and 18 normal controls. Although MCS, CS, and CFS patients had significantly higher rates of current psychiatric disorders and unexplained symptoms than controls, 74 percent of the MCS and 61 percent of the CFS patients did not meet criteria for any major (Axis I) psychiatric disorder, but 69 percent of the CS patients did. In other words, only one-fourth of MCS patients whose illness began after a particular exposure met criteria for a major psychiatric disorder, whereas two-thirds of MCS patients who recalled no specific initiating event met criteria for a major psychiatric disorder. This latter finding underscores the importance of exposure-driven studies for obtaining more homogeneous study populations. There were no significant differences between any of the groups on neuropsychological testing except for one complex visual memory task. Thus, standardized neuropsychological tests did not objectively verify the cognitive difficulties reported by MCS, CS, and CFS patients. The authors acknowledged that no chemical exposure challenges were administered prior to cognitive testing and that this may have led to their negative cognitive findings. In future studies of persons with MCS, exposure challenges will be essential. MCS patients clearly state that their cognitive and mood symptoms occur with exposures. Testing patients for chemical sensitivity without exposing them to chemicals may be analogous to evaluating patients for exercise-induced angina without a treadmill test.

A team of Australian researchers (Dunstan et al. 1995) compared serum organochlorine levels in 22 chronic fatigue syndrome (CFS) patients, 17 patients with CFS-like symptoms and a history of exposure to toxic chemicals, and 34 matched non-CFS controls. DDE (a product of DDT metabolism) and hexachlorobenzene (HCB) comprised 90 percent of total organochlorines measured in each of the three groups. Detectable HCB (> 2.0 ppb) was present in the sera of 45 percent of CFS patients ($p < 0.05$) and 47 percent of the toxic exposure group (not statistically significant), but only 21 percent of controls. The CFS group also had significantly higher mean total organochlorine and DDE levels compared to controls. Although total organochlorines and DDE were also higher in the toxic exposure group than in controls, the differences were not statistically significant. Notably, three CFS patients with the highest organochlorine levels (greater than 30 ppb) also reported hypersensitivity to chemicals. The authors concluded that "The results of this preliminary investigation—that levels of recalcitrant organochlorines are higher in CFS patients compared with controls, and that serum organochlorine concentrations in CFS patients with and without a history of toxic chemical exposure are not significantly different—suggest that these chemicals may have an etiological role in chronic fatigue syndrome . . . ."

Chronic neuropsychological symptoms resembling those in MCS have been reported in British sheep dippers exposed to organophosphate pesticides (Monk 1996; Sharp 1986). Chaudhuri et al. 1997 compared ten individuals with well-documented chronic exposure to organophosphate pesticides, who suffered from incapacitating fatigue, with ten healthy controls with no known exposure to organophosphates. Five of the pesticide-exposed subjects had increased liver function tests, and two had developed non-Hodgkin's lymphoma. Neuroendocrine responses for pesticide-exposed individuals were similar to those of chronic fatigue syndrome patients in several respects: increased prolactin release following buspirone administration, increased growth hormone release one hour after pyridostigmine administration; and reduced growth hormone suppression after dexamethasone. The authors surmise that the clinical similarities between chronic fatigue syndrome and chronic illness following organophosphate exposure, coupled with their similar neuroendocrine responses, suggest that the two conditions share a common pathogenesis.

While evaluating persons affected in three apparent outbreaks of sick building syndrome (SBS) in the United States, Chester and Levine (1994) found that from 10 percent to 90 percent of individuals in the buildings developed persistent health problems reminiscent of chronic fatigue syndrome (CFS). The authors noted that "the agents responsible for the traditional symptoms of SBS may also trigger CFS" and that "CFS can occur in the setting of SBS." The highest attack rate occurred in nine of ten California high school teachers, all of whom used a single small conference room and sequentially became ill. There was no fresh air supply to the conference room, which housed a spirit (solvent-based) duplicator machine. Each of the nine teachers who became ill took a leave of absence, and two retired. Eight were still sick five years later. The teacher who was unaffected had spent less time in the room than the others and often worked outdoors. The study provides no information with respect to whether any of the teachers developed MCS-like symptoms.

In addition to its major overlaps with chronic fatigue syndrome and fibromyalgia, MCS shares features in common with asthma, especially occupational asthma. As discussed in the first edition of this book, reactive airways dysfunction syndrome (RADS) is an asthmalike condition that begins following a major chemical exposure, for example, a chemical spill, release, or fire. Thereafter, shortness of breath, wheezing, or chest tightness may be triggered by low levels of many common irritants, including cigarette smoke, fragrances, and solvents. Meggs et al. (1996a) studied 13 workers with MCS who had been exposed to chlorine dioxide released from a ruptured pipe five years previously. Seven of eight who underwent pulmonary function testing showed evidence of airway obstruction and had positive methacholine challenges. In contrast, in another study, Meggs and Cleveland (1993) found that a group of MCS patients had normal pulmonary function tests. In these and other papers (Meggs 1994), the authors reason that respiratory irritants may induce asthma and/or rhinitis. They label the latter condition "RUDS" for "reactive upper airways dysfunction syndrome," hypothesizing that neurogenic inflammation caused by chemical exposures may underlie both upper and lower airway hyperresponsiveness. Further discussion of the role of inflammation in MCS appears in a later section of this chapter under the heading "Mechanisms."

Kipen et al. (1995) administered a questionnaire to 705 MCS, asthma, and other clinic patients, inquiring about which of 122 common substances caused symptoms. Total scores for 39 patients with MCS and 43 with asthma were significantly higher than those of other patients surveyed, and totals for patients with MCS were higher than those for patients with asthma. Out of the 122 substances, the mean numbers (in parentheses) endorsed as causing symptoms (by men/women) were: MCS (34/42), asthma (19/33), medical clinic patients (9/12), occupational clinic referrals (8/14), and surveillance (healthy) patients (4/7). The authors suggest that a score of 23 or greater affords adequate sensitivity (69 percent) and specificity (89 percent) for differentiating MCS from non-MCS patients. Sixtynine percent of MCS patients and 54 percent of asthma patients met or exceeded this score, whereas only 15 to 20 percent of clinic patients and 4 percent of surveillance patients met or exceeded a score of 23.

Magnitude of the Problem

How prevalent is multiple chemical sensitivity? When the first edition of this book was published, there were no data with which to address this question. However, in a recent telephone random digit dialing survey conducted on 4,046 households throughout California, 16 percent of respondents reported sensitivities to everyday chemicals, and a surprising 6.3 percent answered "yes" to the question, "Have you ever been told by a doctor that you had environmental illness or chemical sensitivity?" (Kruetzer 1996). It may be argued that results from a California-based survey do not fairly represent the entire nation, particularly because many clinical ecologists practice in that state. However, studies in other states suggest similar rates. For example, 4 percent of a sample of nearly 200 retired elderly persons living in Arizona reported that they had extreme chemical intolerances that had been diagnosed by a physician (Bell et al. 1994b), and 3.9

percent of over 1,000 randomly selected rural North Carolinians attested to symptoms of chemical sensitivity that occurred daily (Meggs et al. 1996b). Although these surveys framed their questions somewhat differently, the results are remarkably close, that is, 5 percent plus or minus 1 percent.

Several large surveys suggest that between 15 percent and 34 percent or up to one-third of Americans consider themselves especially sensitive, allergic, or unusually sensitive to certain chemicals and chemical odors, depending on the sample and how the question is worded. Studies of nearly 4,000 EPA office workers in Washington, D.C., several hundred Arizona college students and retirees, and over 1,000 rural North Carolinians suggest that one-fourth to one-third of Americans consider themselves especially sensitive to certain chemicals (Table 8-2). Notably, most of the participants in these surveys who reported that certain odors made them feel ill (i.e., people with chemical sensitivities) were neither chronically sick nor disabled. Thus, although a sizable percentage of Americans report that certain odors make them feel sick, the majority of these people apparently are not greatly incapacitated and thus differ substantially from MCS patients who report suffering from their sensitivities almost daily. Whether—with sufficient exposure—any, some, most, or all of these chemically sensitive individuals would develop MCS (disabling multiorgan symptoms triggered by multiple incitants) cannot be determined. In the future, consideration should be given to incorporating key questions related to chemical, food, and drug intolerances and any history of disabling chemical exposures into national health surveys.

Origins of Chemical Sensitivity

Little has changed since the first edition of this book appeared in terms of the kinds of exposures that seem to trigger symptoms in MCS patients. Had we rewritten this book from scratch, we might have tried to differentiate between those exposures that appear to initiate MCS, most notably certain pesticides and solvents (Cullen et al. 1992; Davidoff and Keye 1996; Lax and Henneberger 1995; Miller and Mitzel 1995) and those that appear to trigger symptoms once the illness has a foothold, for example, fragrances, tobacco smoke, or wearing dry-cleaned clothing. In the first edition (Chapter 3), we used the term "origins" in discussing both initiators and triggers. Here we narrow our discussion of origins to initiators, as we now have a clearer picture of the phenomenology of MCS. Cumulative observations by clinicians and researchers worldwide point to a two-stage process: initiation and triggering.

Patients may accidentally, albeit unintentionally, mistake a trigger for an

TABLE 8-2. Frequency of Chemical/Odor Sensitivity in Selected Populations

Population	n	Question posed	% Answering affirmatively
EPA office workers (EPA 1989)	3,955	Do you consider yourself especially sensitive to . . . [various indoor air contaminants]?	31%
Arizona[a] elderly living in planned retirement community (Bell et al. 1994b)	192	Do you consider yourself especially sensitive to certain chemicals?	34%
University of Arizona[a] college students in introductory psychology class (Bell et al. 1996a)	809	Do you consider yourself especially sensitive to certain chemicals?	28%
Rural North Carolinians (Meggs et al. 1996b)	1,027	Some people get sick after smelling chemical odors like those of perfume, pesticides, fresh paint, cigarette smoke, new carpet, or car exhaust. Other people don't get sick after smelling odors like these. Do any chemical odors make you sick?	33% (39% of women; 24% of men)
California residents (Kruetzer 1996)[b]— Random digit dial telephone survey	4,046	Do you consider yourself allergic or unusually sensitive to everyday chemicals like those in household cleaning supplies, paints, soaps, perfumes, detergents, garden sprays, or things like that?	16% (15–30% in communities having recently experienced a chemical spill or release)

[a]A haven for pollen-allergy sufferers in the past, Arizona is thought to have the highest percentage of atopic individuals of any state.
[b]Health Investigations Branch, Department of Health Services, State of California.

initiator. For example, suppose that someone whose home has been exterminated monthly suddenly experiences shortness of breath and confusion while driving behind a smoky diesel truck. Subsequently, other intolerances develop. To such an individual it may appear that diesel exhaust initiated the MCS, when in fact the exhaust may have been only the first robust trigger of the person's symptoms. At such an early stage in our understanding of MCS, it seems prudent to make causal attributions, especially in individual cases, only with great care. On the other hand, there is accumulating evidence that exposures to organophosphate pesticides, volatile organic chemicals in sick buildings, and various solvents may initiate MCS, based upon observations by independent scientists looking at different groups of individuals. Near-simultaneous onset of MCS in a group of individuals following an identifiable exposure event strongly suggests causation. Such outbreaks provide the ideal setting for exposure-driven studies that ultimately will help clarify the origins of MCS.

In recent years, we have observed a tendency to name MCS-like conditions after suspected initiating events, for example, pentachlorophenol syndrome, toxic carpet syndrome, darkroom disease, and Gulf War syndrome, or after prominent symptoms, for example, reactive airways dysfunction syndrome (RADS), reactive upper airways dysfunction syndrome (RUDS), chronic fatigue syndrome, and fibromyalgia. This reductionist approach helps foster a belief that each of these named entities is a distinct and isolated syndrome. This may cloak a larger view—that is, an underlying, unifying mechanism, for example, that some people lose tolerance following certain chemical exposures, and that thereafter their symptoms are triggered (and their illness is perpetuated) by common, low-level exposures (toxicant-induced loss of tolerance). Mary Lamielle, founder and Director of the National Center for Environmental Health Strategies in New Jersey, a national nonprofit organization dedicated to finding creative solutions for environmental health problems, lamented this trend: "Refusal to look at the larger issue makes it impossible to understand the parts" (Lamielle 1992).

Pesticides

There are now several studies linking chronic, multisystem symptoms to organophosphate or carbamate pesticide exposure in groups of individuals (see the "Epidemiological Studies" section of this chapter). These agents have been implicated in multisystem illnesses and new-onset chemical intolerances in pesticide-exposed casino workers (Cone and Sult 1992) and an attorney whose home was exterminated (Rosenthal and Cameron 1991), as well as other persons exposed to organophosphates (Sherman 1995). A study involving nine European countries (see Chapter 7) revealed reports of MCS-like cases following exposure to various pesticides in six countries (Ashford et al. 1995). Sheep dippers in the United Kingdom exposed to organophosphate pesticide report MCS-like illnesses (Monk 1996). In a controlled study, Stephens et al. (1995) compared 143 farmers exposed to organophosphate sheep dips and an equal number of quarry workers and found that the farmers reported more psychological symptoms and scored significantly worse on three of eight computer-administered psychological tests: symbol-digit substitution, syntactic reasoning, and simple reaction time. Although the authors did not look for chemical intolerances in these farmers, 30 years ago, Tabershaw and Cooper (1966) described a group of 114 agricultural workers with acute organophosphate pesticide poisoning, some of whom developed persistent, MCS-like symptoms. Three years following their initial acute exposure, 22 workers (19 percent) reported that even a "whiff" of pesticide made them feel ill. Sixteen of them quit working with pesticides for this reason, whereas six continued farmwork but avoided pesticide exposure as much as possible. It is unknown how many of 61 workers who could not be located for follow-up for this study may have left because of chronic illness or chemical intolerance. Estimates as to what percentage of occupational organophosphate poisonings results in delayed or persistent neurological and psychiatric effects include 5 percent (WHO 1990) and 4 to 9 percent (OTA 1990), although neither includes MCS in its analysis. In 1961 Spiegelberg described persistent, multisystem symptoms among Germans who had worked in chemical weapons (including organophosphate nerve agent) production for the Wehrmacht during World War II. Notably, he also described multisystem symptoms and new-onset intolerances for alcohol, nicotine, and medications among these workers—hallmarks of MCS—more than 30 years ago.

Recently, the U.S.E.P.A. specifically addressed MCS in its overall assessment of the health impact of an organophosphate insecticide (chlorpyrifos) (Blondell and Dobozy 1997). An EPA memorandum dated January 14, 1997, states "In addition to acute poisoning, chlorpyrifos and other organophosphate insecticides have been reported to be associated with chronic neurobehavioral effects and the reported development of sensitivity to chemicals previously tolerated which is associated with a wide variety of symptoms." The memorandum also discusses the controversy surrounding MCS. During a six-year period ending in 1990, the EPA-funded National Pesticide Telecommunications Network received some 1,022 calls from consumers reporting unusual sensitivity to pesticides. Chlorpyrifos was the pesticide most frequently cited in these complaints. DowElanco, which produces chlorpyrifos, has pledged to the EPA that it will support further scientific studies on the health effects of chlorpyrifos; withdraw chlorpyrifos from use in several United States markets, including indoor broadcast flea control, indoor total release foggers, paint additives, and pet shampoos, dips, and sprays; and improve labeling and pesticide control operator training. In a letter to the EPA dated January 16, 1997, the President and CEO of DowElanco clarified the proposed actions: "However, we must state unequivocally that our proposed initiatives are not prompted in any way by a conclusion that any current label uses create exposures capable of causing human injury, and any attempt to portray them in this light would only make difficult their timely and effective implementation." (Hagman 1997).

Thus there is accumulating evidence dating back over several decades linking organophosphate-type compounds with chronic illness and new-onset intolerances in a subset of exposed persons. This unusual complaint of new-onset chemical, food, and drug intolerances is a sentinel symptom of MCS, one that should alert practitioners to explore possible chemical initiators in their patients. It would be difficult to imagine that so many people with prior identifiable chemical exposures would invent such bizarre and inconvenient intolerances: many now avoid fragrances they formerly enjoyed, no longer drive because traffic exhaust makes them feel ill, and have given up favorite foods such as pizza or chocolate because they feel sick when they eat them.

Indoor Air Pollutants

Indoor air contaminants appear to be among the most potent initiators and triggers of chemical intolerances. Complex mixtures containing low levels of hundreds of different volatile organic chemicals occur indoors and may have synergistic effects not currently well understood. Although the role of indoor air pollutants in MCS was discussed in detail in the first edition, two specific areas require further attention: X-ray developing chemicals, including glutaraldehyde, and new carpet.

A growing number of radiologic technicians and some radiologists have reported onset of MCS-like illnesses, sometimes referred to as "darkroom disease" or "processing room fever," following exposure to darkroom film-developing and fixing chemicals, which may include glutaraldehyde, glutathione, hydroquinone, sodium sulfite, phenol, ethynyl, ammonium thiosulfate, diethylene glycol, potassium hydroxide, nitroindazole, sulfur dioxide, acetic acid, and aluminum chloride or sulfate (Gordon 1987). As early as 1978, Rea reported the case of a 38-year-old physician who experienced gastrointestinal distress, urinary urgency, shortness of breath, chest tightness, peripheral arterial spasm, and cardiac arrhythmias (premature ventricular contractions) when exposed to X-ray developer emissions. When the physician left the environment, his arrhythmias stopped; on at least 20 separate occasions, reexposure was followed by recurrence of his arrhythmia.

In X-ray film developing, small amounts of developer solution are carried over to a volatile fixing solution that is kept at temperatures of $30°C$ or higher. After a brief rinse, the film is fed into a hot dryer. Historically, darkrooms have been relegated to tiny rooms where light could be sealed out (like closets) and where little or no fresh air ventilation was provided.

Airborne emissions from the developer, fixing solution and hot dryer may include glutaraldehyde, acetaldehyde, acetic acid, formaldehyde, and sulfur dioxide. Glutaraldehyde, a known skin sensitizer, has been used in rapid film processing since the late 1960s to prevent the softening of the film emulsion that occurs with the higher temperatures required. Since that time, illnesses attributed to darkroom chemicals have increasingly occurred, often associated with inadequate ventilation and the failure to properly exhaust emissions outside the building (Gordon 1989).

A New Zealand survey revealed that 80 percent of affected radiographers were women, more than 90 percent of whom exclusively used automatic processors (Gordon 1995). One ill radiographer lamented that her greatest battle was not with the photographic companies, health authorities, or the medical profession, but with unaffected radiographers who displayed indifference and antagonism. Recent technological developments in X-ray film developing, such as closed-system mixing of chemicals and laser imaging that completely eliminates the use of processing chemicals, may eventually supplant open chemistry film developing systems (Kuntz 1992).

Exposure to glutaraldehyde has also been implicated as causing MCS by some medical workers exposed in clinics, laboratories, and hospitals while performing cold-sterilization of fiberoptic instruments (e.g., bronchoscopes, cystoscopes) and anesthetic equipment and while working in renal dialysis units.

Carpets

In this revised edition of the book, we have intentionally emphasized the carpet question because we think that it is among the more clearly articulated, ongoing debates over the effects of low-level chemical exposures, particularly complex mixtures, on health. The technical, political, and philosophical issues that surround it are archetypical of those that can be expected with other low-level exposures.

Many MCS patients report that their illness began while they were working or living in a new or remodeled office or home (Lax and Henneberger 1995; Miller and Mitzel 1995). Miller and Mitzel found that among 112 MCS patients who attributed onset of their condition to either pesticide or remodeling exposures, 18 percent reported new carpeting as their current, single most troublesome exposure. Among the 75 patients who had become ill following remodeling of their homes or workplaces, 57 percent mentioned new carpeting in narrative descriptions of the remodeling event.

The Carpet and Rug Institute (CRI), the trade association for the carpet industry, estimates the market for carpet to be over 1.5 billion square yards annually (CRI 1995). In one of its technical bulletins, CRI cites a review by

Dr. Alan Hedge, a consultant to CRI and professor of Environmental Analysis at Cornell University, who states that "[c]oncentrations of VOCs in carpet emissions are substantially below any known thresholds for toxicity effects—orders of magnitude lower than those known to produce effects— a hundred, a thousand, ten thousand times lower than any known effects. New carpet emissions should not create health problems for people—any people." (CRI 1996).

Nevertheless, health problems associated with carpet emissions were described by Randolph in the United States as early as 1962 (Randolph 1962). In 1991, the attorneys general from 26 states petitioned the Consumer Product Safety Commission for a health warning label on carpets. Since 1988, the Consumer Product Safety Commission has received nearly 800 complaints about adverse health effects following carpet installation (Schaeffer 1996). Most recently, the rate has been 30 to 50 complaints per year, with complaints increasing transiently whenever media attention is given to this subject. Reported symptoms include fatigue, sinus infections, muscle and joint pain, nervous system changes, respiratory difficulties, worsening asthma, rashes, and multiple chemical sensitivity. Notably, studies of carpet and textile workers or carpet layers have revealed an increased incidence of central nervous system problems, lymphocytic leukemia, and testicular, ovarian, and large bowel cancers (Huebner et al. 1992; O'Brien and Decloufé 1988; Vobecky et al. 1984).

The air inside buildings or homes where new carpeting has been installed may contain hundreds of volatile organic compounds (VOCs) at levels far below occupational exposure limits (Fig. 3-2). New carpet padding and adhesives also release a spectrum of VOCs indoors. Lack of fresh outside air for dilution ventilation results in higher indoor levels of contaminants. VOCs are more likely to pose a health problem in tightly sealed, energy-efficient office buildings built since the mid-1970s or in homes that typically have no provision for fresh outside make-up air other than leakage via cracks, doors, and windows.

Air contaminants associated with use of carpet, carpet pads, and adhesives indoors pose a challenging analytical task. Conventional chemical assays, for example, gas chromatography (GC) or GC-mass spectrometry, may help identify and quantify certain constituents (not all contaminants are trapped by sampling media or detected by analytical equipment), but this approach provides no information about potential toxicity of the VOC mixture as a whole. Additive or synergistic toxic effects between components could occur, but conventional assays do not provide this kind of information.

Early in the 1990s, Dr. Rosalind Anderson, who obtained her Ph.D. from Yale University School of Medicine and now operates a commercial testing laboratory (which has recently relocated from Boston to Vermont), began testing carpet samples sent to her by persons who claimed they had become ill in their homes or offices after new carpet had been laid. For this work, Anderson's laboratory selected the American Society for Testing and Material's Standard Test Method for Estimating Sensory Irritancy of Airborne Chemicals [ASTM designation E981-84] (ASTM 1984), a biological assay developed for the United States Army by Dr. Yves Alarie of the University of Pittsburgh (Anderson had been Alarie's student). This test method has been used to predict certain rapid-onset responses of humans who might be exposed to substances like riot-control agents, volatile organic hydrocarbons, or pesticides.

In the ASTM assay, mice are positioned in a glass exposure chamber with only their heads exposed to the air being tested. If chemicals in the air cause sensory irritation in the mice, a reflexive change in their breathing pattern occurs culminating in a concentration-dependent decrease in respiratory rate with a characteristic pause at the beginning of expiration due to stimulation of nerve endings in the nasal mucosa (Muller and Black 1995). According to the ASTM protocol, a 12 to 20 percent decrease in respiratory rate for at least three minutes accompanied by this characteristic pause is considered a slight effect; a 20 to 50 percent decrease is considered moderate, and over 50 percent, severe. If the air mixture causes pulmonary irritation, a characteristic pause occurs at the end of expiration.

The mice used in this standardized assay are an outbred Swiss-Webster hybrid selected for their biological variability and intermediate sensitivity to chemical irritants. Alarie previously reported good agreement between results obtained from this ASTM method and Threshold Limit Values set for occupational exposures (Alarie 1973; Schaper 1993).

For testing carpet emissions, Anderson collects the test atmosphere in a 40-liter glass aquarium containing the desired (or available) amount of carpet. The quantity of carpet varies between 1-square inch and 9-square feet, but generally 1-square foot is used. In some experiments, the air temperature has been heated to 37°C. In other experiments, the system is never heated. Animals breathe the test atmosphere for one hour, twice a day for two days, a total of four hours. According to Anderson, tests with clean air only (sham controls) have shown insignificant responses, if any.

Among 125 carpets submitted to Anderson's laboratory for evaluation between January 1993 and June 1994, only 10 percent showed no toxic effects in mice (Anderson 1995). Sixty percent of the carpets were associated with three or more neurotoxic effects in at least one of the four mice. In another series of 12 carpets submitted in 1992, 8 of the 12 samples caused one or more deaths within 48 hours of testing. In a few cases, all four mice died. Log dose–response curves for sensory irritation, pulmonary irritation, neurobehavioral scores, and deaths showed a relationship between the quantity of carpet tested and the severity of effect. Interpreting her data on carpets, Anderson attests, "[w]e are sure that this change in rate and pattern of breathing are predictive of the severity of the human reaction . . . . No other toxicology test has ever been so carefully compared to the human reaction . . . . It is our good luck that mice are somewhat less sensitive to these irritants than are humans so that there is virtually no chance of false positives: If the mouse reacts, [some] human[s] will also." (Anderson 1992).

In addition to the ASTM respiratory measures, Anderson uses a battery of neurobehavioral tests, scoring the mice for paralysis, falling, body tone, freeze, vocalization, lacrimation, bleeding, convulsions, repetitive motions, reach reflex, tremors, gait, activity level, cyanosis, exophthalmos, twitch, attack, foot placement, grip strength, righting reflex, tension, responsiveness, piloerection, face swelling, isolation, and coma. Mice are observed for up to 48 hours post-exposure and any deaths noted. Anderson reports that her laboratory has conducted more than 500 tests with four mice per test using over 300 carpets sent to her and more than a dozen carpet samples purchased locally. Using the four endpoints of sensory irritation, pulmonary irritation, neurobehavioral changes, and death, she observes that "each carpet appeared to have its own mixture of toxic effects, presumably reflecting its own complex mixture of toxic emissions" (Anderson 1995). Carpets submitted to her laboratory for testing were not significantly different from purchased carpets in terms of their effects on mice, suggesting that effects were not due to contamination of older, used carpets by cleaning solutions, pesticides, or other substances.

Concerned with Anderson's findings, the EPA sought to replicate her results independently. A double-blind collaborative study on samples from two carpets collected by the CPSC and sent to both the EPA and Anderson laboratories was arranged. Monsanto and DuPont conducted their own tests on samples from the same carpets. Carpet samples for testing were selected by CPSC from carpets previously tested by Anderson. The EPA states that both carpets had been reported by Anderson to produce toxicity in some test animals and death in others (Tepper et al. 1995). However, the EPA investigators found no evidence that the carpets caused respiratory irritation, neurotoxicity, or pathology. In fact they noted as many adverse effects in control mice as in the exposed mice, which suggested to them that the exposure procedure itself (for example, restraining the mice) may have caused significant health effects unrelated to carpet emissions. Similarly, neither Monsanto nor DuPont found significant toxicity in the carpet samples. On the other hand, Anderson Laboratories reported severe respiratory irritation during the carpet exposures and health effects including death in mice during a one-week post-exposure observation period.

The EPA also characterized the chemical emissions of the test carpets. Notably, the EPA found more than 200 VOCs emitted from carpet during these studies, 15 percent of which were identified and confirmed, 70 percent tentatively identified, and 15 percent remained unknown. Formaldehyde levels in the carpet chambers were 52 and 25 $\mu g/M^3$, compared to $5\mu g/M^3$ in the empty control chamber. Initially, concentrations of the more volatile compounds (containing less than eight carbon atoms) reached their highest levels in the chamber. After the second exposure, concentrations of less volatile compounds (containing more than eight carbon atoms) were higher than previously. A confounding factor was a significant level of the pesticide chlorpyrifos ($32,000$ to $45,000$ $\mu g/M^2$ or 32 to 45 mg/M^2) detected in one of the carpets.

The EPA used certified clean bottled air and humidified it to 25 percent *prior* to passing it through the chamber containing the carpet, whereas Anderson used room air with whatever water vapor was present from background sources. With regard to this, the EPA researchers state that they "do not believe that any difference [in humidification] can account for the disparate findings" (Tepper et al. 1995). The EPA states that it has conducted numerous toxicity studies on carpet with no evidence of the adverse effects reported by Anderson. Anderson feels that humidity may have been a crucial difference between the EPA's studies and her own. She reports that she tried to duplicate the EPA's humidification approach and also found no effects in the mice. Subsequently, instead of exposing the mice to air humidified in this manner, she passed it through a water trap before it reached the mice. When mice were exposed to an aerosol of water vapor collected in this way, neurological effects were again seen, suggesting that key VOCs had been scrubbed out of the air stream by the water vapor. The EPA says it tried this too but still found no toxicity (Tepper et al. 1995).

Yves Alarie of the University of Pittsburgh reports that he replicated Anderson's findings. He asserts that his is the only laboratory that duplicated her work exactly and that he found the same neurological effects she reported. Anderson had her exposure system delivered to Alarie's laboratory so that he might replicate every part of it. When Alarie tested a piece of carpet that Anderson had previously found to cause severe neurotoxic effects in mice, he observed the same response. Alarie, who has worked with mice for 30 years, recalls that the behavior of the mice was "very abnormal." He says they clearly were affected. Some were even ataxic. He reports that he videotaped their responses and reproduced the effects twice. Alarie states he also received samples of the same carpet that CPSC sent to the EPA and Anderson Laboratories for collaborative study. Like the EPA, Alarie did not find toxicity in these samples. However, he says there were several potentially crucial differences between the EPA and Anderson tests. He recalls an industry finding that there could be as much as a ten-fold difference in VOC emissions from equal-sized samples taken from the same piece of carpet. Alarie also points out that during carpet manufacture, styrene-butadiene "glue" is unevenly applied to the carpet backing. If the glue does not cure or polymerize adequately, the styrene-butadiene mixture may outgas for extended periods. Alarie considers the split carpet study, "the most stupid experiment ever planned—sending people such different pieces of carpet and expecting them to get the same results" (Alarie 1996). Furthermore, he recalls that the carpet was one that Anderson had previously found to have minimal toxicity, and that the EPA carpet emission system differed from Anderson's.

Peer reviewers of the Anderson Laboratories and EPA collaborative study identified a variety of possible causes for their divergent findings. These included: problems with blinding; differences in humidity control; the fact that EPA sacrificed the mice after the last exposure and performed post-mortem examinations whereas Anderson Laboratories monitored mice for seven days post-exposure and discarded them without performing a post-mortem examination; possible unidentified pathogens or health problems affecting mice at Anderson Laboratories, which does not follow the same procedures as the EPA to maintain a disease-free environment; use of Swiss-Webster mice from different suppliers; the fact that behavioral examinations on the mice were conducted after each of the four exposures at Anderson Laboratories but only after the second and fourth exposures at EPA; differences in restraining the animals (collar restraints, used by both laboratories, can cause stress and injuries); variability in carpet samples sent to the laboratories for testing; and differences in data interpretation. Unfortunately, the cause for the disparate results was never confirmed (U.S. Environmental Protection Agency, 1993).

Which animal testing approach provides the best gauge of potential toxicity in humans is still the subject of debate. In toxicology, exposing animals to high doses of a toxic agent is considered "a necessary and valid method of discovering possible hazards in humans" (Casarett and Doull 1996): "Obtaining statistically valid results from such small groups of animals requires the use of relatively large doses so that the effect will occur frequently enough to be detected." For example, detecting a serious toxic effect, such as cancer, with a 0.01 percent incidence (representing 20,000 people in a population of 200 million, an unacceptably high number), would require a minimum of 30,000 animals. Consequently, there is no choice but to give large doses to relatively small numbers of animals and extrapolate the findings. People generally are exposed to new carpet for much longer periods than the four hours allotted in the ASTM method. Increasing the exposure duration or raising the exposure concentration via heating might be some ways to reasonably compensate for this difference. As long as the carpet temperature is kept below 100°C, Alarie states, thermal decomposition products are not a concern. However, the chemical composition of carpet emissions changes greatly between 23°C (more typical of ambient air temperatures) and 70°C, raising questions about the practical significance of irritation occurring in rodents at higher temperatures (Muller and Black 1995). Anderson has responded to this concern about heat by now working at room temperature. CPSC has proposed the testing of defined, synthesized mixtures of carpet VOCs at higher concentrations in order to achieve adequate exposures (Muller and Schaeffer 1996). One problem, however, is that such mixtures may omit key VOCs inadvertently.

Researchers at Lawrence Berkeley Laboratory have quantified VOC emissions from several new carpets collected directly from carpet mills and have also measured VOCs emitted by new carpet under simulated indoor air conditions over a seven-week period (Hodgson et al. 1993). Two carpets with styrene-butadiene rubber latex adhesive backing emitted primarily 4-phenylcyclohexene (4-PC) and styrene. 4-PC concentrations in the simulated indoor air experiment ranged from 2 to 5 ppb. This compound, which is the source of "new carpet odor," has an olfactory threshold estimated at less than 0.5 ppb. A carpet with polyvinyl chloride backing emitted vinyl acetate, propanediol, and lesser amounts of formaldehyde, iso-octane, and 2-ethyl-1-hexanol. A polyurethane-backed carpet emitted primarily butylated hydroxytoluene (BHT). The authors note that whereas many carpet VOCs decline rapidly during the first 24 hours after airing out begins, some compounds like formaldehyde and 4-PC outgas more slowly and may be more important in terms of health consequences. Other than for formaldehyde, these authors say, ". . . little is known about the health effects of these VOCs at low concentrations."

The SB (styrene-butadiene) Latex Council, whose corporate members include BASF Corporation, The Dow Chemical Company, and Goodyear Tire and Rubber Company, states that over 90 percent of all carpets made in the United States use styrene-butadiene latex as a bonding agent to hold the carpet yarn and backings together (SB Latex Council 1996). Styrene-butadiene latex is not the same as natural latex, which has been associated with severe allergic reactions in sensitive persons. The latter, which is plant-derived, differs chemically from SB latex.

Following a series of meetings with the EPA, known as "the Carpet Dialogue," the Carpet and Rug Institute developed a voluntary program under which carpet manufacturers have sought ways to reduce emissions from carpet. More than 50 carpet manufacturers participate in this program. The SB Latex Council reports that SB latex manufacturers have voluntarily found ways to reduce 4-PC emissions from their product by more than 60 percent since 1989. CRI (1996) reports that since the inception of the Indoor Air Quality (IAQ) testing program, the industry overall has reduced total VOC emissions by 20 percent and 4-PC emissions by 50 percent. Currently, carpet manufacturers are permitted to affix the CRI IAQ Carpet Testing Program label to a particular product type if a single representative test result does not exceed specified emission criteria for total volatile organic compounds, formaldehyde, 4-PC, and styrene. According to CRI, formaldehyde has not been used in making carpet in more than 12 years, but it is included in the testing battery simply to prove its absence. A single exemplar carpet is retested quarterly to ensure that the product continues to adhere to these requirements. Critics of the program comment that there is no indication that the testing program measures critical emissions, and that a single test does not adequately represent an entire product line.

The Carpet Testing Program label states, "Some people experience allergic or flulike symptoms, headaches, or respiratory problems, which they associate with the installation, cleaning, or removal of carpet or other interior renovation materials," and recommends that "Persons who are allergy-prone or sensitive to odors or chemicals should avoid the area or leave the premises when these materials are being installed or removed." (CRI 1994). In addition, CRI has established a testing program for carpet adhesives. Adhesive products meeting emission criteria for total VOCs, formaldehyde, and 2-ethyl-1-hexanol may display the program's logo.

The carpet itself is not the only potential problem—carpet adhesives and padding also contribute to indoor air VOCs. In 1962, Randolph cautioned that the air of living quarters could be fouled by the odors of rubber rug pads and rubber or plastic backing of rugs or carpets. Now CPSC researchers report detecting more than 100 VOCs, spanning a broad range of chemical classes, from 17 carpet cushion (pad) samples they screened (Schaeffer et al. 1996). Bonded urethane and prime urethane pads account for 90 percent of combined residential and commercial markets. Synthetic fibers, rubberized jute, and sponge rubber cushions together represent less than 10 percent market share. As a group, the synthetic fiber padding samples released the lowest quantities of VOCs.

For consumers, perhaps the best advice is caveat emptor (let the buyer beware). Alarie cautions that although he has worked in inhalation toxicology for 36 years, he knows "nothing about" 50 percent of the roughly 200 chemicals released by new carpet. "Am I going to tell a consumer 'Don't worry about it?' No way. No one on earth can say it's safe." He recommends that if some people are bothered by carpet, they should get rid of it. He says he has advised carpet manufacturers to replace a carpet if its purchaser complains about it. For consumers, he suggests using a "smell test" to determine how irritating a carpet is prior to purchase: Place a four-by-six-inch piece of carpet in a glass jar and seal it. After 24 hours, open the jar and sniff. If the odor is disagreeable, then don't buy the carpet, or air it out longer.

When carpet is manufactured, it comes out of the baking ovens, is cut, rolled, and wrapped, thus sealing in VOCs. Purchasers can ask carpet installers to unroll the carpet and air it out prior to installation. Adhesive and padding are other possible sources of VOCs that should be pretested by consumers. Adhesive may be avoided all together—most carpet can be nailed or tack-stripped to the floor. Alarie thinks that the carpet and rug industry can continue to improve its quality control and put out a better product, one that if properly cured should outgas within a few days.

Although homeowners can opt not to purchase or to remove problem carpet, that choice is usually not available to employees working in an office or to teachers or children in a school where carpet is to be laid. In such situations, more sensitive individuals may be at risk but feel powerless either to prevent carpet from being installed or to seek its removal once it has been installed, and they develop symptoms. They may suspect that their health problems are due to the carpet (or other remodeling exposures), but be unable to substantiate their suspicions. If a child is involved, the parents may vacillate between wanting to pull the child out of school and away from possible harm, wondering whether the carpet truly is affecting their child, and not wanting the school to think they are crazy. This is a dilemma no parent should have to face. Yet many do. Consumers are understandably confused.

To this day, it remains a mystery, even among scientists who were close to the situation, why Anderson and Alarie found toxic effects from carpets that EPA did not. What actually happened and why results from the various laboratories differed likely never will be resolved to everyone's satisfaction. Notwithstanding, there remains a need for a biological assay that does correlate with symptoms reported by affected consumers—an assay that uses relevant exposure durations (more representative of human exposures and therefore much longer than the one to four hours employed to date), employs adequate concentrations of the VOCs generated by the carpets (requiring careful selection of air flow, temperature and humidity), and assesses both neurological and respiratory effects.

Gulf War Exposures

Otto Dix, a German painter turned machine gunner, wrote about his impressions of World War I: "Lice, rats, barbed wire, fleas, shells, bombs, underground caves, corpses, blood, liquor, mice, cats, artillery, filth, bullets, mortars, fire, steel: That is what war is" (Boggett 1996). In contrast, a soldier's description of the Persian Gulf War might be: "Sand flies, filth flies, smoke from oil well fires, pesticides, pyridostigmine bromide, anthrax vaccine, botulinum vaccine, depleted uranium, diesel engine exhaust, smoky tent heaters, burning human waste, CARC paint, lindane, permethrin, DEET, blowing sand, jet fuel, diesel fuel, and maybe sarin: that is what war is." The Gulf War may have involved a greater diversity of chemical exposures than any prior war.

Nearly 700,000 United States troops were deployed to the Persian Gulf during Operations Desert Shield and Desert Storm between August 1990 and June 1991. Approximately 10 percent have undergone special examinations by the Department of Veterans Affairs (DVA) or the Department of Defense (DOD) because of health concerns related to their tour of duty in the Gulf. A 1994 National Institutes of Health Technology Assessment Workshop on the Persian Gulf Experience and Health concluded that "no single disease or syndrome is apparent, but rather multiple illnesses with overlapping symptoms and causes" (NIH 1994). Several studies have demonstrated an excess of self-reported symptoms among Gulf War veterans compared to nondeployed veterans of the same era (CDC 1995; Kaiser et al. 1995).

Miller (1996b, 1996c) reported on the first 59 consecutive Gulf veterans seen at the DVA's Regional Referral Center in Houston, Texas. Exposures that the veterans said caused acute symptoms while they were in the Gulf included: pyridostigmine bromide (a carbamate drug related to organophosphate pesticides and nerve agents) used to protect against possible nerve agent exposure (41 percent); smoke from oil well fires (17 percent); smoke from fuel used in tent heaters (14 percent); vaccines (12 percent); exhaust from diesel vehicles or generators (10 percent); vapors from CARC (chemical agent resistant coating) paint applied to vehicles (9 percent); and vapors from fuels (9 percent). Some also reported illness after exposure to pesticides, debris, or mist from SCUD missiles that exploded nearby, water contaminated with fuel oil, smoke from burning human waste, lindane used for delousing prisoners, and fine dust from charcoal absorbent used in protective clothing.

In 1996, the DOD acknowledged that chemical warfare agents had been found in some Iraqi bunkers that were destroyed by U.S. troops. The U.S. Army is investigating whether soldiers may have been exposed to sarin when smoke plumes from detonated bunkers at Kamisiyah drifted over troop areas. As far as is known, these troops did not manifest classical, acute symptoms associated with organophosphate exposure (such as salivation, lacrimation, incontinence, blurred vision, and pinpoint pupils). However, MCS patients who attribute onset of their illness to organophosphate or carbamate pesticide exposure frequently report only flulike symptoms at the outset (Miller 1996c).

Symptoms reported by the 59 Gulf veterans were strikingly similar to those of civilians exposed to indoor air pollutants in a new or remodeled building or to pesticides (Miller 1996b, 1996c; Miller and Mitzel 1995). All three groups complained of fatigue, musculoskeletal pain, memory and concentration difficulties, and mood changes such as irritability or depression. Although such symptoms often occur in the general population, half or more of these exposure groups reported severe fatigue versus only 3 percent of controls. About one-third or more of those exposed reported severe

TABLE 8-3. Frequency of Selected Symptoms Reported as Severe[a] by Gulf Veterans, MCS Patients, and Controls

Symptom	Gulf veterans (n = 59)	MCS pesticide exposed (n = 37)	MCS remodeling exposed (n = 75)	Control[b] (n = 112)
Fatigue	78%	68%	52%	3%
Depression	29	49	33	6
Headaches	53	38	31	5
Shortness of breath	38	43	31	2
Asthma or wheezing	12	27	15	0

[a]Participants rated their symptoms on a 0–3 scale: 0, not at all a problem; 1, mild; 2, moderate; 3, severe. Frequencies listed in this table reflect ratings of severe (3) only.

[b]Matched for age, sex, and educational level to the two MCS groups (37 + 75 = 112).

depression, headaches, and/or shortness of breath versus 2 to 6 percent of controls (Table 8-3).

Notably, 78 percent of the Gulf veterans reported new-onset chemical intolerances since the Gulf War. For example, mechanics who once liked the smell of engine exhaust or said they used to "bathe" in solvents with no associated symptoms, reported severe symptoms with these exposures since the war. One mechanic related that before the war his idea of the perfect perfume was WD-40; but now the odor of WD-40 and many other low-level chemical exposures make him feel ill. Seventy-eight percent of the veterans reported new food intolerances or feeling ill after meals; 40 percent had experienced one or more adverse reactions to medications since the war; 66 percent of those who used alcoholic beverages reported that even a small amount, such as one can of beer, made them feel ill; 25 percent of those who used caffeine reported feeling ill if they drank coffee or another caffeinated beverage; and 74 percent of those who smoked reported that smoking an extra cigarette or borrowing someone else's stronger brand made them feel ill. Many of the veterans reported confusion or concentration difficulties while driving, yet neither they nor their doctors had entertained the notion that their symptoms might be triggered by exposure to traffic exhaust. (Notably, a recent mortality study found Gulf veterans to be at increased risk for accidental deaths, particularly motor vehicle deaths [31 percent excess], compared with nondeployed veterans of the same era [Kang and Bullman 1996]). More than half of the Gulf veterans reported intolerances in each of three categories—chemical inhalants, foods, and drugs or food/drug combinations.

The veterans in the study conducted by Miller were inpatients in a VA tertiary referral hospital. In contrast, Fiedler et al. (1996a) mailed a pilot questionnaire to about 400 Gulf veterans on the VA registry in the Northeast.

Although the response rate was not reported, 39 percent of veterans who were listed on the VA registry as having fatigue, as well as 30 percent not initially screened for fatigue, considered themselves especially sensitive to certain chemicals, with car exhaust and perfume leading the list.

Fiedler et al. (1996a) have proposed using phenylethyl alcohol (rose oil) to challenge veterans separately via two routes—inhalation and skin absorption. Because Gulf War veterans are a relatively masked group of patients, such an approach may be prone to yield both false negative and false positive results (Miller 1994a, 1996a, 1996b, 1996c). The importance of unmasking patients prior to challenges so that they are at a clean baseline for testing is discussed in detail in Chapter 10 under "Current Reflections on MCS."

Without carefully conducted double-blind, placebo-controlled challenge studies using (1) salient exposures, (2) in a controlled environment, and (3) after an adequate period of unmasking, that is, removal from low-level background chemical exposures, questions concerning the role of everyday, low-level chemical exposures in perpetuating the veterans' symptoms are unlikely to be resolved. Although research using an Environmental Medical Unit (EMU) has been proposed to the Department of Defense, the Department of Veterans Affairs, and the National Institute of Environmental Health Sciences (NIEHS), studies of this kind have yet to be funded. Congress authorized partial funding for such a project in 1994, and the Department of Defense agreed to provide the remaining sum, but an EMU still has not been constructed. Although the U.S. House of Representatives Veterans Affairs Committee understood the need for studies in an EMU, somehow during the sequential stages of appropriations, issuance of a request for proposals, and the scientific review of applications, the original congressional intent was altered so that what ultimately emerged and was funded bore no resemblance to an EMU. Unfortunately, until this tool is made available to physicians, Gulf War veterans and MCS patients are likely to remain in their current Catch-22 of being required to show objective evidence of their disability and having no means by which to do so.

Implants

Some physicians have implicated implanted devices as causing multisystem illness that closely resembles chronic fatigue syndrome and multiple chemical sensitivity (Brautbar et al. 1992). In 1992, the FDA banned the use of silicone breast implants for cosmetic purposes. The FDA moratorium on the use of silicone gel implants was based on insufficient safety data, not on proven toxicity. At the time, FDA Director David Kessler mused, "We know more about the life of a tire than a breast implant" (Kessler 1992).

An estimated two million American women have received silicone breast implants, 80 percent for cosmetic purposes. The most common type of implant consists of silicone gel inside a polyurethane-coated membrane. Also an unknown number of U.S. patients have received temporomandibular jaw-joint (TMJ) implants for TMJ dysfunction and for other reasons. Materials used in these implants include silicone rubber and Teflon film laminated to plastic composite. Friction in the jaw-joint, due to the enormous forces developed during chewing, can lead to the release of microscopic implant fragments. As with most MCS exposure groups, only a subset of those who have received breast, TMJ, or other implants appear to develop central nervous system, pulmonary, skin, and other symptoms. Brautbar et al. (1992) suggested that silicone breast implant patients could serve as an excellent model for chemical sensitivity because the initiating chemical—silicone—is in the body. Although some patients recover fully following removal of their implants, others report persistent, multisystem symptoms.

Among 300 patients who became systemically ill following mammoplasty, implant rupture was not the initiating event. Indeed, rupture preceded systemic disease in only 3 percent of patients (Brawer 1996). Onset of illness began two weeks to eighteen years after the implant, with 90 percent of the entire group becoming symptomatic after six years. Although implant rupture (which occurred in 214 out of the 300 cases) exacerbated preexisting symptoms, it did not impact the rate of disease progression as compared to those who did not experience rupture. According to some authors, silicone may slowly ooze through intact implant membranes (Brautbar and Campbell 1995). An average of 30 symptoms and signs occurred in each patient, with more than half reporting the following symptoms: fatigue, arthritis, chest pain, hair loss, dry eyes/mouth, morning stiffness, myalgias, skin rash, paresthesias, cognitive problems, telangiectasias, and skin pigment changes.

Among 23 consecutive women with silicone breast implants referred for neurocognitive evaluation, the presenting symptoms (moderate to severe) for the majority included: fatigue (100 percent), short-term memory problems (91 percent), slowed thinking (91 percent), sleep disturbance (87 percent), irritability (78 percent), sensitivity to noise or light (74 percent), distractibility (70 percent), headaches (70 percent), dizziness (70 percent), anxiety (70 percent), depression (65 percent), confusion (65 percent), anger (65 percent), shortened attention span (61 percent), forgetfulness (61 percent), impatience (61 percent), labile emotions (57 percent), trouble finding the right word or word reversals (52 percent), and blurred vision (52 percent) (Hoffman et al. 1995).

Some investigators have suggested that silicone is not inert but may lead to inflammatory and immunological responses as it migrates to various organs via the reticuloendothelial system (Brautbar 1994; Vojdani et al.

1992a). Lappé (1995) points out that reversibility of autoimmune symptoms after withdrawal of a suspected drug incitant (e.g., penicillamine or captopril) is generally construed as evidence of a causal link. With silicone implants, such improvement, including normalization of immune measures, reportedly occurs in 40 to 60 percent of patients following implant removal, yet the same sort of causal role is not assigned to silicone by most physicians (Campbell et al. 1994). Other authors remain unconvinced that silicone is immunogenic or can serve as an adjuvant, and point to fundamental flaws in methodology and interpretation by those who make such claims. As is the case for MCS, publications in this area do not always reveal authors' potential conflicts of interest, e.g., serving as an expert witness for one side (Marcus 1996).

Childhood Abuse

Childhood physical and sexual abuse has been cited as a potential causal factor underlying some cases of MCS. Staudenmayer et al. (1993b) compared the sexual and physical abuse histories of 63 "universal reactors" with self-reported sensitivity to multiple chemicals with those of 64 controls chosen on the basis of having chronic symptomatology, an "identifiable psychologic disorder on Axis I of the DSM-IIIR," and "complaints not attributed by the patient to multiple chemicals or foods." Sexual abuse was defined as actual intercourse, and physical abuse as severe trauma with life-threatening intent as perceived by the child. The patients were seen in a private clinic that has advocated the use of psychological deprogramming (see Chapter 4, in section "Possible Psychogenic Mechanisms") to treat MCS patients. Thus, it is likely that there was a referral bias operating in this study, as physicians who view MCS as psychogenic might be inclined to refer patients for such deprogramming. The paper also fails to mention how many of the "universal reactors" attributed onset of their condition to an initial chemical exposure event. Fiedler et al. (1996b) found that only one-fourth of MCS patients whose illness began after a defined exposure met criteria for a major psychiatric disorder, whereas over two-thirds of MCS patients who recalled no specific initiating exposure qualified for a psychiatric diagnosis. Thus patients with an identifiable antecedent exposure seem to exhibit markedly less psychopathology than those with no specific initiating exposure.

Further selection bias likely occurred in the study by Staudenmayer et al. when only about half of the universal reactors and half of the controls agreed to undergo psychotherapy following their initial evaluation. There were no statistically significant differences with respect to a history of physical or sexual abuse between universal reactors and controls who elected not to undergo psychotherapy, nor were there any differences between male universal reactors and male controls who agreed to psychotherapy. However, 50 percent of female universal reactors ($n = 10$) versus 12 percent of female controls ($n = 3$) who agreed to psychotherapy reported a history of physical abuse alone; and 60 percent ($n = 12$) versus 25 percent ($n = 6$) reported a history of sexual abuse. According to the authors, the memory of sexual abuse was "repressed" in 30 percent of universal reactors ($n = 7$) and 12 percent of controls ($n = 3$), but this difference was not statistically significant. Thus, out of the 45 women who agreed to entered psychotherapy, only 5 of 20 universal reactors and 3 of 25 controls had recollections of sexual abuse that were not repressed—that is, had memories of abuse that did not have to be elicited by the therapist. This is not a large difference.

Perhaps it would be more enlightening to look at this study's findings from the reverse perspective: A substantial number of universal reactors appeared not to have any history of childhood physical or sexual abuse. What accounts for illness in those individuals? Bell (1994) suggests that olfactory stimuli such as environmental chemicals and/or emotionally charged events could sensitize limbic pathways in the brain: "[S]ome patients could have initiated their limbic instability with genetic factors, others with life stressors, others with chemicals, and still others with various combinations of genetics, stressors, and chemicals." A history of childhood abuse in some patients does not rule out the presence of chemical intolerances in those patients or their presence in other patients; nor does it rule out the possibility that childhood abuse could have enhanced limbic vulnerability, resulting in hyperresponsiveness to chemical exposures or an enhanced tendency to become sensitized to chemicals later in life. Teicher et al. (1993) studied limbic vulnerability among psychiatric patients and found that it was increased by 38 percent in patients with a history of physical abuse, by 49 percent in those with a history of sexual abuse, and by 113 percent in those with a history of both forms of abuse.

The rate of sexual and physical abuse in the general population and the appropriateness and ethics of various psychotherapeutic techniques for uncovering so-called repressed memories are subjects of heated controversy that cannot be explored in depth in this book. The prevalence rates for childhood sexual abuse from various studies range from 6 percent to 62 percent for females and from 3 percent to 16 percent for males, depending upon how sexual abuse is defined (contact versus noncontact) and other study differences (Cosentino and Collins 1996). According to these authors, "Conservative estimates, based upon the most methodologically sophisticated studies, indicate that one in three to four girls and one in ten boys have been sexually victimized before the age of 18."

Pope and Hudson (1995) surveyed the literature for studies that have tried to test the question of whether memories of childhood sexual abuse can in fact be repressed. They found only four pertinent studies. None of the four provided both clear confirmation of abuse and adequate evidence of amnesia in patients. The authors conclude that the available data do not support the idea that individuals can repress memories of childhood sexual abuse. The American Psychiatric Association (APA 1994) has cautioned that

> . . . repeated questioning may lead individuals to report "memories" of events that never occurred. It is not known what proportion of adults who report memories of sexual abuse were actually abused . . . . A strong prior belief by the psychiatrist that sexual abuse, or other factors, are or are not the cause of the patient's problems is likely to interfere with appropriate assessment and treatment.

Even if childhood physical or sexual abuse were more frequent among MCS patients (which remains to be determined), one could not conclude that such abuse played a causal role. The "addiction-like" cravings and withdrawal symptoms reported by some MCS patients could be the consequence of an inherited brain chemistry that predisposes them toward addictive behaviors. There is some evidence for increased rates of drug addiction in families of persons with chemical intolerances. Bell et al. (1995b) found that twice as many chemically intolerant versus chemically tolerant young adults reported family histories of drug abuse. Despite increased family histories of drug problems, these chemically sensitive young adults reported the lowest rates of current smoking and personal drug abuse and the highest frequency of illness from drinking a small amount of alcohol. Child abuse, whether physical or sexual, might be expected to be more frequent in families whose members have alcohol or other drug addictions. In such cases, MCS might be due to an inherited diathesis rather than childhood abuse. Another study (Bell et al. 1995a) showed no differences in family histories of alcohol or drug problems for MCS patients ($n = 28$) versus controls ($n = 20$). More recently, Bell (1997a) found that persons who had made life-style changes because of their chemical intolerances (a group that resembled MCS patients) ($n = 10$) were more likely to report having blood relatives with physician-diagnosed alcohol problems (60 percent) or drug problems (20 percent) than were chemically intolerant persons without life-style changes ($n = 8$; 25 percent and 12.5 percent, respectively), or normals ($n = 12$; 8.3 percent and 0 percent). Likewise, emotional or physical abuse was reported by 80 percent of chemically intolerant persons with life-style changes, 25 percent of chemically intolerant persons without life-style changes, and 42 percent of normals in the same study. Despite the small sample size, differences between the chemically intolerant with lifestyle changes and normals for childhood abuse were statistically significant, and approached significance for having a blood relative with an alcohol problem. Mean scores on the

McLean Limbic Checklist (Teicher et al. 1993) were twice as high for chemically intolerant patients with and without lifestyle changes (28.2 and 27.3, respectively) versus controls (13.4).

All of the initiating exposures associated with MCS in other industrialized nations have been reported in the United States. A recent review of both published and "gray" literature in nine European countries revealed the following exposures or events as possible initiators of chemical sensitivity (Ashford et al. 1995): organic solvents (all nine of the countries); pesticides including organophosphates and pyrethroids (six countries); amalgam/mercury (four); formaldehyde (four); paint/lacquers (four); stress/psychosocial factors (four); hairdressing chemicals (three); indoor environment (three); new/renovated buildings (three); pentachlorophenol/wood preservatives (three); carpets and glue (two); methylmethacrylate (two); pharmaceuticals (two); printed material (two); anesthetic agents (one); diesel exhaust (one); industrial degreasers (one). Repeated or continuous low-level exposures, rather than a single event, characterized most of the experience (see Table 7-1).

Mechanisms

A profusion of mechanistic hypotheses for MCS have surfaced since the first edition of this book, including olfactory-limbic kindling, other forms of neural sensitization, neurogenic inflammation, genetically based or chemically induced cholinergic supersensitivity, individual differences in the ability to metabolize xenobiotics, such as decreased sulfation capacity, and disorders of porphyrin metabolism. Theory papers concerning most of these hypotheses have appeared in the scientific literature. In a few cases, animal and human studies are under way. Numerous papers also have appeared addressing possible psychological mechanisms for MCS such as odor conditioning, iatrogenic (doctor-induced) belief systems, panic disorder, toxic agoraphobia, post-traumatic stress disorder (e.g., MCS developing after a chemical spill or as a consequence of childhood sexual abuse), somatoform disorder, or depression (Binkley and Kutcher 1997; Göthe et al. 1995; Gots 1995, Guglielmi et al. 1994; Kurt 1995; Pennebaker 1994; Simon 1994a; Sparks et al. 1994a, 1994b; Spyker 1995; Staudenmayer et al. 1993b, 1997).

Staudenmayer and Camazine (1989) described MCS patients, whom they call "universal reactors," as "the latest individuals in a long list of those who historically manifested psychosomatic illness." They further suggest that "Universal Reactors project their problems onto the environment 'since they substitute for an internal instinctual danger an external perceptual one' (Freud 1936)." Norman Rosenthal, psychiatrist and Chief of the Clinical Psychobiology Branch at the National Institute for Mental Health, takes issue with their explanation for MCS: "The problem with such psychodynamic formulations is that they are easier to construct than to test. Even in studies that show patients with MCS to have a tendency to projective defenses, such as Staudenmayer and Camazine have reported, who can say whether these tendencies are a cause, a result, or an epiphenomenon of MCS?" (Rosenthal 1994).

In a critique of studies viewed as supportive of a psychogenic hypothesis for MCS, Davidoff and Fogarty (1994) pointed out the frequently overlooked fact that psychological symptoms are not necessarily psychogenic:

> According to consensus within the American Psychiatric Association, psychiatric diagnoses are descriptive entities that subsume signs and symptoms without explaining them. In other words, psychiatric symptoms and diagnoses are "nonspecific" in terms of etiology; these phenomena may have diverse causes. Consider as an illustration what is commonly called "depression": a constellation of negative affects (such as hopelessness, fatigue), negative cognition (such as pessimism), passivity, and anhedonia. Depression may arise because of a psychological loss, trauma, conflict, or another type of psychosocial stress. In addition, depression may arise in the absence of psychosocial stress because of a structural brain lesion (e.g., stroke, brain tumor); metabolic or endocrine dysfunction (e.g., hypothyroidism, hypoadrenocorticolism, B12 deficiency); medication (e.g., steroids, sedatives, estrogens, analgesics, antihypertensives); or toxic exposure (e.g., lead, organic solvents). Physicians need to understand that neither tests nor clinicians can distinguish origin when presented with a symptom complex alone.

Drawing attention to the fact that psychological diagnoses explain nothing about MCS feels a bit like announcing, "The emperor has no clothes!" in the midst of an otherwise adulatory crowd honoring a beloved sovereign. There are about 37,000 psychiatrists and 241,000 psychologists in the United States (Roback et al. 1994; *Statistical Abstract of the United States 1994*). Any theory of disease so bold as to suggest that depression, anxiety, panic attacks, fatigue, or MCS might be caused by chemical exposures should expect a less than enthusiastic reception. Yet, all physicians are taught that organic bases for illness should be ruled out before psychological explanations are invoked.

Kuhn (1996) observed that the emergence of a new paradigm is characteristically accompanied by pronounced professional insecurity caused by "the persistent failure of the puzzles of normal science to come out as they should." This failure sparks the search for new rules. MCS currently may be in what Kuhn described as the pre-paradigm period, a time "regularly marked by frequent and deep debates over legitimate methods, problems, and standards of solution, though these serve rather to define schools than to produce agreement" (see further discussion on the stages of a paradigm shift in Chapter 9).

A yet-to-be-proven mechanism for disease is a theory of disease. There are two primary schools of thought or fundamental theories that drive and divide MCS researchers: (1) MCS is a psychogenic condition, and (2) MCS is an organic condition. A few researchers, in particular those exploring olfactory-limbic sensitization, have adopted a biopsychosocial approach to MCS and argue that chemicals and stress, together or independently, may alter neurochemistry and produce the illness. All diseases and illnesses lend themselves to a biopsychosocial approach. On the other hand, it would be disastrous if physicians were to presume that cancer was caused primarily by stress or was treatable primarily by psychological interventions. The role of psychological factors in MCS is as yet undetermined, but should not be presumed as major until the role of chemical exposures has been elucidated in a thorough, scientific manner—that is, via double-blind placebo-controlled challenges in a controlled environment to ascertain physical causes of triggering in already sensitive individuals and via investigations of exposures preceding onset of the condition. It is our view that until these direct approaches to the problem have been explored, those who continue to promote untested and untestable psychogenic theories for MCS are part of the problem. Their lobbying of policymakers and others in this regard has contributed to widespread governmental inertia on this issue, making it near impossible to obtain funding for essential studies specifically directed toward MCS. Many of those who advocate psychological explanations in government-sponsored meetings and in the scientific literature are paid corporate spokespersons or consultants with financial conflicts of interest. Yet these conflicts generally are not revealed when these individuals appear in scientific meetings, author scientific articles, serve on official panels or boards, or serve as reviewers of grant proposals. Policymakers and publishers of scholarly journals need to recognize and remedy this appalling injustice. When a grant proposal on MCS is being reviewed, remarks by a single reviewer who is "against" MCS (or who believes or has heard that it is psychogenic) can have a chilling effect on the rest of the panel and effectively kill that proposal. Only an outspoken scientist on the panel who is knowledgeable about MCS, skeptical yet open-minded, could counter such remarks.

It is hoped that the recent emergence of tough, testable theories for MCS involving chemical causation and the availability of relevant animal models will persuade policymakers and grant reviewers that fundamental questions concerning the role of low-level chemical exposures in MCS urgently need to be addressed, and that answers *can* be found by using first-rate science.

In this section, we highlight developments related to some of the physiological mechanisms that have been proposed to explain MCS, describing the limited pilot studies that have been undertaken in this area. We will not address psychogenic hypotheses further here. As discussed above, psychiatric diagnoses are by their very nature descriptive and do not rule out chemical causes. Considering the fact that more than 6 percent of Californians report having been diagnosed with MCS (Kruetzer and Neutra 1996), there has been only a modicum of research into the possibility that physiological mechanisms may underlie this illness. Sadly, we can readily summarize the relevant research on physiological mechanisms for MCS in the past five years in a few pages.

Olfactory-Limbic Sensitization

Originally articulated in the first edition of this book (in Chapter 4), this theory has since then been expanded, and its relationships with kindling, partial kindling, time-dependent sensitization (TDS), and learning are being explored (Bell et al. 1992; Gilbert 1994; Rossi 1996; Sorg 1996a, 1996b). All of these processes are forms of neural sensitization. They may be thought of as a continuum, from very weak sensitization (learning) up to the induction of seizures, known as kindling. The olfactory pathways, especially the olfactory bulbs, are sensitive to both electrical and chemical kindling. Several researchers now are using rodent models to explore the question of whether common chemicals such as chlorinated pesticides, formaldehyde, and toluene, whose classical toxic effects have been widely studied, might affect neural activity in novel ways if given in smaller, repeated doses at varying time intervals (Gilbert and Mack 1995; Kay 1996; Sorg 1996b). Individually, such small doses would cause no problem, but over time might "kindle" a greater response.

Gilbert (1994) has studied the effects of various chlorinated pesticides, including lindane (a pesticide used topically to treat head lice) and endosulfan, on kindling in rats. Kindling is a model for seizure induction whereby repeated low intensity electrical stimuli, any of which alone would not cause seizures, give rise to seizures over time. Persistent biochemical and physiological changes occur in the brains of kindled animals. Particularly sensitive to kindling are the limbic regions of the brain that govern mood (amygdala) and short-term memory (hippocampus). Gilbert (1995) found that although a single oral convulsive dose of lindane did not cause kindling, half as much lindane (a subconvulsive dose) given three times a week for ten weeks led to increasing responsiveness to the pesticide and more frequent myoclonic jerks and clonic seizures over the course of the study. These effects persisted two to four weeks after the last dose was given. Recordings from the amygdala showed abnormal electrical activity, including rhythmic bursts and spike and wave discharge in the absence of observable seizure behavior.

Gilbert and Mack (1995) demonstrated that the doses of lindane and endosulfan required to produce myoclonic jerks in 50 percent of their animals was decreased by more than 60 percent in rats kindled by repeated electrical stimuli to the amygdala (part of the limbic system).

Conversely, endosulfan reduced the level of electrical stimulation necessary to evoke seizures in amygdala-kindled rats. Thus, increased sensitivity after kindling by one agent can transfer to other, unrelated agents. Rats kindled via electrical stimulation of the amygdala became permanently more vulnerable to seizures induced by alcohol withdrawal (Pinel and Van Oot 1975). Likewise, rats kindled by repeated low doses of drugs that cause seizures such as lidocaine or carbachol required fewer stimulations to elicit convulsions when kindled electrically (Gilbert 1994). Partial kindling (kindling in the absence of overt seizure behavior) in cats can lead to animals with a labile affect that exhibit explosive defensive behavior following slight provocation (Adamec and Stark-Adamec 1983).

Time-dependent sensitization (TDS) refers to the ability of intermittent exposures, whether drugs, environmental chemicals, or stress, to induce physiological and behavioral effects that grow with the passage of time between exposures. Many features of TDS are remarkably similar to those of MCS (see first section of Chapter 4 of this book; also see Antelman 1994, Bell 1994, Bell et al. 1996c):

1. Involvement of multiple organ systems either singly or simultaneously. For TDS this includes the nervous system, the immune system, and the cardiovascular system.

2. Initiation by a wide array of chemically dissimilar agents and environmental/physical stressors. For TDS, inducers include such chemically diverse drugs as antidepressants, amphetamines, ethanol, corticosterone, estrogen, interleukin-2, and nicotine, as well as electroconvulsive stressors.

3. Spreading or generalization of sensitivity to chemically different stimuli, referred to as cross-sensitization.

4. Food and alcohol intolerances. Antelman speculates that the food and alcohol intolerances accompanying MCS could be related to alterations in the effect of ethanol on plasma adrenocorticotropic hormone (ACTH) and glucose that have been observed in TDS.

5. Triggering by a provocative stimulus, with symptoms being absent between provocations.

6. Persistence. MCS reportedly can persist for decades in humans. In rats, TDS has lasted at least two months following induction, the longest period tested thus far. The normal life span for a rat is about two years. Persistence of amygdaloid kindling, even after 12 months without further electrical stimulation, has been observed in cats (Wada et al. 1974) and in baboons (Wada and Osawa 1976). These findings have "led most investigators to conclude that kindling is due to permanent changes in the brain" (Corcoran 1988).

7. Bipolarity of response. Stimulatory and withdrawal responses in MCS have parallels with TDS. "Lower intensity" stressors sensitize stimulatory responses and "higher intensity" stressors elicit inhibitory responses.

8. Thresholds for induction. A certain level of initiating exposure appears necessary for the development of MCS—a level or a threshold that appears to vary from person to person. Similarly, a minimum level of stress, evidenced by acute changes in plasma corticosterone levels, may be necessary for the induction of TDS.

9. An apparent increased vulnerability of females for both MCS and TDS.

10. Induction by any of various exposure routes. TDS may be induced by injecting a drug, by placing it in an animal's drinking water, or by psychological stressors.

These similarities, as described by Antelman (1994), offer face validity for TDS in animals as a model of MCS for humans.

Sorg (1996a) hypothesized that MCS might be initiated through a mechanism similar to that involved in neural sensitization by cocaine. She exposed female rats to formalin vapors (11 ppm) or water vapor (controls) for 1 hour daily for 7 days. On the eighth day, a saline (placebo) injection was given, and 24 hours later a cocaine injection was administered. Cocaine and amphetamines are known to sensitize certain brain pathways, for example, the mesolimbic dopamine system, causing animals to move from place to place more frequently (increased locomotor activity). The rats previously exposed to formalin for 7 days showed increased locomotor activity following their cocaine injection, suggesting that formaldehyde had sensitized the same pathways involved in cocaine sensitization. After 3 to 4 weeks with no further exposures, cocaine injection enhanced locomotor activity in only a subset of the formalin-treated animals. These data suggest that repeated formalin exposure can cause temporary sensitization in most animals. According to Sorg, those animals remaining sensitized weeks later could be analogous to patients who develop MCS following chemical exposure.

The levels of formaldehyde used in this study were relatively high, 11 ppm versus the occupational exposure limit (Threshold Limit Value) for formaldehyde of 0.3 ppm for a 40-hour workweek. More recently, Sorg exposed rats to either 1 ppm formaldehyde or plain air for 1 hour per day for 7 days or for 20 days (5 days per week for 4 weeks). No differences in cocaine-induced locomotor activity between exposed and control rats were noted after 7 days of exposure. However, after 20 days of formaldehyde exposure, vertical locomotor activity was significantly increased both 2 to 4 days and 4 to 6 weeks after the last exposure to formaldehyde (Sorg 1996b). These results suggest that formaldehyde sensitization can affect the same neural pathways as those involved in cocaine sensitization. Sensitization occurs after long-term (20-day), but not short-term (7-day), low-level exposure. These findings are consistent with a limbic sensitization model for MCS.

Kay (1996) explored the question of whether inhaled toluene could disrupt electrophysiological activity in the olfactory-limbic area. Five male rats implanted with electrodes in the olfactory bulb, prepyriform cortex, entorhinal cortex, and dentate gyrus were exposed to food odors (mint, almond, orange, vanilla) and toluene vapors. Substances were presented on a saturated cotton swab held in front of each animal's nose for six seconds, one minute apart for 20 presentations. Mint and toluene are known to stimulate the trigeminal nerve. Over multiple exposures, toluene and, to a lesser degree, mint evoked narrow band, high amplitude oscillations in the 15 to 30 Hz range in the olfactory-limbic tract (originating in most cases in the prepyriform cortex). None of the other food odors had this effect. Following toluene sniffing, one animal exhibited long-lasting spontaneous seizure activity in the absence of further exposure, and on at least one occasion exhibited behavior consistent with an absence seizure. In this case, epileptiform activity originated in the hippocampus and spread to the olfactory areas. The limbic oscillations produced by mint are of potential concern, given some investigators' use of peppermint to mask odors and as a sham exposure during challenges of chemically sensitive patients (Staudenmayer et al. 1993a).

The limbic and mesolimbic areas of the brain are particularly sensitizable. Sensitization appears to involve excitatory amino acids, essential neurotransmitters present in central nervous system pathways involving pain reception, olfaction, learning, and memory. Bell (1996) has begun to probe the relationship between limbic sensitization and MCS in a series of clinical surveys. Using the McLean Limbic System Checklist (Teicher et al. 1993), a questionnaire based on symptoms that occur with temporal lobe seizures (seizures that often originate in the amygdala within the limbic system), Bell found that both college students and middle-aged women with chemical intolerances scored higher on this checklist than control subjects did (Bell 1996). Interestingly, women with temporal lobe epilepsy have greatly increased rates of polycystic ovary disease. This is thought to be because the amygdala helps regulate reproductive hormone release by the hypothalamus. If women with MCS have abnormal limbic activation, then their reproductive hormones also may be affected. Indeed, Bell et al. (1995a) found that women with MCS reported significantly increased rates of ovarian cysts and menstrual disorders, compared with controls. Bell (1996) concludes that studies from a variety of perspectives provide face, construct, and criterion-related validity for an olfactory-limbic neural sensitization model for MCS. If, as the above studies in humans and animals suggest, the primary target for toxicants in MCS is the limbic effector system, then MCS may best be regarded as a toxic encephalopathy (Kilburn

1993b). Recently, Bell et al. (1996d; 1997a, b, c) have reported a series of studies consistent with laboratory-based sensitization of chemically intolerant individuals compared with controls. Outcome measures that have demonstrated sensitization include heart rate, blood pressure, plasma beta-endorphin, and resting EEG alpha activity.

Other Hypotheses Involving the Olfactory System

A group of German investigators studied the chemosensory event related potentials (CSERPs) of 23 MCS patients in response to olfactory (hydrogen sulfide) and trigeminal (carbon dioxide) stimuli (Hummel et al. 1996). Room air or 2-propanol, a commonly used solvent, was administered to the nostrils of patients in a double-blind manner. Olfactory "performance" was then assessed via measurement of CSERPs in response to hydrogen sulfide (H_2S) and carbon dioxide (CO_2), odor discrimination ability (using eight odorant pairs), odor detection thresholds (using phenylethyl alcohol), and acoustic rhinometry. Approximately 20 percent of the MCS patients reported symptoms regardless of the type of challenge, leading the authors to speculate that such patients may be susceptible to experimental manipulations, and double-blind studies are essential. The juxtaposition of so many different chemical exposures in a relatively short period of time (an aspect of masking; see Chapter 10, section on "Current Reflections on MCS") may have contributed to this finding. In addition, if patients were not fully unmasked (at a clean baseline) prior to challenge, background chemical exposures extraneous to the study could have caused false positive responses. Changes occurred in the CSERP latencies (decreased after CO_2, increased after 2-propanol), indicative of changes in the processing of both olfactory and trigeminal stimuli. The authors suggested that these changes could be due to shifts in the orientation of cortical generators, that is, neuronal populations involved in processing chemosensory input. The volume of the nasal cavity decreased regardless of the type of exposure administered; however, these changes were not statistically significant. Olfactory thresholds remained unchanged pre- and post-exposure, consistent with the findings of other researchers (Doty et al. 1988). On the other hand, subjects' ability to discriminate odors decreased more after exposure to room air than after exposure to 2-propanol. The authors surmised that the MCS patients' relatively increased ability to discriminate odors after low levels of 2-propanol (concentration near the olfactory threshold) versus room air might increase their ability (relative to normals) to detect odorants and could make them more prone to develop "pathological responses" to odorants. As the study did not include a control group, it remains to be determined whether similar responses would also occur in healthy age- and gender-matched subjects.

Cholinergic and Other Receptor Supersensitivity

A number of investigators have observed that MCS may be initiated by exposure to cholinesterase-inhibiting pesticides (Ashford et al. 1995; Cone and Sult 1992; Miller and Mitzel 1995; Rosenthal and Cameron 1991; Sherman 1995) (see "Origins" section of this chapter). As a group, MCS patients whose illness began following exposure to an organophosphate or carbamate pesticide report more severe symptoms than those exposed to low levels of mixed solvents (Miller and Mitzel 1995). It is noteworthy that organic solvents also inhibit acetylcholinesterase activity in the membrane of human red blood cells in vitro and, by analogy, are thought to act on nerve cell membranes (Korpela and Tahti 1986a, 1986b). These findings, coupled with the integral role of the cholinergic system in virtually every system of the body, have suggested to some investigators that cholinergic hypersensitivity could provide a unifying explanation for MCS.

Overstreet et al. (1996) observed that rats selectively bred for sensitivity to the organophosphate diisopropylfluorophosphate (DFP), the "Flinders sensitive rat," share many features in common with MCS patients. These specially bred rats have increased numbers of cholinergic receptors (by about 20 percent) in certain limbic areas, including the hippocampus and the striatum. When exposed to cholinergic agonists, the rats become hypothermic, move around less (decreased locomotor activity), and will not as readily press a bar for a water reward. They are also more sensitive to nicotine, serotonin agonists, dopamine antagonists, diazepam, and ethanol. Intraperitoneal sensitization to the antigen ovalbumin in these animals led to a greater increase in gut permeability than in control rats (Djuric et al. 1995). In another study, ad libitum sucrose solution for 30 days resulted in depression, gauged by decreased performance on a swim test (Djuric et al. 1996). Notably, adult female Flinders sensitive rats are more sensitive to cholinergic stimulant drugs than their male counterparts (Netherton and Overstreet 1983).

Because Flinders sensitive rats exhibit decreased appetite and activity and increased REM sleep compared to controls, they originally were used as a genetic animal model for depression. Depressed individuals are supersensitive to anticholinesterases and cholinergic agonists (Janowsky et al. 1994). Further, there are known interactions between depression and addictions, including smoking and alcoholism. Imbalances between the noradrenergic or dopaminergic systems and the cholinergic system conceivably could account for heightened sensitivity. Alternatively, disrupted second messengers inside cells, such as changes in G proteins, cyclic AMP, or other second messenger systems, might be involved. This hypothesis might more readily accommodate the diverse classes of chemicals implicated in triggering MCS than a hypothesis involving alterations in a single neurotransmitter system. Individual differences in second messengers could be inborn or induced by chemical exposures. Future research, involving, for example, exposing the Flinders sensitive line of rats to solvents or other chemicals that seem to trigger symptoms in MCS patients, may help elucidate mechanisms and could yield novel diagnostic and treatment approaches for MCS. In 1996, the Department of Defense funded David Overstreet and colleagues to examine the effect of pyridostigmine bromide, a cholinesterase-inhibiting carbamate drug, on Flinders sensitive rats. Many soldiers who served in the Persian Gulf took pyridostigmine bromide to prevent adverse effects from possible nerve agent exposure, and some associate onset of their illness with its use.

Corrigan et al. (1994) described several patients with MCS-like symptoms and prominent fatigue whose illness developed following exposure to various organochlorine insecticides, including lindane. The hypothesis is offered that lethargy, impaired concentration, and intolerance for alcoholic beverages in these cases could be explained by alteration of GABAa receptor sensitivity in the central nervous system. Some of the patients described in this paper had been exposed to pyrethroids, which also act through the picrotoxin site of the GABAa receptor; others had been exposed to organophosphate pesticides, which impact the cholinergic system. The authors suggest that there is a close connection between the cholinergic and GABAergic systems in the limbic region (hippocampus), where their interaction produces theta-like electrophysiologic activity (Konopacki et al. 1993).

Neurogenic Inflammation

Drawing upon her studies of the effects of tobacco smoke on the upper airways, Bascom (1991) hypothesized that differences in human responses to tobacco smoke are the result of different responses in C fiber neurons unmyelinated nerve cells in mucosal tissue whose fibers terminate near blood vessels and glands and in intraepithelial spaces. Stimulation of these nerve endings leads to release of neuropeptides, including Substance P and calcitonin-gene–related peptide, resulting in vasodilation, vascular extravasation, and bronchoconstriction. Currently, little is known about how the functioning of these nerves may be changed in disease in humans. Cells lining the airways can release various cytokines, such as interleukin-1, as can scavenger cells located in the airways (alveolar macrophages). Neutral endopeptidase, an enzyme in airway epithelium, inactivates Substance P, thus reducing nerve stimulation. Smoke decreases the amount of neutral endopeptidase in the airway mucosa. Bascom suggests that MCS patients could be intolerant of VOCs because they have less of this enzyme or less active enzyme in their airway epithelia. Tobacco smoke is a complex mixture of irritating substances, including oxidizers and aldehydes. Tobacco smoke vapors stimulate chemosensitive neurons. When tobacco smoke–sensitive and nonsensitive persons were exposed to tobacco smoke, symptoms and nasal congestion (demonstrated by posterior rhinomanometry) were significantly greater in the historically sensitive subjects. Even smoke vapors filtered to remove particulates caused congestion in sensitive subjects.

Meggs et al. (1996a) similarly hypothesized that chemical irritants may cause release of Substance P and other mediators from sensory nerves in the airway (Meggs 1994), giving rise to reactive airways dysfunction syndrome (RADS) and reactive upper-airways dysfunction syndrome (RUDS) (respectively, chronic asthma and chronic rhinitis stemming from irritant exposure). To test the hypothesis that neurogenic inflammation might explain both RUDS and MCS, Meggs performed fiberoptic laryngoscopy on ten subjects (six males, four females) who met the Cullen case definition for MCS and who ascribed onset of their illness to a chemical exposure. The most robust findings were the cobblestone appearance of pharyngeal mucosa in six patients, suggestive of lymphoid hyperplasia, and focal areas of pale mucosa seen with fiberoptic endoscopy in eight patients.

Subsequently, Meggs et al. (1996a) examined nasal biopsies of 13 patients, all of whom reported chronic illness and new-onset chemical intolerances following exposure to chlorine dioxide, and three control subjects. Significantly more inflammation and increased numbers of nerve fibers were present in the biopsy specimens of the patients. Some specimens revealed abnormal mucosal epithelium with detachment of cells from the basement membrane and defects in the junctions between epithelial cells. No increase in eosinophils was seen as would be expected in allergic rhinitis. Unfortunately, tissue stains for Substance P and other inflammatory mediators were not helpful because of a high background of nonspecific staining of the tissue. Thus, further studies will be needed to confirm this hypothesis.

Metabolic Mechanisms

Inherited or chemically induced variations in the metabolism and/or excretion of toxicants are other proposed explanations for MCS. The ability to detoxify foreign chemicals (xenobiotics) varies greatly within populations. Some individuals appear to have an increased capacity to detoxify certain organophosphates but not others, depending upon their genotypic and phenotypic expression of paraoxonase, the enzyme that breaks down (hydrolyzes) organophosphates (Davies et al. 1996). Thus far, the detoxification capacities of MCS patients have not been explored in depth. It is important to emphasize that not all xenobiotic metabolic reactions decrease toxicity; some increase toxicity by producing intermediate compounds that are more toxic than the parent compound. Thus, MCS patients conceivably could have either a deficiency or an excess of certain metabolic enzymes underlying their enhanced susceptibility.

Sulfation of phenolics is a key step in the detoxification of various phenolic and aromatic drugs and xenobiotics. This process depends upon the presence of a limited supply of inorganic sulfate, which in turn appears to be produced primarily via sulfoxidation of the amino acid cysteine. Studies in humans have shown enormous variation in sulfation capacity (McFadden 1996). There is a wide range of ability to metabolize the drug S-carboxymethyl-cysteine (SCMC), with 2.5 percent of the population appearing to be nonmetabolizers. Decreased SCMC metabolism has been reported in persons with neurodegenerative diseases such as Alzheimer's disease, Parkinson's disease, and motor neuron disease, and with rheumatoid arthritis and delayed food sensitivity. Some investigators are exploring the role of decreased metabolism of SCMC and impaired sulfation in MCS (McFadden 1996).

Porphyria

Because MCS patients exhibit unusual multisystem symptoms triggered by a wide variety of substances, some authors have hypothesized that MCS may be related to the porphyrias, a group of rare diseases whose manifestations include multisystem symptoms and multiple drug intolerance. The classical porphyrias are a group of at least seven diseases, involving primarily skin, neurological and/or psychological manifestations, caused by disturbances of the heme-forming system (Ellefson and Ford 1996). Porphyrins are chemical intermediates produced via a cascade of enzymatically catalyzed biochemical steps leading to the production of heme, the essential iron-containing protein in blood that binds oxygen for transport to tissues, and of cytochromes, e.g., the cytochrome P450 enzymes that constitute the primary detoxification pathway for thousands of diverse xenobiotics. At least half of the enzymes in the cascade can be directly attacked by various environmental pollutants (Kappas 1987). Patients' symptoms depend upon which enzyme(s) is (are) deficient and which precursors accumulate in which tissues, but may include chemical and drug intolerances, photosensitivity (burning, blistering or scarring of sun-exposed areas), abdominal pain, musculoskeletal pain, fatigue, neuropsychological problems, and pink to dark red urine ("porphyria" is derived from the Greek word for purple). Porphyrias may involve any or all of the five "P's": (1) onset after puberty,

(2) psychiatric abnormalities, (3) pain, (4) polyneuropathy, and (5) photo-sensitivity. Symptoms may be absent between attacks, but acute episodes can be precipitated by the four "M's": (1) medications (including estrogens and ethanol), (2) menstrual or premenstrual periods, (3) malnutrition (fasting or low carbohydrate diet), and (4) medical illnesses (particularly infections) (Perlroth 1988). In the classical hereditary porphyrias, enzyme levels usually are 50 percent or more below normal and the porphyrins that accumulate in the urine and the stool are several times normal levels during attacks. In between attacks, porphyrin levels may be normal. Mayo Laboratories has developed tests for five of the eight enzymes in the heme synthesis pathways (Mayo 1995). One of these tests, erythrocyte coproporphyrinogen oxidase, has been criticized as "fundamentally flawed" and its use therefore unjustified (Hahn and Bonkovsky 1997).

Doss (1987) distinguishes between the inherited porphyrias (inborn errors of metabolism) and secondary porphyrinopathies, that is, increased porphyrins in the urine (porphyrinuria) or in the blood (porphyrinemia) induced by other diseases or certain drugs or chemicals. The latter conditions he terms "non-specific disorders without immediate clinical significance," in contrast to porphyrinurias or porphyrinemias related to a porphyria disease. Whereas patients with inherited (primary) porphyrias can produce and excrete heme precursors at high rates during active disease, secondary porphyrinopathies are associated with low-grade or moderate production and excretion and do not have the same dramatic clinical presentation (Doss 1987, Hahn and Bonkovsky 1997). A wide array of unrelated medical conditions may result in secondary porphyrinopathy, including lead poisoning, various drugs, chronic renal failure, malignancies, alcoholism, hepatitis C, cirrhosis, various anemias, hyperbilirubinemia, systemic lupus, and diabetes mellitus, as well as pregnancy and fasting (Doss 1987). Chemical exposures most frequently cited as causing secondary porphyrinopathies include metals (lead, arsenic, mercury) and halogenated hydrocarbons (hexachlorobenzene, polyhalogenated biphenyls, dioxins (TCDD), vinyl chloride, and carbon tetrachloride) (Doss 1987). Some of these substances directly impact the nervous system, thus adding to their neurotoxicity (Silbergeld 1987). The most studied and best understood porphyrinogenic toxicant is lead. The most consistent neuropathological finding in porphyrinopathies is demyelination of central and peripheral nerve fibers (Silbergeld 1987). Slowed nerve conduction velocity also has been reported. Neuropsychiatric symptoms are generally of the so-called soft variety, and rarely have objective clinical signs been associated with them. Changes in personality and mood have been described (Silbergeld 1987).

The hallmark symptom of MCS—adverse multisystem responses to multiple chemicals and drugs—is also a hallmark of the porphyrias. Further, avoidance of problem exposures is accepted as the primary treatment for porphyria. Downey (1994) described 62 patients from his oral pathology practice with a wide array of oral symptoms, including possible adverse reactions to dental restorations or materials, taste abnormalities, burning mucosa or tongue, paresthesias, blisters, temporomandibular joint dysfunction, and pain. He found that about 90 percent of these patients had one or more abnormalities on a test battery of four porphyrin enzymes. Subsequently, several physicians who see individuals with MCS began testing their patients for porphyrin abnormalities, and reported that 60 to 90 percent of their MCS patients exhibited porphyrin and/or porphyrin enzyme abnormalities (Morton 1995, Wilson 1996c).

Ziem (1996) observed that a number of her MCS patients reported dark brown or red urine, especially during exacerbations of their illness. Likewise, 8 of 13 MCS patients she tested had elevated porphyrins in their urine, and 3 of 11 stool tests showed elevated porphyrin levels. Twelve of 14 patients whose blood was tested for porphyrin-related enzymes had one or more outside normal limits. Decreased coproporphyrinogen oxidase activity was most common, occurring in nine patients. Porphyrin enzymes were completely normal in only two cases; four patients had one enzyme, four patients had two enzymes, and four had three enzymes outside normal limits. Ziem noted that the porphyrin accumulations in urine and stool and the enzyme deficiencies seen in this uncontrolled study of MCS patients were milder than those reported for the congenital porphyrias. She hypothesized that chemical exposures may have a broad spectrum of effects and that the heterogeneous patterns of porphyrinopathy in her MCS patients may reflect their own unique chemical exposure histories.

The elevations in porphyrins and decreases in porphyrin enzymes seen in MCS patients generally have been slight. Advocates of a porphyria hypothesis for MCS claim that MCS patients have abnormal results on these assays much more frequently than the general population even though the findings are often only slightly outside normal ranges, and that because MCS patients generally avoid chemical, food, and drug triggers, their laboratory findings may not be as pronounced as if they were still exposed to incitants. On this basis, they advocate that whether the porphyrin abnormalities are the cause or an effect of MCS, they should be diagnosed in their own right as a "disorder of porphyrin metabolism" (Ziem and Donnay 1996).

As a consequence of over 40 work-related MCS claims being filed on the basis of porphyrin abnormalities, the State of Washington's Department of Labor and Industries issued a memorandum on March 8, 1995, detailing its policy concerning the diagnosis of porphyrinopathies, stating that "[t]his diagnostic problem is best handled with a systematic approach incorporating historical, physical, and laboratory information as opposed to dependence upon one or two simple screening tests." The department cautioned against overreliance upon enzymatic results because of their low specificity for diagnosing porphyrias and their vulnerability to various errors. It advised that if a patient's enzyme levels are found to be low, the test needs to be repeated. Temperature variations, delays in processing, and other factors can reduce the quantity of enzyme present in a specimen. In its memorandum, the department laid out three clinical and three laboratory criteria for porphyria. At least one criterion in each category would need to be met in order to confirm a diagnosis of porphyria:

Clinical criteria:
1. Evidence of autonomic neuropathy, for example, abdominal pain accompanied by ileus, diarrhea, tachycardia, or hypertension. The abdominal pain of porphyria is commonly accompanied by pain in the extremities.
2. Symptoms of peripheral neuropathy, especially if accompanied by weakness, paralysis, hypoesthesia, and/or absent deep tendon reflexes.
3. Evidence of psychosis.

Laboratory criteria:
1. A 24-hour urinary uroporphyrin or coproporphyrin excretion that is more than double the upper limit of normal.
2. A 24-hour urinary porphobilinogen excretion that is above the upper limit of normal, especially when collected during an acute symptomatic episode.
3. A fecal coproporphyrin excretion that is more than double the upper limit of normal.

The memorandum further stated that abnormal laboratory findings should be confirmed with a second test. In addition, an exposure criterion must be met if exacerbation by environmental exposure is claimed, that is, confirmed exposure to hepatotoxic chemicals or known porphyrinogenic drugs.

Ellefson and Ford (1996) of the Mayo Laboratories, which perform porphyrin analyses, acknowledge that "[a] few individuals who represent a small percentage of persons with chemical sensitivity disorder (CSD) can be expected to have active porphyria." They recommend that evaluation of suspected cases include analysis of the excretion of δ-ALA, porphobilinogen, and porphyrins, and the levels of porphyrins in red blood cells, but caution that marginally high values are not diagnostic, although it may be appropriate to look for potential causes for the increase, eliminate them, and remeasure periodically at four- to six-month intervals until levels return to normal.

Immunological Mechanisms

To date, no consistent immunologic abnormalities have been found in MCS patients (see discussion of "Immunologic Mechanisms" in Chapter 4). Nevertheless, immune alteration is an appealing hypothesis. Only a few university-based studies in this area have been conducted since the first edition of this book. Fiedler et al. (1992) performed immunological and neuropsychological evaluations of 13 patients meeting Cullen's MCS criteria who had no psychiatric history and had been in good health prior to a chemical exposure. No significant immunological abnormalities were found. Simon et al. (1993) compared 41 patients who had a recorded diagnosis of MCS with 34 patients with musculoskeletal/back injuries. No significant differences were seen between patients and controls on blinded specimens analyzed for T cells (total, CD4, CD8), B cells, CD25 (IL-2R$^+$), or autoantibodies against parietal cells, mitochondria, smooth muscle, brush border, or nuclear components. Lower interleukin-1 (IL-1) generation among patients was interpreted as probably being due to laboratory error. Subsequently, Simon acknowledged that unpublished data on sequential samples from the same patients submitted to test the reliability of the measurements showed that the data was not reproducible and "basically random," casting doubt upon the validity of the study's reported results (Simon 1994b).

McConnachie and Zahalsky reported that halogenated aromatic hydrocarbons, including chlordane, heptachlor, and pentachlorophenol, may cause T-cell activation and depressed lymphocyte responses to mitogens (McConnachie and Zahalsky 1991, 1992). Some abnormalities were detectable up to ten years after exposure to chlordane. Among 12 persons exposed to the pesticide chlorpyrifos, elevated CD26 (activated T) cells ($p < 0.01$) and increased numbers of various autoimmune antibodies were seen, compared to two control groups (Thrasher et al. 1993). The authors described those exposed to chlorpyrifos as having a "high rate of atopy and antibiotic sensitivities."

In a series of 289 individuals exposed to a variety of chemicals in computer manufacturing operations, including phthalic anhydride, formaldehyde, isocyanide, trimellitic anhydride, and aliphatic and aromatic hydrocarbons, 23.1 percent (versus 6.7 percent of controls) had T-lymphocyte helper/suppressor ratios less than 1.0, and 14.9 percent (versus 3.3 percent) had ratios greater than 2.5 (Vojdani et al. 1992b). Further testing in those with abnormal helper/suppressor ratios revealed lymphocyte activation, decreased lymphocyte responses to T- and B-cell mitogens, and increased autoantibodies directed against formaldehyde, trimellitic anhydride, phthalic anhydride, and benzene (haptens), compared to controls. The pathological significance of such antibodies, if any, is unclear. For example, workers exposed to toluene diisocyanate who have no evidence of disease nevertheless may have antibodies against toluene diisocyanate. In contrast, a number of workplace chemicals (colophony, plicatic acid) are widely recognized as causing specific respiratory sensitization despite the fact that no immunological mechanism has been demonstrated (Bernstein 1996).

Cullen and Redlich (1995) summarized the concerns over immunological testing in MCS: "So-called immunological tests, such as serum antibodies or patterns of leukocyte surface markers, remain of undetermined relation to human health, as are concentrations of most nutrients and antioxidant enzymes. Such tests may be used as biomarkers of MCS or similar idiosyncratic disorder only when clear distinction is shown between affected and unaffected people."

Finally, it deserves repeating that one of the reasons that no consistent immunologic abnormalities (or markers) are found in MCS patients may be the fact that MCS is a class of disease, the subsets of which may have different origins or pathophysiology.

CHAPTER 9

Reviews, Commentaries, and Polemics

Aside from the original research since the first edition that has advanced our understanding of the origins, mechanisms, and patient characteristics relevant to MCS (reviewed above), a number of interpretive commentaries have appeared, some without peer review. We discuss them in this chapter. Not reviewed here are the many articles on MCS that have appeared in the popular press that reflect the full range of opinion concerning MCS.

Scientific investigation related to chemical sensitivity is being stymied by scientists and physicians with financial conflicts of interest (e.g., those working for the chemical industry and those acting as defense expert witnesses in legal cases on MCS) who serve on government panels, editorial review boards, and grant review committees. These conflicts generally remain undisclosed; yet, from an ethical standpoint, they merit no less scrutiny than situations in which physicians obtain money from pharmaceutical companies for testing their products. In the latter instances, those presenting results at professional and governmental meetings, publishing in journals, or participating in grant proposal review committees generally must disclose potential conflicts, which in some cases would restrict participation. No such protections against conflict of interest have been afforded to MCS patients or researchers of chemical sensitivity. However, for both original scientific work and interpretative commentaries, whether published in scientific peer-reviewed journals or in other venues, there has been an increased call for fuller disclosure of possible bias on the part of the authors (Landrigan 1995). Sources of possible bias are not limited to financial support for the published work, but include financial ties in general, involvement with litigation, and disciplinary orientation, to name just a few situations that provide incentives that could possibly bias the work (Ashford 1988 and 1995). We argue that readers are entitled to information regarding possible bias so that they can appropriately judge for themselves the implications and the conclusions reached and thus put the work into proper perspective.

Published Work in Scientific Journals

The observations of, and theories posited for, low-level chemical sensitivity have not fared well when viewed from the perspective of traditional toxicology. A publication entitled "The Science of Toxicology and Its Relevance to MCS" (Waddell 1993) is worth reviewing. The article begins by referring to Paracelsus who, in the early sixteenth century, argued that the degree of adverse effect is proportional to the dose of the chemical. This is the origin of the "dose–response" concept, sacrosanct in modern toxicology, and the concept that gives rise to the oft-quoted axiom that "the dose makes the poison." Waddell endorses the use of double-blind, placebo-controlled challenges to test the "cause" of MCS, but acknowledges the difficulty of performing them with substances that have strong taste or odor. In doing so, he overlooks the subtleties of the word "cause" with respect to MCS. We, and others, have sought to distinguish possible "initiators" of the condition from potential "triggers" in an already affected individual. When the author bemoans the assertion that everything under the sun is reported to "cause" MCS, he is referring to the triggering of symptoms in persons already chemically sensitive.

Waddell further states that the operational definition of chemical sensitivity that we offered in our first edition (on page 29) "may appear on the surface to be more reasonable," but then he incorrectly implies that we say that the symptoms that appear on repeated, blinded, placebo-controlled challenges can be different in response to the same incitant. Our expectations are in fact that the same incitant in an appropriately unmasked subject would elicit the same symptoms upon each rechallenge. However, if the subject is not unmasked or deadapted, this may not be observed. We do believe that the dose makes the poison, however, we would refine Paracelsus's axiom as "the dose plus the host makes the poison," allowing for the observation that, as with allergy, not everyone is equally susceptible to every substance.

The field of allergy focuses on a subset of individuals who exhibit IgE-mediated responses. These responses often can be muted, if not eliminated, with allergy shots (immunotherapy). With chemical sensitivity, it seems that the processes of adaptation and deadaptation can result in a constantly changing host, such that the subject under scrutiny is a moving target. That this should frustrate traditional toxicologists is understandable. But it in no way establishes that chemical sensitivity "is contradictory to the fundamental principles of toxicology" as Waddell and other MCS critics claim (see also Gots 1995).

In the same year that Waddell published his view that MCS fails to conform with toxicological principles, two editorials were published recommending a variety of approaches for characterizing and understanding patients with MCS (Kilburn 1993b; Richter 1993). Both called for properly designed and blinded challenge studies to objectively test for chemical sensitivity in specific patients.

One reason MCS has been said to violate accepted principles of toxicology is that it does not appear to follow a conventional dose–response relationship. For this and other reasons, critics allege that there is no plausible causal relationship between chemical exposures and the patients' symptoms. Waddell correctly observed that "humans have a desire to assign a cause for everything," and that the history of humankind is filled with examples of humans' attempts to assign cause to every event, particularly to illness, misfortune, and death. A key question is, how do we distinguish between a chance association and true cause-and-effect? In an article entitled "Chemical Sensitivity: Symptom, Syndrome or Mechanism for Disease?" Miller (1996a) recounted the nine criteria, offered by Sir Austin Bradford Hill (1965), that have been widely used by epidemiologists to help them make this distinction:

1. Strength of the association, that is, between the exposure and the illness. For example, in 1775 Percival Pott observed an enormous, perhaps 200-fold, increase in scrotal cancer among chimney sweeps versus workers not exposed to tar or mineral oils, a strong association indeed. However, Hill cautions, we should not be too ready to dismiss a cause-and-effect hypothesis merely on the grounds that the observed association appears slight because there are many instances in medicine in which this occurs, yet a cause-and-effect relationship exists. For example, only small percentage of those who harbor meningococcus develop meningitis from it, and only a tiny minority of those who are stung by bees develop anaphylaxis. Analogously, only a small percentage of those exposed to certain pesticides or a sick building appear to develop MCS.

2. Consistency. Have different people in different places and times observed the association? Hill considers this especially important for rare hazards or conditions. With regard to chemical sensitivity, a num-

ber of observers have independently described chemical sensitivity arising in persons exposed to organophosphate pesticides (Cone and Sult 1992; Miller and Mitzel 1995; Rosenthal and Cameron 1991, Sherman 1995).

3. Specificity of the association. The more the association is limited to specific exposures and/or to specific types of disease, the clearer the case for causation. Research on inducing exposures for MCS might reveal strong, specific associations. With respect to triggering, at first blush there might appear to be a lack of specificity in terms of both exposures and symptoms. However, individual MCS patients report specific symptoms with specific exposures. Unlike cases of cancer or heart disease, cause-and-effect for symptom triggering in MCS can be tested experimentally in humans, providing direct experimental measurement of the specificity of the association (if it exists), the strongest form of evidence possible for an environmentally related illness.

4. Temporality. Does the exposure precede the illness? Some critics have noted that for many MCS patients depression or somatoform tendencies preceded their "initiating" exposure event. However, large numbers of MCS patients show no evidence of prior psychopathology. Perhaps the strongest evidence for temporality is the temporal cohesiveness between exposure and onset of symptoms observed in large exposure groups, for example, the Environmental Protection Agency's sick building occupants and sick Gulf War veterans, many of whom report new-onset intolerances and have no evidence of psychiatric problems predating their exposure.

5. Biological gradient. An association that follows a biological gradient or dose–response curve strongly suggests causality. Hill acknowledges that it is frequently difficult to obtain a satisfactory measure of exposure. However, a dose–response relationship similar to that inferred for allergic conditions (Waddell 1993) may also pertain to chemical sensitivity: There is a dose–response relationship for the first, sensitizing exposure in a susceptible individual; with subsequent exposures, the now sensitized person also responds in proportion to the dose, but at a much lower dose level than most people. Also suggestive of a dose–response relationship is the observation by many MCS patients that the longer they remain in an exposure situation, the more severe their symptoms become, and the longer they persist. Again, in contrast to cancer or other environmentally related disease, the triggering phase of chemical sensitivity lends itself to direct human testing of a dose–response relationship, thus obviating the need for speculation about a biological gradient.

6. Plausibility. Hill comments that it is helpful if the causation we suspect

is biologically plausible, but that what is plausible depends upon the biological knowledge of the time: "In short, the association we observe may be new to science or medicine and we must not dismiss it too light-heartedly as just too odd." In fact, there are some medical conditions that have features strikingly similar to MCS and are well-accepted, for example, reactive airways dysfunction syndrome (discussed previously) and multiple drug allergy syndrome (Sullivan 1991). These parallel clinical observations may be signs pointing in the direction of biological plausibility for MCS.

7. Coherence. The cause-and-effect relationship under scrutiny should not conflict with other generally known facts about the disease, for example, the pathology or biochemistry of the illness. As little research on MCS has been done, so far this has not been a problem.

8. Experiment. Experimental evidence can provide the strongest support for a cause-and-effect relationship. Perhaps one of the reasons why MCS patients are so dogged in their insistence that chemicals are causing their symptoms is the strength of the experimental evidence they perceive when they deliberately avoid and then are reexposed to incitants. Part of the appeal of MCS, at least to some environmental scientists, is that it poses an experimentally testable hypothesis, via direct human challenge studies, in contrast with most other environmentally related illnesses of major concern, such as environmentally induced cancers. Of course, experimental conditions must be optimized, that is, there must be unmasking in an environmentally controlled hospital unit, if the most robust effect is to be seen. Currently, the only obstacle to the undertaking of these studies is lack of funding.

9. Analogy. Under certain circumstances, cause-and-effect can be inferred by analogy. The sensitivities reported by MCS patients are reminiscent of the heightened sensitivity to tobacco smoke reported by those who have recently quit smoking. Likewise, there are close parallels between MCS and addiction, in which caffeine and food cravings, as well as binging, are also reported. MCS patients describe "going through withdrawal" or "detox," during which the symptoms they report are reminiscent of those reported by drug abusers, yet most MCS patients systematically avoid even mildly addictive substances. Other possible analogues to MCS are reactive airways dysfunction syndrome (RADS) and toluene diisocyanate (TDI) sensitivity, particularly the former, in which a single major exposure may lead to airway hyperresponsiveness to multiple, chemically unrelated inhalants. We must ask ourselves if the airways can develop heightened sensitivity to multiple chemicals in this way, by analogy, why could not the central nervous system do so as well?

To Hill's criteria, Miller would add a tenth criterion, one that would apply to symptoms (or illnesses) that are primarily subjective in nature:

10. Unique symptomatology. The more obscure or unique a symptom is, particularly if it is reported by several independent exposure groups (e.g., industrial workers, white collar professionals, Gulf War veterans), the greater the likelihood of causation. For MCS, it would be difficult to imagine that the curious symptom of odor intolerance, which has been reported by demographically diverse groups following various exposure events, could be "invented" by all of them. Equally unexpected and counterintuitive are MCS patients' practices of avoiding fragrances, foods, alcoholic beverages, and other substances that they formerly relished. Why would people who really liked pizza, chocolate, and beer give these things up unless they made the subjects ill? Why would a mechanic who loved his job, and used to think that WD-40 would make a wonderful perfume, suddenly report that odors at work made him ill if, in fact, they did not? Why would doctors, lawyers, teachers, and others say they quit their professions because of severe mental confusion around fragrances and engine exhaust if this were not the case? Scientifically, it would be absurd to dismiss such eccentric behaviors in otherwise sane individuals without searching exhaustively for a plausible biological basis.

Hill suggests that his criteria should be used to "study association before we cry causation." He further cautions that none of the criteria indisputably revokes a cause-and-effect hypothesis, and none is a sine qua non. In the aggregate, these criteria assist us in determining causation. As discussed earlier, MCS appears to involve two steps: (1) induction by a major or repeated exposure, and (2) subsequent triggering of symptoms by chemically unrelated, low-level exposures. Each of these two steps requires causal validation. Validation of the second step lends itself to direct experimental testing. Validation of the first may rest upon epidemiological investigations and animal studies.

We earlier emphasized the difficulties and limitations in using traditional epidemiologic approaches for clarifying chemical sensitivity (see discussions in Chapters 1 and 2). [See also "Some Preliminary Thoughts on the Potential Contribution of Epidemiology to the Question of Multiple Chemical Sensitivity" (Neutra 1994).] Some of these problems become less serious when observations focus on event- or exposure-driven research, but they are not all resolved. In a thoughtful commentary reviewing three published Canadian "outbreaks of concern" involving nonspecific symptoms or hazardous exposures in a community or a workplace (Cole et al. 1996), the authors observe that "[t]he nature of environmental and occupational health outcomes precludes their fitting into traditional medical models of disease causation." Cole et al. state:

> We advocate expansion of the disciplinary base, the research methods, and the contextual framing of studies of communities or work places in which health concerns exist. We understand that trade-offs may result between acceptance of such evidence along traditional epidemiological lines, but we maintain that traditional approaches are inadequate for the task of understanding, and they may forestall better conflict resolution in outbreaks of concerns.

These authors recognize the need for taking preventive action even in the face of inconclusive scientific findings. [See also the accompanying editorial by the editor-in-chief of the Archives of Environmental Health, which, although generally supportive of this view, also argues that subjective reporting of symptoms and "feeling" states can and should be verified by neurobehavioral performance tests in the absence of objective physical signs (Kilburn 1996).]

Because MCS does not appear to be readily explained by, or consistent with, simplified views of traditional toxicology, epidemiology, and disease causation, much attention has focused on the possibility that the condition is psychogenic. A particular useful critical review of the post-1980 literature on MCS, focusing on human studies with original data that address psychogenic origins implicitly or explicitly, was provided by Davidoff and Fogarty (1994). Ten papers were reviewed: Black et al. 1990; Brodsky 1983; Doty et al. 1988; Rosenberg et al. 1990; Schottenfeld and Cullen 1985; Simon et al. 1990; Staudenmayer and Selner 1990; Stewart and Raskin 1985; Terr 1986, 1989. Six of the ten studies used a descriptive case history design and no control group, three studies used a case-control design, and one study used a quasi-experimental design. According to Davidoff and Fogarty (1994), there were many problems with these studies related to sample selection, measurement, and study design:

> Five sample selection problems were identified, including small sample size, and nonrepresentative sample sources (those with obvious biases, such as psychiatric referrals or workers' compensation claimants) . . . . Seven measurement problems were identified, including no control for investigator bias and insufficient information about the assessment instruments . . . . Three study design problems were identified including inadequate controls for a chronic illness explanation of symptoms and conclusions about cause-and-effect in the absence of both a testable hypothesis about cause-and-effect and a study design capable of elucidating cause-and-effect relationships . . . . Only one study reviewed (Doty et al. 1988) was judged to have fewer than eight methodologic problems . . . .

The identical test findings used by Simon et al. 1990 to support a psychogenic hypothesis could have been used to support a biogenic hypothesis. The tests are unable to distinguish between the two hypotheses . . . . Study design problems were judged to be prominent in 9 of the 10 articles reviewed, the exception being Doty et al. 1988. [A]lthough Black et al. 1990 and Simon et al. 1990 reported their goals were to explore associations between MCS syndrome and psychiatric disorders, the conclusions of both groups suggested the study data supported the thesis that MCS syndrome was attributable to "psychological vulnerability" or psychopathology. [*Authors' note:* In a later publication, Simon et al. 1993, greater care is taken not to confuse psychological symptoms with psychogenic origin of MCS. See the discussion of this work in Chapter 8.]

This review suggests that existing research studies considered widely supportive of a psychogenic origin for MCS syndrome have serious methodological flaws . . . . [T]he nature, timing, and origin of these [psychiatric symptoms] . . . can be explained plausibly by at least six competing hypotheses . . . . [S]tudies must be designed to differentiate between [these] competing hypotheses . . . .

The review by Davidoff and Fogarty is notable for its clarity of thinking and analysis, absent in many other commentaries. A journal recognized for its predominantly industrial orientation, *The Journal of Occupational and Environmental Medicine,* chose this article to feature in its "Selected Review of the Literature" forum, and gave the paper an extraordinarily positive endorsement: "[t]he authors . . . convey full awareness of widespread skepticism directed towards patients exhibiting characteristics of multiple chemical sensitivities syndrome (MCS) and to supporters of MCS as a distinct diagnostic entity. Their rigorous methodology, application of diagnostic criteria, and careful analyses may serve to convert some hardened nonbelievers at least into hardy agnostics" (Wittmer 1996).

In the same year, a less analytic two-part review of MCS continued some of the confused thinking characteristic of earlier papers (Sparks et al. 1994a and b). The first part, entitled "I. Case Definition, Theories of Pathogenesis, and Research Needs," reviewed most of the recent work relevant to both biological/physical and psychogenic theories for MCS (except the review by Davidoff and Fogarty). Despite the title, this review treats the issue of case definitions superficially. At first, the authors are appropriately circumspect about drawing definitive conclusions from the limited data available on MCS. For example:

The available evidence shows that patients diagnosed with MCS are very heterogeneous and that more than one causal mechanism may be operative in different cases.

The illness belief/behavior theory of MCS arises from historical and sociologic evidence and clinical observations but has not been subjected to controlled investigations.

However, the authors use the ambiguous term "psychophysiologic" to describe the possibility that MCS could have a biological or physical origin yet manifest itself in psychological symptoms. Rather than adding clarity, this term is likely to confuse the uninitiated reader, who could easily interpret the prefix "psycho" as referring to causes rather than symptoms. Furthermore, as in other publications, the word "cause" is used ambiguously, sometimes to refer to the initiation of chemical sensitivity and sometimes to the triggering of symptoms. That this ambiguity should have continued in 1994 is disappointing.

The authors are appropriately open minded about possible neuroinflammatory mechanisms, limbic system kindling, time-dependent sensitization, and odor conditioning. However, there is a tug of war throughout the first part of this paper between not overstating the evidence for any particular mechanism and where the authors ultimately end up in the second part of their paper—indicating a strong preference for psychogenic origins (Sparks et al. 1994b). The inconsistency between the two parts of the paper is evident in the statement made in Part One that "[b]ecause none of the above views of etiology is universally accepted on the basis of substantial scientific evidence, dogmatic adherence to one of them is unwise as a basis for managing patients with an MCS diagnosis." Yet the authors violate their own admonition in Part Two of this paper (see Miller 1995) by recommending psychiatric or psychological interventions as the only appropriate course of treatment.

Although the authors agree with the recommendation made in the first edition of this book, and endorsed in the NAS workshop, the AOEC workshop, and the New Jersey workshop, that an environmental medical unit be used to conduct research using double-blinded, placebo-controlled challenges, they introduce some confusion by calling the unit a "specially constructed environmental challenge chamber" (see the discussion of the differences between an environmental medical unit and a conventional challenge chamber in Chapter 10).

Worth analyzing for its flawed arguments is a paper by Ronald Gots (1996). Gots is affiliated with the International Center for Toxicology and Medicine, the Environmental Sensitivities Research Institute,[1] a so-called

---

[1]ESRI is a corporate-supported entity with an "Enterprise Membership" fee of $10,000 per year. Board members include DowElanco; Monsanto; Procter and Gamble; the Cosmetic, Toiletry and Fragrance Association; and other companies and trade associations involved in the manufacture of pharmaceuticals, pesticides, and other chemicals. An ESRI advertorial, published in newspapers around the country, is reproduced below:

Multiple Chemical Sensitivity: Fear of Risk or Fact of Life?

(NU)—Scientists are increasingly concerned that a doubtful new diagnosis—supposedly caused by everything "man made" in the environment—is unnecessarily making thousands of Americans miserable each year. One of these so called "modern diseases" is called MCS, for Multiple Chemical Sensitivities. Many established sci-

(*continued*)

independent entity that is funded by the chemical industry and trade associations, and the National Medical Advisory Service, which provides medical experts for corporate defendants. Although he correctly concludes in the body of his paper that psychogenic causes of disease should be considered when no plausible physiologic cause is known, Gots takes the logic one unjustifiable step further in the abstract, where he asserts that "[e]verything that is known about MCS to date strongly suggests behavioral and psychogenic explanations for symptoms." He relegates research on limbic kindling, time-dependent sensitization, and respiratory tract or olfactory sensitivity to the status of scientific curiosities that require behavioral mechanisms to tie them to symptomatology. Even if he were correct about the absence of physiological evidence (and he is not), the presence of psychological problems in patients is not proof of psychological causation. The work of Fiedler et al. (1992) and that of Simon et al. (1990, 1993) amply demonstrate that there are MCS patients with no premorbid or subsequent psychological problems.

Gots dismisses, as entirely unreliable, patients' reports of symptoms that "provide a completely subjective catalogue of chemicals and doses, one which is ever changing, one which cannot be uncovered or tested, and one which therefore has no limits." There is, in fact, statistically significant evidence of a strong correlation between patients' subjective symptoms and objective measures in a population suffering in sick buildings (Wyon 1992). Of course, research using double-blinded challenges in an EMU would settle the question once and for all for MCS patients.

Gots leaves one small window open, admitting that patients could possibly be responding to the odors of chemicals. Ignoring the mounting scholarship on effects of low levels of volatile chemical substances on the brain, he asserts that "[t]his responsiveness to smell, of course, speaks to the psychogenic nature of this phenomenon, since there is no relationship between odor and actual toxicity." He quotes Barsky and Borus (1995), who view "functional somatic syndromes" as including "chronic fatigue syndrome, total allergy syndrome [MCS], food hypersensitivity, reactive hypoglycemia, systemic yeast infection . . . , Gulf War syndrome, fibromyalgia, sick building syndrome, and mitral valve prolapse." Sick building syndrome, for one, is now generally accepted as a physical disorder (Mølhave et al. 1993; NRC 1992b) although the mechanisms still are not well understood. Notably it was once called "mass psychogenic illness" by the National Institute for Occupational Safety and Health (Baker 1989;

---

entists and physicians doubt MCS actually exists; it exists only because a patient believes it does and because a doctor validates that belief. For information on MCS, write the Environmental Sensitivities Research Institute, 6001 Montrose Road, Suite 400, North Bethesda, MD 20852. (News USA 1996)

Kreiss 1989) and still is by some researchers in the United Kingdom (Wessely 1992).

In order to justify his criticism of chemical avoidance practices, Gots lumps together as "deleterious" both "avoidance techniques and sauna treatment." This journalistic device needs to be appreciated for what it is: a joining of potentially useful and questionable practices under one label, similar to lumping all unexplained disorders under the heading of "functional somatic syndromes" (Barsky and Borus 1995). We take issue with such criticism, which would bar the use of judicious trials of avoidance in selected patients, especially in the face of the mounting, though not yet definitive, evidence for a physiological origin for MCS—and in the absence of convincing evidence that all MCS is psychogenic. The papers published since the first edition of this book that purport to "prove" psychological causes either simply document psychological symptoms (which are not necessarily psychogenic) in some, but not all, MCS patients (Simon et al. 1990, 1993), make unsupported claims concerning the efficacy of psychological interventions (Staudenmayer et al. 1993b) (see discussion of "Psychological Therapies" in Chapter 10), or are, for the most part, recycled opinion (Staudenmayer 1996, 1997; Gots 1995; 1996).

Both Gots and Staudenmayer appear dismissive of many medical conditions such as sick building syndrome that are as yet unexplained by dominant medical or scientific theory. In contrast, in 1992, the National Research Council (NRC) Subcommittee on Immunotoxicity observed (NRC 1992b):

> Some authorities suggest that SBS [sick building syndrome] is of psychologic origin or that it represents an anxiety state, mass hysteria, or a conditioned reflex. Investigators at the Karolinska Institutet in Sweden have demonstrated, however, that blinded passers-by exposed to a mobile breathing chamber linked to the air supply of a "sick building" experienced the same reactions to building air as had the building's inhabitants. [Reference was to Berglund et al. 1984 in NRC, p. 132]
>
> The sick building syndrome is a real phenomenon, in which susceptible occupants of closed buildings have symptoms of headaches, eye and nasal irritation, mucous membrane irritation, lethargy, and difficulty with concentration. A role for VOCs [volatile organic compounds] in the etiology is suggested, and the hypothesis that this syndrome is solely of psychologic origin is not consistent with existing data. (NRC, 1992b p. 138)

Other Writing and Publications

Already documented in earlier parts of this book is the battle waged by warring factions of the medical community, principally the traditional allergists and the clinical ecologists. The American Medical Association (AMA) initially was heavily influenced by the position paper of the American College of Physicians (ACOP) on Clinical Ecology (1989), and it briefly weighed in on the side of the allergists. However, a few months later, after becoming aware of the NAS workshop on multiple chemical sensitivity, a Canadian MCS workshop, and the first edition of this book, as well as recent review articles, the AMA retracted its earlier critical assessment and in December 1991 retreated to a more agnostic position, stating:

> The fact that the diagnostic tests and therapy recommended by clinical ecologists are largely unproven by controlled clinical trials does not necessarily establish the lack of scientific validity. Well-controlled studies could validate and prove a scientific basis for many of the tests and therapies associated with multiple chemical sensitivity. (Estes et al. 1992)

The AMA went on to say that "[u]ntil such accurate, reproducible and well-controlled studies are available, the AMA Council on Scientific Affairs believes that multiple chemical sensitivity should not be considered a recognized clinical syndrome." The council recommended that "the AMA continue to monitor the published literature on clinical ecology and report as appropriate." The AMA made a subtle, but crucial distinction between the lack of proven efficacy of certain therapeutic approaches and evidence for MCS itself. The American College of Physicians does not have a position statement on MCS, although it issued one on Clinical Ecology in 1989. The California Medical Association withdrew its position statement on Clinical Ecology in 1993.

In contrast to the AMA's cautious position, the American Council on Science and Health, an organization often at odds with environmental scientists and occupational physicians concerned about preventing occupational and environmental disease, especially cancer, produced an informational booklet on MCS (American Council on Science and Health 1994). The booklet essentially equates MCS with "junk science" and is one-sided, quoting selectively and out of context from various documents, including the AMA's published position; and it indicts by innuendo those who advocate research to clarify the nature of the condition. The publication also perpetuates the illogical notion that the current lack of proven therapies proves the nonexistence of the condition. Notably absent is any reference to the neuroscience literature on MCS, ignoring one plausible neurological mechanism for chemical sensitivity.

The 1996 Berlin Workshop

In 1995, the European research team that conducted the nine-country MCS exploratory study described in Chapter 7 urged:

Perhaps the most immediate need is for the convening of a workshop in early 1996 of a small number (30–50) of invited participants from Europe and North America to discuss the experience and evidence related to chemical sensitivity to date, and to make recommendations for further research. Invited participants should include knowledgeable researchers, practitioners, governmental authorities, and policy makers. *Both proponents and critics of the condition should be included.* The workshop format should allow for presentations, discussion, dialogue, and challenge of views in a structured, focused, and constructive way. To the extent possible, the workshop should help participants resolve differences and agree on research priorities. (Ashford et al. 1995, emphasis added)

This recommendation to the European Commission was not acted upon. Instead, with prompting from German scientists and government officials, the International Programme on Chemical Safety (IPCS) held an invited conference in Berlin in February 1996 in collaboration with three German agencies—the German Federal Ministry of Health, the Federal Institute for Health Protection of Consumers and Veterinary Medicine (BGVV), and the Federal Environmental Agency (UBA). IPCS is jointly supported by the United Nations Environmental Programme (UNEP), the International Labour Organization (ILO), and the World Health Organization (WHO). The workshop included presentations by invited participants from Europe, the United States, and Canada, and was also attended by representatives of various so-called nongovernmental organization (NGOs). The conference process was compromised from the beginning, with the four invited NGO representatives all coming from industrial NGOs and none from environmental, consumer, labor, or other publicly based groups. No representatives of MCS patients, environmental groups, or labor unions were present at the workshop. The purposes of the three-day meeting were to review information on MCS to determine whether MCS constitutes a syndrome; to examine relationships with other environmental illnesses; to identify possible etiological factors; and to discuss diagnoses, diagnostic testing, differential diagnoses, and approaches to treatment. Ad hoc panels that included the NGO representatives from the four transnational corporations with potential interests at stake met to review issues and reported back to the workshop as a whole. The four "NGO representatives" were full-time employees of BASF, Bayer, Monsanto, and Coca Cola, the first three of which claimed affiliation with an industry-funded science institute (the European Centre for Environment and Toxicology).

A U.S. industry consultant, Ronald Gots (the director of the corporate-financed Environmental Sensitivities Research Institute mentioned above), who has been vocal in insisting that MCS is only a mental disorder (Gots 1996), not only was a full participant in this international meeting but was also invited to present the "U.S. perspective" on MCS even though he has not published any original peer-reviewed research on MCS. (Ronald Gots is not only the director of an anti-MCS "research institute" but, as noted above, also directs the National Medical Advisory Service, which provides medical experts to corporate defendants involved in litigation over MCS.)

During the meeting, the corporate NGOs and various outside "observers"—from German federal environmental and health ministries— joined invited participants in the discussion, participated in the drafting of recommendations, and were present during the voting on those recommendations (because votes were not recorded there is some lingering uncertainty as to who actually voted). Some of the German federal observers were directly involved in a major lawsuit concerning "wood preservative syndrome" (attributed to pentachlorophenol). (For a discussion of the wood preservative problem, see Ashford et al. 1995.) One "NGO" representative was an employee of Bayer, which is one of the owners of DESOWAG, a defendant in the wood preservative lawsuit. Another "NGO" representative who was present at the meeting has been openly critical of physicians who assist MCS patients.

This hardly impartial group of observers and participants recommended in their report that MCS be renamed "idiopathic environmental intolerance." Initially unaware of the compromised process, the chair of the workshop, Dr. Howard Kipen, as well as other participants and scientists, registered their objections to the process after the close of the meeting (Abrams et al. 1996). Howard Kipen and Claudia Miller, in separate letters to the IPCS, have insisted that their objections to the term "idiopathic environmental intolerance" (IEI) be included in the report if it is ever published. Soon after the Berlin meeting, certain workshop participants reported to the media and at scientific meetings that the "idiopathic" in IEI meant "self-originated," rather than "being of unknown etiology" (a more familiar meaning of the term as it is used in medicine)—and they erroneously proclaimed that IEI had become WHO's official name for the condition, and that the new name should replace the term "MCS." Soon after receiving a letter of protest from 80 prominent United States scientists and physicians concerning the Berlin meeting (see below), IPCS clarified the status of the IEI name by issuing a notice stating that WHO had "neither adopted nor endorsed a policy or scientific opinion on MCS." In response to these objections, IPCS has limited the distribution of the final "unedited" report. In its current form, the report now contains several disclaimers, for example, that the document does not necessarily represent the decisions or stated policy of UNEP, ILO, or WHO; that it does not constitute a formal publication; and that it should not be reviewed, abstracted, or quoted without the written permission of the Director of the IPCS.

Even though review of, abstraction from, or quotation from the report was prohibited by IPCS because of objections registered by Kipen, Miller, and other scientists, the so-called consensus recommendations appeared in an "unauthored" paper in a recent issue of *Regulatory Toxicology and Pharmacology* (Anonymous 1996), further corrupting a process that was flawed from its conception. Notably, this paper was published in a special issue of *Regulatory Toxicology and Pharmacology* along with the proceedings of a "State of the Science Symposium" on MCS that was co-sponsored by Gots' National Medical Advisory Service in 1995—well before the Berlin Workshop. The journal acknowledges that the supplement containing the Berlin recommendations was "made possible" through a grant from [Gots'] Environmental Sensitivities Research Institute, "a charitable, non-profit, scientific, and educational organization dedicated to the open exchange of objective scientific information and data among physicians, scientists, industry, the government, and the general public" (see earlier footnote concerning ESRI). In a letter dated April 17, 1996 (Abrams et al. 1996) to the IPCS conference sponsors (the World Health Organization, the International Labour Organization, and the United Nations Environmental Programme) objecting to the MCS workshop as well as prior IPCS activities involving chrysotile asbestos, 80 prominent and predominantly academic and independent scientists and physicians (including the former directors of the U.S. National Institute of Environmental Health Sciences, the National Institute for Occupational Safety and Health, and the National Cancer Institute), none of them clinical ecologists, urged that IPCS and its U.N. sponsors:

1. Immediately halt work on the IPCS chrysotile asbestos criteria document, and conduct an expert, impartial review of whether it serves any useful purpose to complete it.

2. Immediately halt work toward the issuance of reports on MCS. They should reconstitute the MCS panel exclusively with scientists who have published research (not merely opinion) on chemical sensitivity and related scientific issues and with doctors who have actual experience in managing patients with MCS. The IPCS and its U.N. sponsors should also assign NGO places to legitimate NGOs, and, to the fullest extent possible, should identify and exclude scientists with financial conflicts of interest.

3. Have ICPS report publicly on the complete extent of conflicts of interest of all members of IPCS expert scientific panels (i.e., participants, representatives, and observers) on all projects now under way, pursuant to IPCS's own Programme Advisory Committee recommendation that there be "open declarations of both professional and personal interests related to the issues under evaluation in order to avoid . . . conflicts of interest."

The IPCS Director, joined by UNEP and ILO, responded in a published letter (Mercier et al. 1996) that simply categorically denied improper influence, asserting that "[t]here are no instances where any final reports of the IPCS were biased by industry or other NGOs" and "[t]he fact that there have been no improper influences must be readily apparent to and understood by anyone objectively viewing the process." Obviously, there were persons at the conference, including the chairman, who were clearly of another opinion. Following this exchange of letters, there was a meeting of the IPCS Programme Advisory Committee in Halifax, Nova Scotia, October 23–25, 1996, at which the sponsoring agencies discussed this and other issues with the IPCS. One item raised was the possibility of convening a second workshop in 1997 in the United States. As we go to press, a formal response with regard to such a meeting is yet to come.

Commentary

Thus, as we look back over the past five years, instead of science and accumulated knowledge providing a rational basis for dialogue and consensus, the MCS debate has become even more acrimonious and acerbic—fueled by concerns for liability and regulation. Where five years ago critics of MCS at the NAS workshop nevertheless agreed upon a necessary and clear direction for research to clarify the origins and nature of chemical sensitivity, the most vocal critics now have essentially withdrawn their recommendations for it, arguing that mounting evidence does not justify further research. Fortunately, others, while still appropriately skeptical, increasingly advocate research to clarify the situation.

In the meantime the acrimonious debate has expanded to Europe where, as in North America, concerns for liability and more stringent regulation have spurred some industry representatives to compromise fair and full discussion of chemical sensitivity and to improperly influence international organizations whose charge is to foster honest and open discussion in the interest of protecting public safety, health, and the environment.

Even if these biases were not operating, established scientific and medical paradigms for physiology, biochemistry, and disease do not easily embrace new observations, data, and theory. Thomas Kuhn has written extensively about the painfulness of paradigm shifts in his *Structure of Scientific Revolutions* (Kuhn 1996). (An expansion of the Kuhn view, which has poignant relevance for chemical sensitivity, is provided below.) An example of a recent paradigm shift is illustrated by the now accepted view that peptic ulcers are caused by the bacterium *Helicobacter pylori* and can be effectively treated with antibiotics (Peterson 1991). The longstanding scientific and medical conviction that ulcers had their origin in stress died very hard. It took more than a decade for the new paradigm to topple the old, even though evidence of an infectious etiology has been available for a relatively long time (Gilovich and Savitsky 1996). The proponent of the now accepted view was initially derided and ridiculed, as have been many pioneers before him in medical science. Still, the arrogance of tenacious clinging to outdated views continues in many parts of medical science, aided now by the more politely dismissive label "junk science," replacing the earlier label "fraudulent." Critics of a physiologic basis for MCS accuse some physicians of unjustifiably reinforcing somaticizing patients' "belief systems" (Staudenmayer et al. 1993b), when in fact it is increasingly evident that it is the critics' intractable belief in dominant medical models that retards much-needed research and progress that could help us truly understand the condition. Blind adherence to old paradigms, coupled with vested financial interests, and the reputational consequences of rejecting prior positions are powerful incentives militating against change.

Stages of a Paradigm Shift (adapted from Kuhn 1996)

- Ignore departures from the existing paradigm.
- Deny that an anomaly exists; blame it on faulty observation or testing error. Deride the proponents of the new paradigm.
- Acknowledge the anomaly, but call it "idiopathic."
- Try to explain the anomaly with the existing paradigm, sometimes by making minor adjustments.
- Seek alternative paradigms to contradict or minimize the one proposed.
- Recognize the paradigm as valid, but within a narrow context relegated to "exceptions."
- Accept the new paradigm as offering some explanatory power, but retain the old paradigm too.
- Discredit the old paradigm; deride any attempt to reinstate or rehabilitate the old paradigm.
- Accept the new paradigm with enthusiasm.
- Begin again.

CHAPTER 10

Research and Medical Needs

Current Reflections on MCS

Below, we summarize points made in Chapter 7–9.

Phenomenologically, toxicant-induced chemical sensitivity appears to develop in two stages: (1) loss of tolerance following acute or chronic exposure to various environmental agents, such as pesticides, solvents, or contaminated air in a sick building; and (2) subsequent triggering of symptoms by extremely small quantities of previously tolerated chemicals, drugs, foods, and food/drug combinations. Masking may prevent both patients and their physicians from seeing more than the tip of MCS iceberg (Fig. 10-1). Although sensitivity to chemicals may be one of the consequences of this two-stage process, "chemical sensitivity" may not be the most appropriate term for describing the process itself.

There are two principal reasons for this. First, although "chemical sensitivity" certainly sounds like an inconvenient problem, the words fail to convey the potentially disabling nature of the condition and its postulated origins in a toxic exposure. Although some may balk at using the word "toxic" in this manner, numerous investigators from different geographic regions have published strikingly similar descriptions of individuals who report disabling illness following exposure to recognized environmental contaminants, albeit at levels not generally regarded as toxic (Ashford et al. 1995;

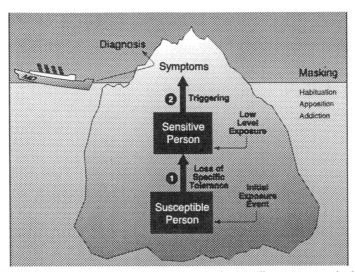

FIGURE 10-1. Phenomenology of toxicant-induced loss of tolerance. Illness appears to develop in two stages (depicted as the portion of the iceberg below the waterline, not visible to the physician): (1) loss of specific tolerance following acute or chronic exposure to various environmental agents, such as pesticides, solvents, or contaminated air in a sick building; and (2) subsequent triggering of symptoms by extremely small quantities of previously tolerated chemicals, drugs, foods, and food/drug combinations (e.g., traffic exhaust, fragrances, caffeine, alcohol). Physicians (in the Titanic) see only the patients' symptoms and formulate a diagnosis based upon them. Because of masking, both physicians and patients may not "see" more than the tip of the MCS iceberg, failing to observe that everyday, low-level exposures may be triggering symptoms. Even when such triggers are recognized, an initial exposure event that may have initiated loss of specific tolerance may go unnoticed or may not be linked to the patient's illness (UTHSCSA © 1996).

Cone and Sult 1992; Cullen 1987; Miller and Mitzel 1995; Rosenthal and Cameron 1991; Ziem 1992). Yet, for the *affected* individuals, the exposure appears to have been toxic.

Paracelsus aptly opined that dose makes the poison. However, as our knowledge has grown, it has become evident that "dose + host" makes the poison. Not everyone exposed in a sick building or to a chemical spill develops MCS. Thus, individual susceptibility, whether physiological (inherited or acquired) in origin or psychogenic, appears to play a role in determining who gets sick. The term "chemical sensitivity" fails to convey the key observation that chemical exposures appear to initiate a process that results in chemical sensitivity.

Another problem with the term "chemical sensitivity" is that it suggests that those afflicted become intolerant of chemical exposures only, when, in fact, caffeine, alcoholic beverages, various drugs, and foods reportedly trigger symptoms in these individuals once the process has been initiated (Bell et al. 1992, 1993c; Miller 1992; Miller and Mitzel 1995; Randolph and Moss 1980). Of course if one views the term "chemical" broadly enough, it could encompass such exposures as foods, alcoholic beverages, and caffeine—and we were careful to include these substances in our discussions earlier in this book—but it is not the popular understanding of the word "chemical." An alternative term, "toxicant-induced loss of tolerance" (TILT), has been proposed (Miller 1997). This term offers several advantages. First, it describes the process as it has been observed by clinicians and patients. Second, it allows for the possibility that various toxicants may initiate the process. Third, it does not limit the resulting intolerance to chemicals. Finally, it sharpens the focus of the current debate over chemical sensitivity by positing a theory of disease that can be subjected to objective testing. The possibility that psychologically traumatic events could also initiate loss of tolerance for various substances is not denied, but, in our review of the literature, the confirmed or anecdotal evidence for this is less robust than that for chemical agents.

Historically, new theories of disease have arisen when physicians observed patterns of illness that did not fit accepted explanations for disease in their time—for example, first the germ theory and subsequently the immune theory of disease. Likewise, the range of illnesses under discussion here does not conform to current, accepted explanations for disease or toxicity. Objections to accepting chemical sensitivity as an organic disease have included concerns that:

1. Too many different chemicals have been said to cause it.[1]
2. Patients report too many symptoms involving any and every organ system.
3. No confirmed physiological mechanism explains it.
4. No biomarker for it has been identified.
5. Total avoidance of chemicals is impractical.

Theories of disease attempt to explain what is going on inside the "black box," the patient, prior to overt illness, as illustrated in Schematic 10-a.

---

[1] The simple and simplifying word "cause" obscures the complex, at least two-step process thought to characterize chemical sensitivity. Initiating causes and triggering causes play very different roles in the process, just as cancer initiators and cancer promoters do in prevailing theories of cancer causation.

Schematic 10-a. Theories of disease. (UTHSCSA © 1996)

A theory of disease is a yet-to-be-established, general mechanism for a class or family of diseases. For the germ theory of disease, the boxes depicting the general mechanism of infection would look something like those in Schematic 10-b.

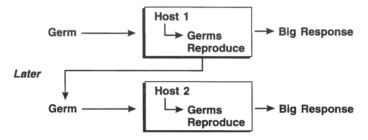

Schematic 10-b. Germ theory of disease. (UTHSCSA © 1996)

Note that:

1. Many different kinds of germs cause responses.
2. There are many different responses involving any and every organ system (skin, respiratory, gastrointestinal, neurological).
3. Specific mechanisms for different members of the disease class vary greatly, for example, cholera versus AIDS versus shingles.
4. There is no single biomarker; identification of specific germs took years.
5. Prevention (avoidance, antiseptics, sanitation, use of gloves) preceded knowledge of specific mechanisms.

For the immune theory of disease, the boxes might look like those in Schematic 10-c.

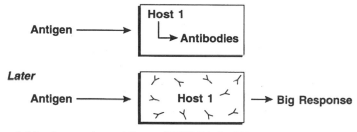

Schematic 10-c. Immune theory of disease. (UTHSCSA © 1996)

Here, as with the germ theory of disease:

1. Many different kinds of antigens cause responses.
2. There are many different responses involving any and every organ system (skin, respiratory, gastrointestinal, neurological).
3. Specific mechanisms for different members of the disease class vary greatly, for example, poison ivy versus allergic rhinitis versus serum sickness.
4. There is no single biomarker; identification of specific antibodies took years.
5. Prevention (avoidance, allergy shots) preceded knowledge of specific mechanisms.

For the two step-theory of carcinogenesis, the boxes might look like those in Schematic 10-d.

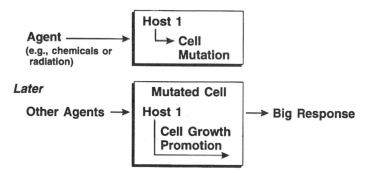

Schematic 10-d. Two-step theory of carcinogenesis. (UTHSCSA © 1996)

Here, as with the germ and immune theories of disease:

1. Many different kinds of agents may cause responses.
2. There are many different responses involving any and every organ system (skin, respiratory, gastrointestinal, neurological).
3. Specific mechanisms for different members of the disease class may vary greatly, vinyl chloride versus polynuclear aromatics versus radiation.
4. There is no single biomarker; biomarkers for different cancers are different.
5. Prevention (diets rich in certain vegetables or nutrients, avoidance of radiation and certain industrial chemicals) preceded knowledge of specific mechanisms.

For toxicant-induced loss of tolerance (TILT), the boxes might look like those in Schematic 10-e.

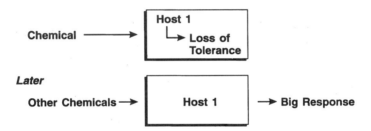

Schematic 10-e. TILT theory of disease. (UTHSCSA © 1996)

For toxicant-induced loss of tolerance, as for the theories of disease discussed above:

1. Many different kinds of chemicals may cause responses.
2. There may be many different responses involving any and every organ system.
3. Specific mechanisms may vary greatly.
4. It is conceivable that there is no single biomarker for response; identification of biomarkers may take years.
5. Prevention (avoidance of initiators or triggers) may precede knowledge of specific mechanisms.

Although the concept "loss of tolerance" may sound vague, in fact it is not. What these individuals report is a loss of specific tolerance to particular chemicals, foods, and drugs (Miller 1997). Note that this theory does not exclude the possibility that toxicant-induced loss of tolerance could turn out to be a special kind of toxicity or a variation on the immune theory of disease, just as allergy and delayed-type hypersensitivity are special cases that fall under the general classification of immunological disorders. Entertaining TILT as a possible new theory of disease facilitates a paradigmatic shift, one which challenges us to view the illnesses associated with chemical intolerance in a new framework—as a class of disorders, parallel to infectious diseases, immunologic diseases, and cancer. The TILT spectrum of disorders may be caused by a broad range of agents (solvents, pesticides, combustion products, implants), leading to a diverse spectrum of diseases (Fig. 10-2), some affecting only single organ systems, like migraine, asthma, and RADS, and others affecting many simultaneously, like MCS, chronic fatigue syndrome, and "Gulf War Syndrome."

Much effort has been devoted to developing a consensus case definition for chemical sensitivity (see discussion later in this chapter), with a singu-

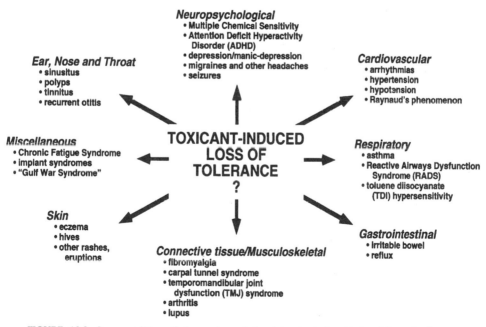

FIGURE 10-2. Some conditions that may have their origins in toxicant-induced loss of tolerance. (UTHSCSA © 1996)

lar lack of success. This lack of success would not be surprising if in fact what we were dealing with was a new class or family of disorders. Certainly, it would not be feasible to develop a single clinical case definition that would embrace all infectious diseases, all immunological diseases, or all cancers. Toxicant-induced loss of tolerance also may not lend itself to a case definition approach for study.

Theories of disease that withstand scientific scrutiny come along infrequently. The past century has witnessed the incorporation of the germ, immune, and cancer theories of disease into medical practice. Equating toxicant-induced loss of tolerance with any of these theories, each of which has been widely corroborated, would of course be premature. On the other hand, toxicant-induced loss of tolerance has certain earmarks of an emerging theory of disease, including vituperative disputes among physicians extending over several decades (Miller 1997).

Experimental Considerations and Approaches to MCS

For ethical reasons, the first stage of TILT (initiation) is more difficult to model in humans than the second stage (triggering). Ultimately, epidemiological studies and animal models may elucidate the first stage. Event-driven studies, for example, the opportunities presented by "natural experiments" such as the Gulf War, a pesticide spill, or a sick building, are enormously important here. Fortunately, the second stage readily lends itself to testing via direct human challenges, providing a potent form of scientific evidence through observation, which, unlike associations uncovered in epidemiological studies, can establish causation. However, in the design of human challenge studies in this area, certain key clinical observations need to be taken into account. First, the commonly reported biphasic (stimulatory and withdrawal-like) pattern of the patients' symptoms, particularly those symptoms involving the central nervous system, must be understood in order for investigators to perform meaningful test challenges on these patients (see the section on "Adaptation" in Chapter 2). Second, masking, described further in the next section, may "hide" responses to low-level chemical challenges and therefore should be minimized prior to testing. Controlling masking may be analogous to controlling background noise in studies on sound. Below we discuss further these clinical features, their incorporation in experimental designs, and how failure to incorporate them may weaken, if not negate, the outcome of challenge studies.

Chemically sensitive patients resemble drug addicts in one sense: members of both groups often report intense cravings and debilitating withdrawal symptoms. However, chemically sensitive patients' responses are not primarily to drugs. These individuals more commonly report addictions to caffeine or certain foods. The stimulatory and withdrawal symptoms reported by chemically sensitive patients are frequently identical to those reported by normal persons exposed to much greater amounts of the same substances. For example, after drinking one cup of coffee, chemically sensitive patients may report feeling "hyper," jittery, talkative, nervous, or anxious, or experiencing paniclike symptoms (stimulatory phase). Hours to days later, they may report withdrawal symptoms such as fatigue, yawning, confusion, indecisiveness, irritability, depression, loss of motivation, blurred vision, headaches, flulike symptoms, hot or cold spells, or heaviness in their arms and legs (withdrawal phase) (cf. Table 2-2). Similar symptoms occur during caffeine withdrawal among some low-to-moderate caffeine users in the general population (Silverman et al. 1992). Large numbers of chemically sensitive patients and many Gulf War veterans with unexplained illnesses report that one drink of an alcoholic beverage causes inebriation or a severe hangover (Miller 1992, 1994a; Miller and Mitzel 1995). These augmented responses suggest that those afflicted have lost their prior natural or native tolerance for low levels of many common environmental substances.

Early in their illness, prior to eliminating caffeine from their diets, many chemically sensitive patients report having consumed chocolate, coffee, tea, or cola addictively, often in large quantities (Miller 1992). Some carried huge cups of coffee or tea around with them wherever they went. Many report later having stopped use of all caffeine and xanthines, generally on the advice of a friend or a physician, and subsequently experiencing several days of intense withdrawal symptoms. Frequently they report that it was only after eliminating all xanthines from their diets that they were able to discern the effects of consuming a single cup of coffee or a chocolate bar. Most report becoming aware of the unpleasant effects of caffeine only after a trial of partial or complete caffeine avoidance. In this regard, chemically sensitive patients resemble certain reformed smokers or alcoholics who after quitting their addictants report extreme sensitivity to minute amounts of them. Terms such as "addiction," "withdrawal," and "detox" pepper the vocabulary of chemically sensitive patients. One described the condition as being "like drug abuse without any of the fun." These parallels to addiction provide perspective; they may help explain why the mechanisms that underlie chemical sensitivity have been difficult to define, and why biological markers have proved elusive.

Drug addiction and toxicant-induced loss of tolerance share a number of features in common. TILT also has features reminiscent of toxicity and allergy (Table 10-1). However, it is its resemblance to addiction that is perhaps most striking and has escaped the attention of many physicians and researchers.

If toxicant-induced loss of tolerance were a mechanism underlying certain cases of chronic fatigue, migraine, asthma, or depression, it might be

TABLE 10-1. Features of Toxicant-Induced Loss of Tolerance Compared with Features of Addiction, Allergy, and Toxicity

Feature	Toxicant-induced loss of tolerance[1]	Addiction[1]	Allergy[1]	Toxicity[1]
Chemical/drug intolerance	+	+	+	+
Ambient air incitants	+		+	+
Food intolerance	+		+	
Alcohol intolerance	+	+		
Caffeine intolerance	+	+		
Withdrawal symptoms	+	+		
Craving, binging	+ (foods)	+ (drugs)		
Sensitization	+		+[2]	
Induction by chemicals	+		+	+
Induction by biologicals	? molds[3]		+	
Multisystem symptoms	+	+	+	+
Frequent CNS symptomatology	+	+		+
Well-defined mechanism(s)			+	+
Genetic susceptibility	+	+	+	+
Dose/response relationship	+[4]		+[4]	+

[1]Categories are not "pure" and may overlap in a given host, e.g., haptenation of a chemical toxicant may initiate an immunologic response; brain and liver toxicity may accompany alcohol addiction.

[2]Low molecular weight chemicals may combine with tissue proteins producing "haptens" that evoke immune responses.

[3]There are anecdotal cases of MCS arising following exposure to molds. Notably molds release not only allergenic spores and fragments (when disrupted), but also characteristic spectra of volatile organic chemicals.

[4]Dose–response does occur for allergens: With the first, sensitizing exposure in a susceptible individual, there is a dose–response relationship; with subsequent exposures, the sensitized person also responds in proportion to dose, but at a much lower dose level (Waddell 1993). The same kind of dose–response relationship may pertain for TILT, but has not been rigorously demonstrated. For the initiating event, dose information is sparse. Once affected, chemically sensitive individuals generally report increasingly severe symptoms the longer they remain in an exposure situation, an observation that suggests a dose–response relationship.

reasonable to ask if these patients also report chemical intolerances. In fact, some, but not all, of these patients report them (Buchwald and Garrity 1994; Kipen et al. 1995) (see Chapter 8, section on "Overlaps with Other Illnesses"). Many chemically sensitive patients carrying these same diagnoses relate that it was not until they accidentally or intentionally avoided a sufficient number of their problem incitants that they noticed feeling better. Then, when they reencountered one of those incitants, robust symptoms occurred. As they repeated the iterative process of avoidance and reexposure, they noticed that particular symptoms occurred with particular exposures. Most indicate that had they not avoided many chemicals and foods simultaneously, or "unmasked" themselves, they would not have figured out what was making them sick. This process of sustained observation and discerning cause-and-effect relationships requires a certain degree of intellect and insight. Perhaps this explains why self-identified chemically sensitive persons have, on average, high educational levels. Others may fail to make the connection between chemical exposures and symptoms and thus be "selected out" of the MCS pool.

"Masking" and "unmasking" are colorful lay terms for which there is no scientific equivalent. Nevertheless, investigators' ability to understand masking and unmasking and manipulate these variables knowledgeably may determine the success of studies in this area. When chemically sensitive patients follow a diet free of their problem foods and live in a relatively chemical-free home in the hills of central Texas where there are no major agricultural or industrial operations or air contaminants, they say they are in an "unmasked" state. Under these circumstances, if a diesel truck were to drive by, they claim they could identify specific symptoms due to the diesel exhaust, perhaps irritability, headache, or nausea.

On the other hand, the patients say that when they travel to a large city such as Houston or New York City, stay in a hotel room, and eat in restaurants, they become "masked." In the presence of many concurrent exposures (exhaust, fragrances, volatiles offgassing from building interiors, various foods) in New York City they feel chronically ill as if they had flu. If a diesel truck were to drive by them under these circumstances, they say they would not be able to attribute any particular symptoms to the exhaust. There would be too much "background noise" from overlapping symptoms occurring as a consequence of overlapping or successive exposures. In theory, such background noise or masking would hide the effect of individual exposures. Responses would blur together (also see Chapter 2, under "Adaptation").

Masking appears to involve at least three interrelated components, any of which could interfere with the outcome of low-level chemical challenges in these individuals: (1) habituation, (2) apposition, and (3) addiction. In real life, these three components operate concurrently although here they will be considered individually.

There is some notation that can be used to help depict these components. In the addiction literature, responses to addictive drugs are often illustrated graphically by using a biphasic curve or sine wave (Fig. 10-3). The portion of the sine wave above the horizontal axis represents symptoms with *onset* of exposure, often called "stimulatory" symptoms; the portion below represents symptoms with "*offset*" or cessation of exposure, often referred to as "withdrawal" symptoms. The height or the amplitude of the sine wave in either direction is proportional to the severity of the response. For a person who is not particularly sensitive to the substance, the curve would be much flatter, with zero or low amplitude in either direction. The length of the biphasic curve represents the duration of symptoms following an exposure, reportedly ranging from minutes up to several days, depending upon the exposure and the individual. Of course, the particular nature of the symptoms would vary from one sensitive subject to the next and from

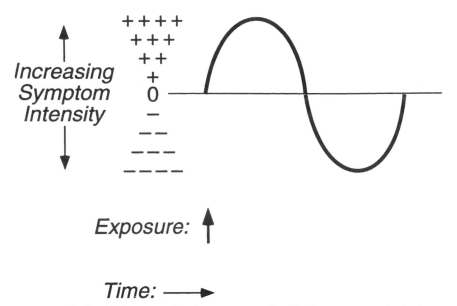

FIGURE 10-3. Graphical representation of symptom progression following exposure to a single substance in a person sensitive to that substance (e.g., caffeine, a solvent, alcohol, nicotine). The portion of the biphasic curve above the line represents symptoms with onset of exposure (stimulatory symptoms) and the portion below, symptoms with offset of exposure (withdrawal symptoms). Amplitude is proportional to symptom severity. The length of the curve (duration of symptoms) may range from minutes to days. (UTHSCSA © 1996)

substance to substance.

Suppose that some researchers wished to test a putatively sensitive subject by exposing him to a low concentration of xylene. Xylene is a common indoor air contaminant and a component of Mølhave's mixture (Mølhave et al. 1986), which has been used in human inhalation challenge studies. It would be important to ensure that their subject was unmasked, that is, responses were at true baseline, prior to challenge. The following components of masking would need to be considered and controlled:

1. Habituation. With continuous or repeated exposure to an environmental stressor, adaptation or acclimatization frequently occurs; that is, symptoms tend to diminish in intensity as exposure continues. Reportedly, chemically sensitive patients' acute symptoms also decrease with sustained exposure. However, in their case, when the exposure ceases, they frequently report marked withdrawal symptoms. Such withdrawal symptoms have not been described with true adaptation or acclimatization. Thus what MCS patients exhibit is more akin to habituation than to true adaptation or acclimatization. Suppose now that a subject who is to be challenged with xylene works in a sick building where he routinely is exposed to low levels of xylene on a regular basis. If he is administered a test exposure to xylene below the odor threshold (0.62 ppm) (AIHA 1989), it may

produce little or no effect if he has been working in that building during the preceding week (see Fig. 10-4). On the other hand, if he were to avoid the building and all other sources of xylene for four to seven days before testing, a more robust response to the xylene challenge would be anticipated under the disease model posited for MCS (Miller 1994a, 1996a).

Thus, a sensitive subject's response to a challenge may range widely in intensity, from none to maximal, depending upon how recently that person has been exposed to the test substance or a chemically related substance. If insufficient time has elapsed, for example, less than four days, the challenge may yield a falsely negative response as a result of habituation. If too much time has elapsed, for example, weeks or months, sensitivity may have waned.

2. Apposition. Suppose next that the research subject is sensitive to multiple substances. On the day when he is scheduled for challenge testing, he gets up in the morning, uses some scented soap or hairspray, cooks breakfast on a gas stove, and drives his car through heavy traffic to get to the laboratory. Inside the laboratory building he rides an elevator where he is exposed to people wearing various colognes. If he were sensitive to several of these exposures, his responses might overlap in time. Such responses reportedly can last for hours or days. If so, they could persist during a placebo challenge, resulting in a false positive response. Thus, apposition or juxtaposition of the effects of closely timed exposures is a second component of masking that must be controlled prior to and during challenge studies (Fig. 10-5).

Masking: Habituation

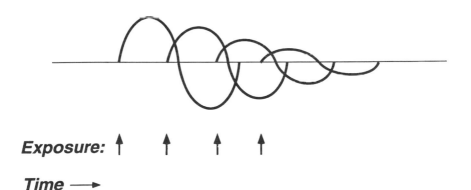

Exposure: ↑ ↑ ↑ ↑

Time ⟶

FIGURE 10-4. Habituation. Symptom severity declines with repeated, closely timed exposures (inhalant or ingestant) to the same substance. Symptoms become less acute and more chronic in nature, with a blurring of the relationship between symptoms and exposure. (UTHSCSA © 1996)

Masking: Apposition

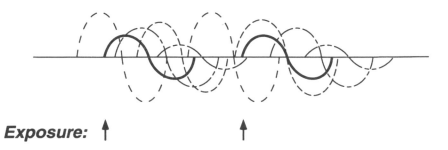

Exposure: ↑   ↑

Time ⟶

FIGURE 10-5. Apposition. If an individual is sensitive to many different substances, then the effects of everyday exposures to chemicals, foods, or drugs may overlap in time. This apposition of effects might yield an individual who feels bad most of the time, but the effect of any single exposure is not apparent to either the individual or his or her physician. Apposition would tend to mask the effect of interest (solid lines) in much the same way that background noise masks a sound of interest. (UTHSCSA © 1996)

3. Addiction. Many of the symptoms that chemical sensitivity patients report mirror those commonly associated with addiction. Addiction itself may be a component of masking, one that clearly is related to the habituation phenomenon described above. However, addiction implies compulsive, more or less conscious use of habit-forming drugs, alcohol, tobacco, and caffeine, whereas habituation to air contaminants in a sick building may occur entirely outside the affected individual's awareness. Addicted individuals consciously or subconsciously time their next "hit" so as to forestall withdrawal symptoms (Fig. 10-6), a phenomenon recognized to occur in alcohol, tobacco, and caffeine addiction. However, food addictions (cravings) also are reported among chemically sensitive patients. Randolph described wheat, eggs, milk, and corn as the most common addictants in his patients (Randolph 1962; Randolph and Moss 1980). Frequently, these individuals report intense cravings and ravenously consume enormous quantities of foods, for example, a pound of chocolate, several bags of popcorn, a dozen doughnuts, or 30 cups of coffee in one day. Patients most often report having experienced this kind of addictive consumption early in their illness, before they practiced avoiding problem exposures.

Foods may contain bioactive constituents such as tyramine, monosodium glutamate, and opiates (Bell et al. 1992). Persons who routinely use tobacco, caffeine, alcohol, or foods containing bioactive substances may need to

Masking: Addiction

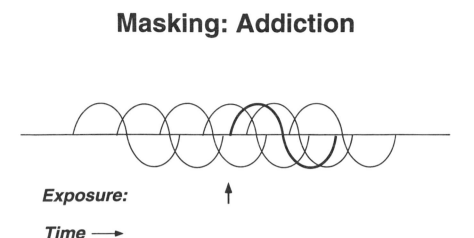

Exposure: ⬆

Time ➡

FIGURE 10-6. Addiction. A sensitive person who is addicted to caffeine, alcohol, nicotine, or another substance may deliberately take that substance at frequent, carefully spaced intervals to avoid unpleasant withdrawal symptoms. Such exposures may also mask the effect of interest (e.g., a challenge test using xylene). (UTHSCSA © 1996)

avoid them prior to challenge testing because the pharmacologic effects of these agents could override or mask the effect of an experimental challenge. Failure to eliminate addictants before testing may result either in false positive challenges due to lingering symptoms from an addictant used in the hours or days preceding a placebo challenge or in false negative challenges due to masking by an addictant.

After the germ theory of disease was introduced in the late 1800s, many overly enthusiastic investigators who were careless in their bacteriological technique announced that they had discovered the causative agents for tuberculosis, yellow fever, and other diseases. These pronouncements and subsequent retractions became so frequent that in 1884 the President of the New York Academy of Medicine lamented that a "bacteriomania" had swept over the medical profession (Warner 1985). In order to prevent future such pseudo-discoveries, Koch, who identified the organisms responsible for tuberculosis, anthrax, and cholera, proposed a set of rules for etiological verification. Koch's postulates required that: (1) the microbe must be present in every case of the disease; (2) it must be isolatable in pure culture; (3) inoculating a healthy animal with the culture must reproduce the disease; and (4) the microbe must be recoverable from the inoculated animal and be able to be grown again.

Just as bacteriomania engulfed the medical profession in the 1880s, "chemomania" is poised to engulf it now. Chemical sensitivity is in need of a set of postulates to ensure that future causal determinations are scientifically based. Figure 10-7 illustrates the following set of postulates, which, if met, would confirm (and if not met, refute) that a person's symptoms were triggered by a particular substance or exposure:

1. When a subject simultaneously avoids all chemical, food, and drug incitants, remission of symptoms occurs (unmasking).

2. A specific constellation of symptoms occurs with reintroduction of a particular incitant.

3. Symptoms resolve when the incitant is again avoided.

4. With reexposure to the same incitant, the same constellation of symptoms reoccurs, provided that the challenge is conducted within an appropriate window of time. Clinical observations suggest that an ideal window is four to seven days following the last exposure to the test incitant.

Chemical Sensitivity: Postulates

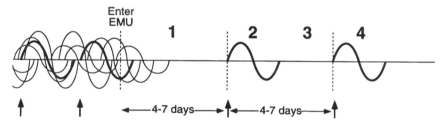

FIGURE 10-7. *Testing chemical sensitivity postulates using an environmental medical unit (EMU). In the left-most portion of the figure, before entering the EMU, a chemically sensitive individual is experiencing symptoms in response to multiple exposures (chemicals, foods, drugs). Effects overlap in time. The effect of any particular exposure cannot be distinguished from the effects of other exposures, and the person's symptoms may appear to wax and wane unpredictably over time.*

- *Postulate 1. When all chemical, food and drug incitants are avoided concurrently, remission of symptoms occurs. Anecdotally, patients report going through "withdrawal" or "detox" for the first several days and experiencing increased irritability, headaches, depression, etc. Anecdotally, after four to seven days most report feeling well and theoretically are at a clean baseline.*
- *Postulate 2. A specific constellation of symptoms occurs with reintroduction of an incitant.*
- *Postulate 3. Symptoms resolve when the incitant is again avoided.*
- *Postulate 4. Reexposure to the same incitant, within an appropriate window of time (estimated to be about four to seven days), produces the same symptoms.*

For research purposes, challenges should be conducted in a double-blind, placebo-controlled manner. (UTHSCSA © 1996)

In order to apply these postulates, the timing of exposures and degree of masking should be rigorously controlled. To accomplish this, a hospital-based clinical research facility called an "environmental medical unit" (EMU) is needed in order to isolate subjects from background exposures (Fig. 10-8) (Miller 1992, 1994a, 1996a, 1997; NRC 1992a) (see also section on "The Environmental Unit" in Chapter 2). We are not advocating that the EMU be used as a routine diagnostic tool. We wish to emphasize that not all patients can or should be evaluated in such a facility; its purpose is to create an environment in which human challenge studies can be conducted without background chemical interferences for research purposes. The EMU would be constructed, furnished, and operated so as to minimize exposure to airborne chemicals. For example, no disinfectants, perfumes, or pesticides would be allowed in the unit. Ventilation would maximize fresh outside air and would provide optimal particulate and gas filtration. Patients would eat chemically less-contaminated foods and water, testing one food per meal in order to determine the effects of individual foods. If symptoms persisted despite this approach, fasting for a few days would be attempted prior to reintroduction of single foods.

The rationale for housing subjects in an environmentally controlled facility for several days prior to challenges is twofold: (1) to prevent extraneous exposures to inhalants or ingestants so that responses to them are not mis-

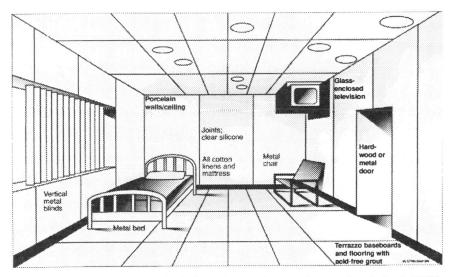

FIGURE 10-8. Preliminary design sketch of a patient room in an environmental medical unit (EMU). Note use of nonoutgassing construction materials and furnishings, and point source control (ventilated television enclosure). (UTHSCSA © 1996)

construed as positive responses when placebo challenges are administered (false positives), and (2) to minimize masking that might blunt or eliminate responses to active challenges (false negatives).

Although the terms "exposure chamber" and "environmental medical unit" sound similar, conceptually they differ in important ways:

1. Patient safety. By definition, an environmental medical unit is in a hospital-like facility where patients can remain 24 hours a day in a clean environment for up to several weeks. Like an intensive care unit or a coronary care unit, the environmental medical unit would be a specialized, dedicated hospital unit. The EMU must be located in a hospital-like setting in order to accommodate very sick patients. An exposure chamber (a small space in which subjects generally are expected to spend minutes to several hours) does not offer a comparable level of care. Because chemical challenges may precipitate bronchoconstriction, mental confusion, severe headaches, depression, and other disabling symptoms, these patients should not be tested in an exposure chamber on an outpatient basis.

2. Control of interfering exposures. Conventional exposure chambers do not reduce background chemical exposures for extended periods (up to several weeks) so that the effects of a number of challenges in a patient can be assessed accurately. This is the central limitation of chambers and the reason why they should not be used to rule in or rule out chemical sensitivity, although they may be perfectly acceptable for studying the effects of air in sick buildings and other exposures in relatively healthy individuals. If subjects are not kept in a chemically clean environment for several days prior to and during challenges, false positive responses due to interfering exposures and false negative responses due to masking may occur. In contrast with an exposure chamber, an environmental medical unit would minimize interfering exposures both before and during challenges, thus maximizing the reliability and the reproducibility of test responses.

Availability of an EMU would allow physicians to refer a wide variety of cases, in which environmental sensitivities were suspected, to the unit for definitive evaluation. There physicians could observe firsthand whether a patient's symptoms improved after several days on a special diet in a clean environment. If improvement occurred, single chemicals at concentrations encountered in normal daily living and single foods could be reintroduced one at a time while the effects of each introduction were observed. Thus the EMU would serve as a tool for ruling in or ruling out environmental sensitivities in the most direct and definitive manner possible. Studying complicated patients such as chronic fatigue sufferers or ill Gulf War veterans in a conventional exposure chamber would provide none of the same information because chambers allow only short-term residence, do not control the entire range of background contaminants, and provide inadequate separation from background exposures prior to challenges.

We have found the following analogy helpful in conveying the importance of controlling exposures for extended periods prior to challenge. If one wished to determine whether headaches in a coffee drinker were due to caffeine, it would not work simply to give the person a cup of coffee and ask her how she felt. It is intuitively obvious that the individual would need to stop using caffeine for a while before a meaningful test of caffeine sensitivity could be performed. In this instance, a false negative challenge would be the most likely consequence of failure to avoid coffee prior to challenge. Similarly, placing a putatively sensitive person in a conventional exposure chamber and exposing him to a low concentration of a chemical might not produce any noticeable effect. On the other hand, if he were to remain in a clean environment such as an EMU for a few days before being tested and his condition improved, one could then perform meaningful challenges.

Placing patients in an EMU would simultaneously control all three components of masking: stopping all exposures several days prior to challenge testing and spacing test exposures four to seven days apart would preclude habituation; eliminating background chemical noise and allowing the effects of each challenge to play out before introducing the next one would control apposition; and excluding drugs, alcohol, nicotine, and caffeine and spacing individual foods four to seven days apart would interrupt any addiction. Individual sensitivity could then be evaluated in the EMU following the previously stated set of postulates for etiological verification (Fig. 10-7).

For research purposes, challenges need to be performed in a double-blind, placebo-controlled manner. Patients with chronic fatigue syndrome, migraines, seizures, depression, asthma, or unexplained illnesses such as the Persian Gulf War veterans' illnesses could also be tested for sensitivities in an EMU using these postulates. Thus, the EMU could be used to determine whether or not particular patients with these diagnoses had a masked form of chemical intolerance.

What evidence is there that unmasking patients in an environmental medical unit and conducting challenges within a four- to seven-day window of time is either useful or necessary? Many credible patients and their physicians (often but not always clinical ecologists) have attempted this maneuver. They report anecdotally that patients' symptoms resolve within a few days after they enter such a facility and that robust symptoms occur when challenges are conducted after several days of avoidance (Randolph

1965; Rea 1992). Although it may rightly be argued that these are anecdotal observations, that is how scientific inquiry usually begins. Other evidence corroborates these anecdotal observations in MCS patients: Withdrawal symptoms of several days' to a week's duration are known to occur in some persons following cessation of exposure to nitroglycerine (dynamite workers' headaches) (Daum 1983), caffeine (Griffiths and Woodson 1988; Silverman et al. 1992), nicotine, and alcohol. Note that these are chemically unrelated substances. In individuals chronically exposed to xylene (Riihimaki and Savolainen 1980) or ozone (Hackney et al. 1977a, 1977b), reexposure following several days' avoidance results in robust symptoms (see discussion in "Adaptation" section of Chapter 2). Foods may require one to several days to navigate the digestive tract before they are eliminated. Taken together, these observations suggest that individuals with sensitivities to multiple incitants might experience effects that could linger as long as several days after avoidance begins. Thus, it may be argued that patients should be removed from their entire background of food and chemical exposures for four to seven days prior to challenges (unmasked), as Randolph first proposed (Randolph 1962; Randolph and Moss 1980).

Although it is conceivable that synergistic or additive chemical combinations may be necessary to reproduce certain symptoms, this is a limitation of any form of challenge testing. Wherever possible, within the bounds of safety and feasibility, chemical combinations believed to precipitate the most robust and measurable responses should be explored. However, 40 years of clinical observations, albeit anecdotal, suggest that single test substances may suffice for the majority of sensitive subjects. Confirmation or refutation of these claims seems a logical first step that should precede testing of complex mixtures. Finally, because isolating patients in a hospital environment such as the EMU may have unanticipated psychological consequences, early studies in this area should also examine the responses of control subjects in the same environment (Miller et al. 1997).

Good pathological and physiological theories provide "a unified, clear, and entirely intelligible meaning for a whole series of anatomical and clinical facts, and for the relevant experiences and discoveries of reliable observers . . . " (Carter 1985). Theories and experiments that overlook salient observations or do not control experimental conditions adequately may lead to erroneous conclusions. During the late nineteenth century, researchers collected sputum from patients with tuberculosis but were unsuccessful in culturing any organism. As a consequence, some concluded that tuberculosis was not an infectious disease. These early investigators did not know that the tubercle bacillus was fastidious and would grow out only after many weeks on a specialized culture medium. Correspondingly, scientists' ability to observe and understand chemical sensitivity may depend upon optimization of experimental conditions, that is, appropriate timing of challenges and use of an environmental medical unit for unmasking patients. To date, studies in this area have failed to unmask patients prior to challenge. When false positive and false negative responses occurred, the investigators concluded that chemical sensitivity was psychogenic in origin (Leznoff 1993, 1997; Staudenmayer et al. 1993a).

In summary, features of toxicant-induced loss of tolerance overlap those of allergy, addiction, and classical toxicity, yet TILT may be distinct from each of these. TILT appears to involve a two-step process (resembling allergy) in which persons lose specific tolerance (resembling addiction) for a wide range of common substances following a chemical exposure event (resembling toxicity). Just as the germ theory describes a class of diseases sharing the general mechanism of infection, the TILT theory of disease posits a class of chemically induced disorders characterized by loss of tolerance to chemicals, foods, drugs, and food/drug combinations. In the same way that fever is a symptom commonly associated with infectious diseases, chemical sensitivity may be a symptom associated with the TILT family of diseases. Although clinical case definitions have been developed that describe particular infectious diseases, no clinical case definition can be applied to the entire class of infectious diseases. The same may be true for TILT disorders. The fact that this phenomenon does not fit already accepted mechanisms for disease is often offered as evidence that the condition does not exist. However, the same criticism would have applied equally well to the germ and immune theories of disease when they first were proposed. *"What is plausible depends upon the biological knowledge of the time"* (Hill 1965, emphasis added).

Looking to the future, carefully conducted epidemiological studies and animal models likely will play an important role in characterizing the initiation stage of TILT, during which tolerance is lost. In the meantime, rigorous testing of the second stage of TILT, that is, the triggering of symptoms by tiny doses of chemicals, foods, drugs, caffeine, or alcohol, is needed if progress in this area is to occur. Adoption of a set of relevant, testable hypotheses for etiological verification will serve to ensure the credibility of those endeavors.

In spite of the scientific advances that suggest a physiological origin for MCS and the availability of several theories offering plausible biologic mechanisms, leading proponents of a psychogenic diathesis insist that "[t]here are no postulates [for a physiologic basis for MCS] that generate testable hypotheses" (Staudenmayer 1996), and "[t]he organic theories of MCS require major paradigm shifts in pathophysiologic understanding of both clinical disease and toxicology. Psychogenic theories do not" (Gots

1996). In some ways, this latter view is consonant with our own—acceptance of a physiological basis for chemical sensitivity would require reformation of the classical disease models, if not a major paradigm shift. As many balanced workshops of scientists and clinicians have shown, however, there is a remarkable and growing consensus as to precisely what kind of research needs to be done in this area. This is discussed in the following section.

Research Needs

Pivotal medical, compensation, litigation, regulatory, and policy questions rest upon a full understanding of chemical sensitivity. Nevertheless, remarkably little funding has been directed toward researching this illness, for a variety of reasons. Because of limited funding, the few studies that have been done have had "shoestring" budgets. Consequently, data are meager. Scientists involved in other rapidly expanding fields may find the relative paucity of research in this area surprising. Economic stakes are high. Insurers, agencies such as the DVA and the DOD that provide medical care and compensation, the chemical industry, and manufacturers of consumer products including carpets, building materials, fragrances, and other goods could be affected greatly by the outcome of research on chemical sensitivity. Norman Rosenthal, Chief of the Psychobiology Branch at the National Institute of Mental Health and an internationally known investigator in the area of seasonal affective disorder (SAD), succinctly summed up the difficulty (Rosenthal 1994): "In trying to research MCS . . . we are in a Catch-22 situation. It is difficult to attract research money for a controversial condition and it is difficult to resolve the controversy without the necessary research." He notes wryly that "[i]n the case of SAD, no one can be blamed or held liable; darkness is a feature of our natural landscape. This is not so in the case of MCS, where someone can be held liable for the injury or injuries."

If, as we are increasingly inclined to believe, chemical sensitivity is a class of diseases rather than a single identifiable clinical entity, this has significant implications both for the direction that research should take and for the problem of constructing a case definition for characterizing patients. We have already indicated that research is needed both on the question of what initiates chemical sensitivity (and through what mechanisms) and on what triggers symptoms in an already affected person (and through what mechanisms). Of course, although knowledge of mechanisms is important and intellectually satisfying, it is not essential for accepting the condition as real if we are otherwise convinced, through careful scientific observation and analysis, of the phenomenologic basis for the condition. Understanding mechanisms is important for convincing us of the origins of a condition where causal connections are highly uncertain, and it is important for devising certain medical and public health preventive interventions. However, again, useful interventions can precede a full understanding of mechanisms (see prior section of this chapter); that is, we do not have to know everything before we do anything.

Questions pertaining to initiation and to triggering are important for different reasons. The first question has special relevance for (1) undertaking primary prevention and (2) providing fair compensation. If we can eliminate the initiating events, we can perhaps avoid creating the problem in the first place, either by eliminating causal exposures or, eventually, by identifying particularly susceptible persons and ensuring that they avoid those exposures. The second question is important for (1) undertaking secondary prevention (avoidance of triggering exposures), including special accommodation in housing and employment, and (2) providing medical treatment if appropriate. For both questions, case definitions would be helpful, and may be essential. However, case definitions designed for one purpose, for example, special accommodation, are not necessarily appropriate for other purposes, for example, compensation or research (Ashford and Miller 1992). An example from occupational medicine is the differing case definitions for hearing impairment and hearing disability. A hearing loss of 25 dBA is the usual threshold for receiving state workers' compensation, but preventive actions are taken for workers with far less hearing loss than this.

Observations from both North America and Europe (Ashford et al. 1995) reveal only a handful of putative initiators, compared to the large number of reported triggers of chemical sensitivity. These different initiators could operate through different specific mechanisms and result in different diseases, sharing some overlapping symptomatology, including chemical, food, and drug intolerances. Prospectively following persons involved in unfortunate "natural experiments," particularly people who have shared relatively homogeneous initiating events, such as Gulf War exposures or a specific chemical spill, is useful. Further, investigating these events reduces the chance that the presence of several diseases that are similar, but also not identical enough to receive the same label, will confuse our ability to notice patterns of symptoms and symptom severity. Exposure- or event-driven studies may also result in improved consistency in biomarkers due to a more homogenous population, which can help further our understanding. Animal experiments with putative initiators are necessary adjuncts to the development of a credible theory of initiation.

What is becoming quite clear is that studying patients who arrive at the condition through a variety of pathways is a slow, arduous way of unraveling the puzzle. Imagine trying to understand the cause of infectious diseases by analyzing a group of patients, all of whom have fever, some with meningitis, some with tuberculosis, some with AIDS, and some with influenza. On the other hand, stratification or selection of patients presenting with "chemical sensitivity," such as those becoming sensitive reportedly as a result of exposure to organophosphate pesticides or household remodeling (Miller and Mitzel 1995), will add clarity to the picture. The kinds of studies that are not very helpful are ones focusing on a patient population referred to a single physician on the basis of insurance claims (Terr 1986), patients with a psychiatric diagnosis (Staudenmayer and Selner 1990; Staudenmayer et al. 1993a, 1993b), or those who have previously seen many physicians without being successfully treated (Rea 1992). Stratification by symptoms or signs, or by social or economic criteria, will not enhance our understanding, if—as we suspect—chemical sensitivity is a class of diseases, rather than a single syndrome. Many of the attempts to understand this condition have resulted in sharply divergent conclusions, usually honestly arrived at, but resulting from the fact that different observers are looking at different parts of the elephant. For example, a recent publication (Staudenmayer 1996) argues that emotional trauma could lead to exquisite sensitivity to chemicals. For the subset of patients observed by a psychologist/psychiatrist, this could well be correct. However, this conclusion does not necessarily apply to all patients.

A major limitation of the published studies on MCS to date is the fact that few have been performed on patients under exposure conditions and none on "unmasked" patients. In order to maximize the opportunity for detecting an abnormality, it may be important to compare markers in patients before, during, and after a salient exposure. Physicians who evaluate individuals with suspected occupational asthma (which may be the consequence of toxicant-induced loss of tolerance) often have patients keep a record of their peak flow readings before, during, and after exposures at work. Some physicians perform a provocative inhalation challenge with the suspected asthma inducer. At baseline or random points in time, patients with occupational asthma may exhibit normal pulmonary function. In parallel fashion, provocative challenges may be key to detecting and diagnosing chemical sensitivity.

Because masking could alter patients' responses, exposure challenges need to be performed after patients have been removed from their usual background of everyday exposures, including the challenge substance itself, for a sufficient period of time that any habituation they may have developed does not interfere with responses during testing. Again, in the case of occupational asthma it is recognized that inhalation challenges should not be conducted either too soon or too many months after removal from the workplace. In the former case, tolerance may have developed, and, in the latter, sensitivity may be waning. Thus, in order to observe the most robust effect of a particular exposure, patients may need to be tested within a prescribed window of time, perhaps four to ten days after their last exposure, and in the absence of background exposures that may trigger extraneous symptoms. For this purpose, we have proposed that patients be housed in an environmentally controlled hospital unit (an EMU) prior to challenges. Because patients in an unmasked state conceivably could have very robust responses upon challenge, these challenges should take place in settings where appropriate emergency medical care is available.

At the conclusion of the NRC workshop on chemical sensitivity, participants unanimously endorsed human challenge studies using a controlled environment, assigning this approach their highest priority for research in this area (NRC 1992a). A federally sponsored meeting of occupational and environmental health physicians (AOEC 1992) and authors of a nine-country exploratory study in Europe also placed an environmental medical unit high on their lists of research recommendations (Ashford et al. 1995).

Future research on MCS also depends upon the development of a case definition for the condition. Six of the most widely cited case definitions (Table 10-2) differ greatly in terms of the minimum number of organ systems that must be affected (one to three); whether patients with other definable clinical or psychological conditions should be excluded; whether provocative challenges are required; and whether the illness has to have been acquired following a documented exposure (Table 10-3). One of these case definitions asserts that symptoms in one organ system are sufficient for diagnosing the condition (NRC) and two do not specify a minimum number of affected organ systems (Ashford/Miller and Nethercott et al.). This distinction is key because if two or more organ systems must be affected in order to meet a case definition, this requirement would exclude from study patients who had a single condition such as asthma, migraine headaches, or irritable bowel as their only health problems. All but two of the case definitions (AOEC and Cullen) agree that other definable clinical conditions such as asthma, arthritis, vasospasm, and seizure disorder should not be excluded; the majority agree that chemical sensitivity could be an etiology for these diagnoses, which themselves are simply descriptive clinical labels. None of the case definitions excludes psychological conditions such as somatization disorder or depression. The AOEC and Cullen definitions exclude other clinical diagnoses, but not psychological ones. Such an approach might tend to bias study populations toward those with psychological problems. Finally, only one case definition (Cullen) requires that the condition be acquired in relation to documentable environmental exposure. The other definitions acknowledge that some patients report

TABLE 10-2. Proposed Case Definitions for Multiple Chemical Sensitivity

Clinical Ecologists (definition appearing in each issue of the journal *Clinical Ecology*):

Ecologic illness is a chronic multi-system disorder, usually polysymptomatic, caused by adverse reactions to environmental incitants, modified by individual susceptibility and specific adaptation. The incitants are present in air, water, food, drugs, and our habitat.

Cullen (1987):

Multiple chemical sensitivities (MCS) is an acquired disorder characterized by recurrent symptoms, referable to multiple organ systems, occurring in response to demonstrable exposure to many chemically unrelated compounds at doses far below those established in the general population to cause harmful effects. No single widely accepted test of physiologic function can be shown to correlate with symptoms.

Ashford and Miller (1989):

The patient with multiple chemical sensitivities can be discovered by removal from the suspected offending agents and by rechallenge, after an appropriate interval, under strictly controlled environmental conditions. Causality is inferred by the clearing of symptoms with removal from the offending environment and recurrence of symptoms with specific challenge.

National Research Council (1992a), Workshop on Multiple Chemical Sensitivities, Working Group on Research Protocol for Clinical Evaluation:

1. Sensitivity to chemicals. By sensitivity we mean symptoms or signs related to chemical exposures at levels tolerated by the population at large that is distinct from such well recognized hypersensitivity phenomena as IgE-mediated immediate hypersensitivity reactions, contact dermatitis, and hypersensitivity pneumonitis.
2. Sensitivity may be expressed as symptoms and signs in one or more organ systems.
3. Symptoms and signs wax and wane with exposures.

It is not necessary to identify a chemical exposure associated with the onset of the condition. Preexistent or concurrent conditions, e.g., asthma, arthritis, somatization disorder, or depression, should not exclude patients from consideration.

Association of Occupational and Environmental Clinics (1992) Workshop on Multiple Chemical Sensitivity, Working Group on Characterizing Patients:

1. A change in health status identified by the patient.
2. Symptoms triggered regularly by multiple stimuli.
3. Symptoms experienced for at least six months.
4. A defined set of symptoms reported by patients.
5. Symptoms that occur in three or more organ systems.
6. Exclusion of patients with other medical conditions (psychiatric conditions are not considered exclusionary).

Nethercott et al. (1993):

1. The symptoms are reproducible with exposure.
2. The condition is chronic.
3. Low-level exposure results in manifestations of syndrome.
4. Symptoms improve or resolve when incitants are removed.
5. Responses occur to multiple, chemically unrelated substances.

TABLE 10-3. Features of Proposed Research Case Definitions for MCS[1]

	Ashford/ Miller	AOEC	Clinical Ecology	Cullen	Nethercott et al.	NRC
Minimum number of organ systems that must be affected	not specified	3	2	2	not specified	1
Excludes other definable clinical conditions such as asthma, arthritis, vasospasm, seizure disorder	No	Yes	No	Yes	No	No
Excludes psychological conditions such as somatization disorder, depression	No	No	No	No	No	No
Provocative challenge required to document	Yes	No	No	No	No	No
Must be acquired in relation to a documentable environmental exposure	No	No	No	Yes	No	No

[1]*Sources:* Ashford and Miller 1991; Association of Occupational and Environmental Clinics (AOEC) 1992; *Clinical Ecology Journal* (definition appears in each issue); Cullen 1987; Nethercott et al. 1993; National Research Council 1992a.

lifelong illness or becoming ill following a series of less well-defined exposures over several years, and recognize that such individuals may be chemically sensitive as well.

The case definition we have proposed requires provocative challenges in a controlled environment to document chemical sensitivity in a patient and, for research purposes, challenges should be double-blind and placebo-controlled. It continues to be our opinion that such an approach is required to define the etiologies of MCS, as well as the etiology of other clinical conditions in which environmental triggers have been alleged by some such as chronic fatigue, headaches, depression, and asthma. In our view, other case definitions prematurely exclude potential cases from study. For example, an unknown but perhaps sizable number of patients with asthma might have bronchoconstriction and inflammation on the basis of low-level chemical exposures. We have urged that such patients not be excluded from study, and that a broader perspective be adopted, as chemical sensitivity may not be a single illness but perhaps is a class of disorders, which share the same general mechanism but may involve different specific mechanisms.

Some argue that challenges in a controlled environment would be costly, and that not everyone could be evaluated in such a specialized unit, especially given the fact that no such research facility currently is available. Our response is that not everyone should or needs to be evaluated in an environmental medical unit. However, enough research must be done to assure the critics that low-level sensitivity can be objectively verified and to develop surrogate measures that correlate well with results obtained in this way, that is, biological markers with high predictive value.

A 1992 Canadian Workshop (Health Canada 1992) proposed a three-tiered definition for MCS: "possible, more probable, and most probable." "Possible" cases need to satisfy the relatively broad U.S. National Research Council definition (Table 10-2). "Probable" MCS cases must meet the criteria for "possible" and demonstrate both improvement in symptoms with reduction or cessation of suspected exposure and recurrence of symptoms with reexposure. "Most probable" MCS cases would be diagnosed via double-blind challenge in a controlled environment after appropriate deadaptation or unmasking. This three-level definition (perhaps substituting the modifiers "possible, probable, and proven") might be appropriate for research, purposes, but thought needs to be given to its use for treatment or compensation.

Research Recommendations

The most recent North American meeting on chemical sensitivity, held in November 1995, in New Jersey (NIEHS, 1997), attempted to address areas and approaches for further research. In the workshop on Empirical Approaches for the Investigation of Toxicant-Induced Loss of Tolerance (Miller et al. 1997), the participants suggested three research(able) questions to test the hypothesis that sensitivity to low-level chemical exposures develops via a two-step mechanism: (1) initiation by an acute or chronic chemical exposure, followed by (2) triggering of symptoms by low levels of previously tolerated chemical inhalants, foods, and/or drugs. (Other working groups developed recommendations for studying specific mechanisms.) The three research questions were:

1. Do some individuals experience sensitivity to chemicals at levels of exposure unexplained by classical toxicological thresholds and dose–response relationships, and outside normally expected variation in the population?

2. Do chemically sensitive subjects exhibit masking that may interfere with the reproducibility of their responses to chemical challenges?
3. Does chemical sensitivity develop as a consequence of acute, intermittent, or continuous exposure to certain substances? If so, what substances are most likely to initiate this process?

Bernard Weiss, one of the workshops participants, had earlier (Weiss 1994) suggested using single-subject designs for experiments involving MCS patients. Using such a design, he had previously tested 22 children whose parents thought they had benefited from the Feingold diet for hyperactivity. Two of the children responded consistently to a challenge using food dyes. Weiss viewed this consistency of response using a repeated-measures approach as more meaningful than one-time observations in larger numbers of people, stating that "[c]onsistency in a single individual may be more informative than significance tests in a large sample." Thus, a study that was able to demonstrate consistent responses in a single MCS patient would be highly significant.

Double-blinded, placebo-controlled challenges performed in an environmental medical unit were seen by the workshop participants as essential for addressing the first two questions, and detailed experimental protocols were developed. The third question relates to the initiation of chemical sensitivity. Here two approaches were identified: (1) begin with a particular exposure history and try to determine whether some of those exposed developed chemical sensitivities, or (2) choose subjects with a particular medical condition, disease, or symptom and look for patterns of prior exposure. These two approaches have been termed "event-driven" studies by a European research team (see below).

A nine-country European exploratory study (Ashford et al. 1995) was undertaken by an international research team with expertise in toxicology, occupational medicine, indoor air, chemistry, environmental and occupational health, law, and sociology. No allergists or clinical ecologists were included. The team's research recommendations included:

> ... the development of protocols for taking a complete occupational and environmental exposure history in patients who report sensitivity to low levels of chemicals. The protocol itself should be developed by a consensus of knowledgeable researchers, physicians, and patients and should give special attention to uncovering and documenting exposure to: 1) known sensitizers and neurotoxic agents; 2) substances often associated with the onset of chemical sensitivity, such as solvents, pesticides, new or renovated buildings, anesthetic agents, and wood preservatives; and 3) stressful or traumatic life events. In addition, protocols for follow-up in terms of changes in signs, symptoms, and disease over appropriate time periods need to be established.

Obvious opportunities for future study include: 1) the follow-up of previously-exposed cohorts of persons most likely to present with or develop chemical sensitivity; 2) the prospective follow-up of populations and persons involved in "natural experiments" that might result in chemical sensitivity, such as chemical spills or relocation to a new or renovated building; 3) the work-up of selected persons in an environmental medical unit (EMU) in which double-blind, placebo-controlled studies are conducted to explore the nature and existence of chemical sensitivity in individual persons; and 4) the exploration of possible animal models that may elucidate mechanisms for chemical sensitivity.

In investigating options (1) and (2), it is important that both an occupational and [an] environmental exposure history be taken and that outcomes (signs, symptoms, and disease) be tracked over a sufficiently long period of time to allow the discovery of chemical sensitivity if it in fact occurs. "Initiating" exposures or events should be distinguished from subsequent triggering agents or incitants. Option (3) is important for investigating whether symptoms resulting from low-level exposures are reproducible on an individual basis. Note that an environmental control unit is not an exposure chamber. It is a specially-designed hospital unit where patients can be housed, removed from possible triggers (in food, water, air, etc.), and re-challenged under carefully controlled conditions. Option (4) is regarded as essential for clarifying the nature of chemical sensitivity. Both human and animal observations have provided important insight as to possible mechanisms for chemical sensitivity. Neurotoxic pathways, in particular, need to be examined. Analysis of use patterns in different countries for pesticides, anesthetic agents, and other possible sensitizers may reveal useful information.

The work-product from the latest North American meeting, described earlier in this section, is consistent with and builds upon the recommendations from previous conferences on MCS. Regarding "environmental exposures and sensitivity syndromes," the Subcommittee on Immunotoxicity of the National Research Council's Committee on Biologic Markers recommended that "[t]here is a need to establish a multidisciplinary team of experts in lung physiology, immunotoxicology, clinical immunology, psychiatry, toxicology, occupational medicine, and industrial hygiene to study patients with purported syndromes. A standard comprehensive panel of clinical procedures should be applied to aid their diagnoses. Blinded challenge tests, using well-defined cohorts with established exposures, might need to be conducted" (NRC 1992b).

Table 10-4 summarizes the recommendations from the two prior U.S. meetings. Of special importance were the following recommendations emerging from three working groups participating in the 1991 National Research Council Workshop on MCS (NRC 1992a):

TABLE 10-4. Summary of Research Recommendations from U.S. Meetings on Multiple Chemical Sensitivity (MCS)

I. National Research Council meeting on Multiple Chemical Sensitivities, March, 1991 (NRC 1992a).
   A. Sponsors: EPA, ATSDR, NIEHS
   B. Participants: Invited clinicians, immunologists, toxicologists, epidemiologists, psychiatrists, psychologists, and others involved in research or clinical activity relevant to MCS
   C. Recommendations (three groups):
     1. Clinical Evaluation Group: Proposed a case definition for research (see Table 10-2, Proposed Case Definitions). Also suggested:
       a. Development of a uniform patient database.
       b. Hypothesis-driven specialized evaluations.
       c. Development of an environmental control unit for study of adaptation/deadaptation hypothesis, control of exposures, and challenging subjects.
       d. Prospective studies of exposure events.
     2. Exposures and Mechanisms Group
       a. Double-blind controlled exposure challenges, examining the possible role of "adaptation" and "deadaptation."
       b. Evaluation of MCS patients in their usual environment, as symptoms and exposures vary over time.
       c. Development of animal models that mimic the human syndrome.
       d. Evaluation of biopsy or necropsy tissue for pathologic changes.
       e. Development of database of chemicals, foods, drugs, and associated symptoms and signs.
     3. Epidemiology Working Group
       a. Improvement of case definition.
       b. Multi-center clinical case-comparison studies using agreed-upon set of criteria and tests.
       c. Use of information from case-comparison study to construct a population-based study to determine the prevalence of MCS.
       d. Follow-up of a defined population subjected to a discrete and sudden exposure to assess the initiation of hypersensitivity and its natural history.
     4. Consensus was reached among all workshop participants that challenging subjects in a well defined environment should have the highest priority for future research.
II. Association of Occupational and Environmental Clinics (AOEC) Meeting on Chemical Sensitivity, September 1991 (AOEC 1992).
   A. Sponsor: ATSDR
   B. Participants: Invited speakers representing divergent views on MCS; members of the AOEC, which includes occupational medicine physicians from academia and private practice.
   C. Recommendations (four work groups):
     1. Group on Characterizing Patients: Proposed a case definition for research (see Table 10-2, Proposed Case Definitions).
     2. Group on Characterizing Events
       a. Assessment of incidence and prevalence of MCS.
       b. Surveys of specific occupational cohorts and cross-cultural studies of "naive" populations, such as pesticide-exposed agricultural workers in the Third World.

(continued)

TABLE 10-4. Summary of Research Recommendations from U.S. Meetings on Multiple Chemical Sensitivity (MCS) (*continued*)

———————————————————————————————————————

      c. Longitudinal studies of populations exposed in "natural" experiments such as a sick building.

      d. Case registries for descriptive and future serologic studies of panels of MCS patients.

      e. Double-blind, placebo-controlled challenge studies.

      f. Studies to determine whether chemical exposures truly can be blinded.

  3. Group on Treatment Methods

      a. A study of the effects of early intervention in an exposed population, such as critical incident counseling.

      b. Randomized, controlled trials of therapies that have some reasonable theoretical basis.

  4. Group on Mechanisms

      a. Challenge studies, including but not limited to chamber studies (the latter should address the issue of adaptation).

      b. Studies of olfactory function and the nasal-olfactory-limbic pathway.

      c. Neuro-imaging studies including the use of pharmacologic probes.

      d. Prospective studies of cohorts of persons sensitive to chemicals.

      e. Studies of families of MCS patients, both medical and psychological.

———————————————————————————————————————

A case-comparison study of patients seen in occupational and environmental medicine should enroll patients who claim to respond to low levels of environmental chemicals. The information from this multi-center study should be used to study the prevalence of MCS in the general population.

Populations with well-defined exposures, such as victims of a toxic spill or workers with uniform occupational exposure, could be studied longitudinally, for the development of chemical sensitivities. Patients with multiple chemical sensitivity will be selected because of symptoms or signs related to chemical exposures at levels tolerated by the population at large . . . . Symptoms must wax and wane with chemical exposures and may occur in one or more organ systems. Although many patients describe the onset of this syndrome with an acute toxic chemical exposure, such an initiating exposure is not required for inclusion. Patients with pre-existent or concurrent diseases such as asthma, arthritis, and psychiatric illnesses are not to be excluded from study, because many believe that chemical exposures play a role in inducing or exacerbating these conditions.

Research units or environmental control units in which MCS patients will be housed in a chemical-free environment are needed. Challenges will then be conducted in a double-blind fashion with attention to adaptation and de-adaptation phenomena. Responses of patients to controlled exposures should be monitored with immunologic, neurologic, endocrinologic, psychologic, and social markers and measures. Dose–response relationships should be studied.

The Association of Occupational and Environmental Clinics, made up mostly of occupational medicine practitioners in university-based or independent clinics, held an invited workshop (AOEC 1992) in September 1991. Some of the important findings (Rest 1992) were:

> Existing "natural experiments" might provide fertile ground for research. For example, new apprentice pesticide applicators could be examined at baseline and followed over time to assess the frequency of developing symptoms consistent with MCS. Populations in other exposure situations, such as sick buildings, could also be followed in this way. Such longitudinal studies might also provide opportunities for applying experimental study designs [e.g., chemical challenges].
>
> The group noted a lack of knowledge about the natural history of MCS and identified a need for and the challenges of designing studies to elucidate its natural history. The group offered several ideas about how such studies might be done but did not necessarily endorse them.
>
> The group identified a need for more formal diagnostic assessment of MCS patients and recommended use of a double-blinded, placebo-controlled, cross-over design.

Participants in this workshop did not achieve consensus on recommending event-driven or exposure-driven studies, in contrast with the earlier NAS panels. Unfortunately, as discussed above, we believe that unless patients are properly stratified, it may be difficult to gain sufficient understanding by restricting studies to clinic populations.

As a result of a grant received from ATSDR, the Environmental Health Investigations Branch (EIIIB) of the California Department of Health Services convened an advisory group to develop a variety of empirical approaches for the study of MCS patients (Kreutzer and Neutra 1996). A subset of the study panel, including Dr. Herman Staudenmayer (who along with Dr. John Selner uses a challenge chamber to test patients in Denver), discussed issues surrounding "isolation challenge studies" in Appendix E of the California report. Particularly disappointing was the implied equivalence of using an exposure chamber and an environmental medical unit. Among alleged "limitations" of challenges conducted in isolation units were:

> [An a]rtificial environment removes complex interactions of the "real world" in which exposures occur (e.g., other chemicals or psychosocial stressors).
>
> Exposures may not be delivered to the subject in the same fashion as that which causes their symptoms. . . .
>
> Some triggers must be masked, leaving open the possibility of confounding (sensitization) by the masker.

In contrast, the first two limitations were considered and not viewed as insurmountable obstacles by the participants in the New Jersey working group that outlined a protocol for an EMU (Miller et al. 1997). Similar concerns apply to many other exposure studies, yet the value of human challenge studies is enormous. Further, clinical observations support the utility of this approach for MCS. The third limitation can be handled by challenge with carefully chosen triggers at concentrations below the odor threshold, or alternative routes of administration (e.g., transdermal route, oral capsules containing caffeine).

In its discussion, the California report created a straw man:

> The expectations that isolation challenges can either document the reality of MCS to skeptics or reveal its non-existence to "true believers" may be overly optimistic . . . . What would be sufficient evidence that MCS does not exist? Every time a case is placed in isolation and fails to demonstrate the alleged sensitivity, there would be much scientific rationalization for the apparent failure: improper case selection, incorrect delivery of the chemical trigger to the subject, incomplete deadaptation, sensitization to something else in the isolation environment, absence of necessary cofactors for adequate triggering (both chemical or psychological).

As in the field of epidemiology, there is no perfect study. It is the accumulation of evidence from good studies that finally leads to acceptance or rejection of an association. It is not the "true believers" that need to be convinced, but rather serious scientists who are looking for answers. To establish whether MCS exists, it is not necessary for every subject to demonstrate chemical sensitivity upon challenge as long as some do. It is theoretically possible that a stratification of chemically sensitive and somaticizing subjects could emerge. Of course patients with and without apparent somatoform disorder could be chosen for study. Finally, if "isolation" means using a chamber, in distinction to properly preparing a subject in an environmental medical unit, then achieving accurate and reproducible results could very well be a problem. Thoughtful researchers need not await equivocal results from chamber challenge studies before they object to a chamber design.

The tone in Appendix E differs greatly from that in the main body of the California report, which states that "we don't believe that pragmatic attempts should be abandoned to develop a 'best possible' protocol [for conducting challenges] that might be generally acceptable to most researchers . . . and we would hope to see funding made available to continue this process" (Kreutzer and Neutra 1996).

In other parts of the report, the California advisory panel addressed epidemiologic approaches for characterizing persons with MCS. Protocols for both retrospective population-based studies and prospective "post-chemical spill" studies were developed. Although the former may be useful in clarifying the prevalence of self-reported chemical sensitivity, we believe that the latter approach is more likely to enhance our understanding of the various pathways that lead to the class of diseases constituting chemical sensitivity (see the discussion in Chapter 9).

Medical and Patient Needs

Although some progress has been made in advancing the understanding of the origins and the nature of chemical sensitivity, many persons suffering from sensitivity to low-level chemical exposures have pressing, unmet needs. While we believe it premature to recommend specific medical treatments at this time, several pathways might be explored to minimize suffering and prevent worsening of the condition. These include appropriate avoidance of problem chemical exposures and foods, low- or no-cost alterations of patients' physical environments, and psychological support as needed (see below).

Although patients often turn to federal agencies, including the EPA, ATSDR, NIEHS, NIOSH, and various state agencies, for assistance, no clear pathway is offered for obtaining balanced information or medical attention from knowledgeable and caring physicians. Canada has two government-sponsored clinics devoted to clinical research and evaluation of persons with MCS (see Chapter 7), but no comparable government-sponsored clinics are available to patients in the United States, nor are we aware of efforts to develop any. No nation has taken steps toward establishing an environmental medical unit for research purposes, despite the high priority assigned to this approach by a number of scientific groups and meetings (NRC 1992a, AOEC 1992, Ashford et al. 1995).

Currently, in an often frantic effort to regain their health, patients often exhaust their financial resources, consulting dozens of specialists and attempting a host of unproven treatments. Forty percent of MCS patients in one study reported having consulted ten or more medical practitioners (Miller and Mitzel 1995). Many have sought help from clinical ecologists, who accept these patients as having a bona fide illness. Others have remained with their family physicians, sought out specialists including occupational medicine physicians, or gone to university-based occupational and environmental medicine clinics.

It is extraordinarily difficult for caring practitioners to know how best to treat these patients. Often, other patients who have gone through this experience have provided help. Various patient support groups offer counsel and referrals. But until research sheds more light upon MCS, what can physicians do? Most physicians in academia whom we know are reluctant to apply the diagnosis of MCS to their patients even when they believe it provides the most parsimonious description of what they are seeing. They may opt instead to apply "piecemeal" but recognized and compensable diagnoses such as asthma, toxic encephalopathy, or migraine headache to these cases. At present, physicians who make a diagnosis of MCS can expect their assessment to be challenged by workers' compensation boards, employers, and others. They are likely to be asked what tangible evidence (clinical signs or laboratory tests) they relied upon to make their diagnosis. Ironically, for many psychiatric conditions, including depression and schizophrenia, no "objective" diagnostic tests are available either, but that does not keep doctors from diagnosing them. With MCS, things are different—the stakes are higher.

"Increasingly, in difficult circumstances, the reasonable trend in medicine is to explain the options and allow the patient to decide" (Vasey 1995). We believe that health care providers should, at a minimum, discuss with patients the possibility of MCS *if symptoms and circumstances warrant it.* Which symptoms? Fatigue, memory and concentration difficulties, mood changes, and multisystem health problems (summarized in Table 8-1) are commonly reported. As more organ systems are involved, MCS especially should be considered (although only one organ system may be affected in some cases). What circumstances? If an identifiable chemical exposure, especially one involving solvents, pesticides, a sick building, remodeling, or new construction, preceded onset of the symptoms; if a major change in the patient's health, as documented by increased health care utilization or absenteeism, occurred; if the patient exhibited clinical signs or abnormal laboratory tests temporally related to the exposure, such as increased liver function tests, a depressed white blood cell count, or decreased cholinesterase level; if anyone else who shared the same experience became ill, particularly with similar problems; and if everyday exposures now reportedly provoke symptoms, then the likelihood of MCS increases. If new-onset intolerances include not only chemical inhalants, but also some foods (or feeling ill after meals), medications, alcohol, or caffeine, then the probability that the practitioner is dealing with MCS increases still further. None of these factors alone "proves" the presence of MCS. On the other hand, the more of these features a patient manifests, the more the practitioner should suspect that MCS could be occurring.

When such circumstances warrant their doing so, practitioners should familiarize themselves with the illness and discuss with their patients the divergence of opinion in the medical community concerning MCS (see boxed text). They should talk with their patients about the wide range of treatments other MCS patients have tried, including psychological therapies and avoidance strategies, the lack of controlled scientific studies confirming the efficacy of these and other treatments, and the need for research to understand mechanisms and find effective interventions. Physicians should avoid chastising or rejecting patients who have tried or are currently using alternative therapies. Many patients have turned to unproven treatments in desperation and do not have the medical background physicians do. Without being dogmatic, caregivers still can convey an earnest desire to see patients improve and can serve as their advocates during a difficult illness. Ample time should be allotted for visits and/or telephone consultations, particularly if travel is a barrier, so patients can discuss their problems and concerns adequately. As Sparks et al. aptly observed, "The evaluation of a patient presenting with MCS may take several hours and it is necessary to allot sufficient time, even if inadequately reimbursed" (Sparks et al. 1994b).

GUIDANCE FOR PHYSICIANS ON MCS

Develop an understanding of MCS. Become acquainted with the symptoms most commonly attributed to it and include it in your differential diagnosis. Be aware that many MCS patients also report various food, alcohol, caffeine, and medication intolerances, and ask about these.

Take a careful exposure history paying particular attention to the time each symptom began and exposures that may have preceded onset of illness, such as a chemical spill, fire, pesticide application, remodeling, or moving to a new home or office.

Explain to the patient the current controversy in the medical profession about MCS, in particular whether it is psychogenic, toxicogenic, or a combination of these. Discuss the fact that more research is needed before the mechanism(s) underlying the condition is understood and before specific therapies targeting that mechanism(s) are tested and shown to be effective.

Determine whether a judicious trial of avoidance and reexposure might help clarify whether exposures at work or home could be causing symptoms. In a patient who has only recently become ill, consider whether ongoing exposures may need to be interrupted to prevent possible long-term disability.

Encourage the patient to become a careful observer by keeping a diary of symptoms and exposures, including both inhalants and ingestants, and noting any consistent relationships between them.

Explain that psychological support sometimes can be a useful therapeutic adjunct, as for any illness. MCS can disrupt career, lifestyle, and relationships, placing enormous stress on these individuals and their families.

Discuss the range of treatments that have been used by patients with MCS and the lack of scientific evidence to support their use. Point out potentially harmful aspects of therapies that are unproven or lack a scientific basis.

Describe treatment options and the potential risks and benefits of each, empowering the patient to choose among acceptable options.

When introducing new medications, be especially watchful for new symptoms that may signal an adverse drug reaction.

Telling a patient to avoid exposures that trigger symptoms may not be enough. On a trial basis, have the patient minimize nonessential exposures to fragrances, cleaners, and other products that release volatile organic chemicals.

Understand that a patient who smokes or uses alcohol or caffeine on a regular basis or who has ongoing chemical exposures may have difficulty discerning the relationships between exposures and symptoms.

The importance of taking a careful exposure history, though it is time-consuming, cannot be overemphasized. Having patients construct a time-line with their symptoms and medical problems stratified across the top half, and lifetime events (such as job changes, changes in residence, military service, surgeries, pregnancies, remodeling, and/or pesticide applications) along the bottom half of the line, may help both patient and physician elucidate temporal relationships between exposures and the onset of symptoms where such relationships exist.

The use of a standardized questionnaire that addresses symptoms, intolerances, and exposures will facilitate history-taking. A questionnaire that permits patients to rate their symptoms and intolerances, both before and since an identified exposure event, if one is implicated, will provide the practitioner with a more quantitative, albeit subjective, sense of how severe the patients' symptoms are in their own eyes. Such a questionnaire can be readministered at intervals to gauge improvement or deterioration in patient health and to assess the impact of any interventions. See Appendix C for one such questionnaire, the Environmental Exposure and Sensitivity Inventory (EESI), which was developed from a study of MCS patients by Miller and Mitzel (1995) and has been used to assess changes pre- and post-exposure (Gammage et al. 1996). Symptom items on the EESI were derived via factor analysis applied to a large number of specific symptoms reported by MCS patients who had experienced a well-defined, initial exposure to a pesticide or low levels of mixed solvents associated with remodeling.

Material safety data sheets (MSDSs) may be useful for identifying exactly what the patients' exposures were or are. Assistance in gathering and interpreting exposure information can be provided by occupational medicine physicians, toxicologists, or industrial hygienists. Insofar as possible, past medical records, which may be voluminous and difficult to obtain, should be examined. Although a comprehensive physical examination emphasizing organ systems relevant to a patient's symptoms is essential, frequently the findings are unremarkable. Baseline laboratory tests such as a complete blood count and chemistry profile are helpful, as well as tests indicated by specific symptoms or findings such as thyroid function tests, pulmonary function tests, peak flow monitoring over time, tests for collagen-vascular disease, and neuropsychological testing in some patients (Weaver 1996). Blood tests for environmental chemicals should be used only if specific exposures are suspected and the substances can reasonably be expected to persist (e.g., chlorinated pesticide, recent organophosphate pesticide, but not most solvent exposures).

Referrals to specialists frequently are necessary, given the complexity of the patients' symptoms, and informed specialists who understand MCS can provide reassurance to both the patient and the referring physician. However, care must be taken to avoid excessive invasive testing and polypharmacy (see below).

Treatment

With regard to treating MCS patients, we are currently at such an early, observational stage in our understanding of this illness (or class of illnesses) that making therapeutic recommendations seems premature. However, it is clear from the history of medicine that preventive interventions based upon well-reasoned guesses as to the causes of diseases have had decisively positive impacts on public health in the past—for example, John Snow's breaking of the handle on the Broad Street pump in London, thus interrupting the cholera epidemic some 30 years before Koch identified the bacterium that causes cholera; the use of gloves and handwashing to prevent iatrogenic spread of organisms causing infection prior to their specific identification; and allergen avoidance and allergy shots to mitigate allergic reactions long before the discovery of IgE antibodies. In the case of MCS, mounting evidence suggests that MCS can be caused, and be exacerbated by, chemical exposures, and that intervening by having patients avoid chemical and other incitants may prevent initiation or worsening of the condition. Such measures might include prohibiting the use of certain pesticides indoors or requiring that new buildings outgas before occupancy.

Earlier we used the term "toxicant-induced loss of tolerance" to describe MCS as representing a possible emerging theory of disease (see earlier section in this chapter entitled "Current Reflections on MCS"). Converging lines of evidence support this theory: (1) the fact that similar reports of multisystem symptoms and new-onset intolerances have been reported by different investigators in different regions among different demographic groups following exposure to many different types of chemicals; (2) the internal consistency of these patients' complaints of intolerances for not only tiny doses of inhaled chemicals but also various foods, caffeine, alcohol, and medications; (3) the degree to which the illness mimics addiction; (4) the identification of plausible anatomical substrates (the olfactory-limbic system, the cholinergic nervous system), whose malfunction might explain many MCS symptoms; and (5) recent animal models that replicate key features of chemical sensitivity.

Given this evidence and the clear consensus among MCS patients that chemical and food avoidance strategies are helpful (see below), increasing numbers of physicians have begun to recommend avoidance strategies for these patients.

Avoidance

There have been a few, anecdotal reports (mostly unpublished) of individuals with MCS in whom the condition apparently was recognized early in its course, who were advised to avoid further exposure, and who recovered (Hileman 1991). Such reports suggest that early intervention (avoidance) could prevent long-term disability for some patients. Our current difficulty in treating MCS once it has become entrenched underscores the importance of early intervention. Physicians, especially primary care doctors, should be alert to cases with MCS-like presentations where there could be ongoing exposure to pesticides, remodeling, solvents, workplace chemicals, or other substances, and the patient could be in the initiation stage of MCS. Such patients might benefit from removal from the exposures on a trial basis, to determine whether improvement occurs, and judicious reexposure, to determine whether symptoms recur. For the vast majority of patients, immediate recovery from initiating exposures is not known to occur, but few persons in sick buildings or whose homes have been exterminated have been recognized by physicians as experiencing possible evolving MCS early in the course of their illness. Supportive treatment as for any chronic illness is required, parallel to that used for chronic fatigue, fibromyalgia, and pain syndromes. Indeed, a multidisciplinary approach similar to that employed for treating patients with chronic pain has been suggested (Weaver 1996). Weaver advises: "Regardless of the treatment chosen, it is important to emphasize that functional improvement and increased patient control, not cure, are the goals. Complete resolution of odor sensitivity may not be possible, given the chronic history of this illness thus far and the common prevalence of such complaints." One of us (CSM) has seen a few Gulf War veterans whose onset of "Gulf War Syndrome" in fact occurred years after the war, and temporally coincided more closely with postwar exposures to pesticides, building remodeling activities, or a toxic waste disposal site than with their wartime exposures. Such ongoing exposures can be interrupted (on a trial basis) only if they are identified during careful history-taking as discussed above.

Avoidance of problem chemicals is consistently the single most helpful intervention reported by MCS patients. In a survey of 206 MCS patients with an average educational level of nearly four years of college, 71 percent rated avoidance of problem chemicals and 54 percent avoidance of problem foods as "very helpful." In contrast, although 52 percent had tried psychological or psychiatric therapies, only 17 percent of those who had tried them rated them as "very helpful" (Miller 1995).

A survey of 305 MCS patients conducted by researchers at DePaul University revealed that at least three-quarters felt that the following interventions had been of "enormous" or "major" help to those who had tried them: avoiding chemicals that cause reactions (93 percent), creating an environmentally safe living space (86 percent), moving to a less polluted area (76 percent), and avoiding foods that provoke reactions (75 percent) (LeRoy et al. 1996).

Similar results were found in a grassroots patient survey of 243 MCS patients, 47 percent of whom were on disability and 5 percent on workers' compensation: 95 percent reporting avoidance of chemical exposures, 79 percent reporting relocation to avoid pollution, and 76 percent reporting avoidance of problem foods to be of "enormous help" or "major help" (Johnson 1996).

Lax and Henneberger (1995) found that MCS patients seen in their occupational medicine clinic who reported avoiding at least half of their self-reported incitants were more likely than nonavoiders to report feeling better at a follow-up interview conducted six months to two and a half years after their initial clinic visit. Notably, in this sample, other potential explanatory variables did not predict either health improvement or worsening, including age, gender, number of doctors consulted, having visited an environmental medicine physician, having applied for workers' compensation or Social Security benefits, or the number of symptoms reported at the initial evaluation.

One form of avoidance that has helped some MCS patients—who say that they are able to work fairly trouble-free, but only if they avoid chemical exposures—is workplace accommodation. Many examples of successful workplace accommodation for these patients exist although relatively few find their way into print. Examples of specific accommodations have included: providing access to fresh air and better ventilation, removal of business machines from the immediate work environment, use of a different work space, removal of carpeting, changes in cleaning agents used by maintenance staff, adoption of integrated pest management, and arranging for telecommuting (home-based work).

After the outbreak of illness precipitated by remodeling and new carpet installation in the EPA headquarters building, some employees were accommodated in a specially designed work space free of common chemical exposures (see Chapters 3 and 7), or were allowed to work from a computer terminal at home (Keplinger 1994). Another example of successful accommodation occurred at Oak Ridge National Laboratory in Oak Ridge, Tennessee, where three of four women with preexisting chemical sensitivities reported exacerbation of their illness after they were relocated to a newly constructed office building. One was moved to another building, and the others controlled their symptoms by modifying their time of occupancy or by using a room air cleaner (Gammage et al. 1996).

Physicians can facilitate such accommodation by making reasonable written requests to employers or schools on their patients' behalf. Although increasing numbers of MCS patients are consulting physicians in university-based occupational and environmental health clinics, who are increasingly identifying the condition, in our experience few such centers offer guidance as to how patients might undertake a trial of avoiding potential chemical, food, and drug incitants and attempt subsequent, judicious reexposure. In contrast, many of these same clinics recommend trials of avoidance for patients suspected of having occupational asthma and the use of serial peak flow readings to document changes in airway resistance related to workplace exposures. A parallel approach could be used for chemical sensitivity, having patients keep a diary and rate individual symptoms on a 0 to 10 scale several times daily.

Over the past five years, one of us (CSM) has been asked to see some 80 sick Gulf War veterans for the Department of Veterans Affairs at its Regional Referral Center in Houston, Texas. In contrast with patients in most studies of MCS, 90 percent of these veterans were males with an average educational level of about a year of college. As discussed earlier in this volume (Chapter 8), the majority reported multisystem symptoms and new-onset intolerances to chemicals, foods, and/or drugs since the war. Many of these veterans would be considered relatively "masked"—that is, they were consuming large amounts of caffeine (up to 30 cups of coffee or tea per day in some cases to ward off fatigue); taking various medications; exhibiting food cravings; living in homes with gas heat/stoves, new carpet, and other exposures; engaging in work or hobbies that exposed them to chemicals; and using fragrances, disinfectants, and fabric softeners. Other than recommending judicious trials of avoidance and having patients keep a diary of symptoms and exposures it is difficult to know what to do to help patients like these. The VA currently does not accept a diagnosis of chemical sensitivity; yet, for some of the veterans, chemical intolerances are their predominant complaint and MCS is the most parsimonious unifying diagnosis that fits their symptoms. There was (and still is) no research environmental medical unit available in which patients could be tested for their chemical intolerances and receive a more definitive diagnosis. Thus, many ill Gulf veterans, like MCS patients, remain in limbo—required to show objective evidence of their chemical intolerances (before a diagnosis of MCS will be recognized by the VA or other official agencies) yet having no means of doing so. As discussed earlier, a 1991 conference of the Association of Occupational and Environmental Clinics, sponsored by ATSDR, concluded that an environmental medical unit was important and that adaptation [masking] was a key issue: "Studies that do not address adaptation will be criticized and will not lead to consensus" (AOEC 1992). In the current absence of a research environmental medical unit to evaluate such patients, what can physicians do to facilitate avoidance and testing for sensitivities?

Although avoidance of potential incitants is a simple concept, it is not a simple task, especially for persons whose ability to concentrate and organize is impaired. Reportedly, avoidance of alcoholic beverages, nicotine, caffeine, medications, and problem foods can precipitate painful withdrawal symptoms in sensitive individuals during the first few days, including severe headaches, disorientation, depression, and malaise. Thus, *we cannot emphasize enough the importance of patients not attempting avoidance regimens on an outpatient basis without the close supervision of a knowledgeable medical practitioner.* Patients with a history of severe problems with asthma, headaches, depression, or other medical conditions, particularly those who ever have required emergency management, preferably should be housed in an environmentally controlled hospital setting during the initial stages of caffeine, alcohol, or food avoidance.

Surveys of individuals with MCS consistently indicate that the majority find avoidance of problem foods to be of major help in controlling their symptoms (Johnson 1996; LeRoy et al. 1996; Miller and Mitzel 1995). The survey by Miller and Mitzel, which focused on persons who had become ill following a well-defined exposure to a pesticide or to mixed, low levels of solvents associated with remodeling, found that 60 percent of those surveyed thought that their diets had been affected "a great deal" by their illness, and only 3 out of 112 respondents reported that food or other ingestants (e.g., chlorinated water, monosodium glutamate) did not make them ill. Whereas occupational medicine doctors may be reluctant to work on dietary issues with patients because it is not part of their medical training, allergists—who are trained to work through food intolerances and allergies with their patients—are unaccustomed to dealing with chemical exposures. Thus, MCS patients "fall in the crack" between the two types of specialists who would seem to be the most likely candidates to address their needs.

A variety of elimination diets have been used by patients, most notably the four-day rotary elimination diet in which no food (or food group) is repeated more than once every four days (Rinkel 1944; Rinkel et al. 1951). Other popularized versions of this diet, following the same general approach have appeared (Mandell and Scanlon 1979). As scientists, we obviously cannot endorse such therapeutic measures without adequate scientific studies to demonstrate their efficacy. However, modifying the timing of food intake so as to uncover possible non-IgE-mediated food intolerances could prove to be a useful intervention, as long as adequate nutrition is maintained in the process. For patients found to have multiple food intolerances, however, achieving adequate nutrition may prove difficult. Here again, research in an environmental medical unit, using blinded food challenges, will be essential for sorting out both food and chemical intolerances and documenting them in a scientifically acceptable manner. Long-standing problems with digestion and/or use of restrictive diets may lead to vitamin, or mineral deficiencies that need to be identified and corrected.

Medications

Adverse reactions to many drugs, whether prescribed or over-the-counter, are reported frequently by MCS patients. Physicians who attempt to treat these individuals with standard doses of drugs often report that usual dosages are not tolerated or that over time the patients develop side effects or idiosyncratic reactions, further complicating their clinical pictures (McLellan 1987). In such situations, the question frequently arises of whether new symptoms in an MCS patient are due to a medication (or one of its constituents, e.g., a food dye, excipient, or diluent), or whether the symptoms should be intensively worked up as a possible new medical problem. The latter approach—exhaustive evaluation of all new symptoms— can become costly and expose the patient to still more possible incitants, such as anesthetic agents used during biopsies, X-ray contrast dye, and methacholine challenges, adding yet another layer of "unexplained" symptoms. Frequently, MCS patients report that their symptoms resolve when drugs are removed, rather than when they are added. Thus practitioners, while needing to be alert to new symptoms that could signal life-threatening problems (e.g., cancer or coronary artery disease), also need to be on the lookout for unexpected and unusual reactions to drugs and other environmental triggers. Some MCS patients report benefiting from much smaller than normal doses of certain medications (e.g., analgesics and sometimes antidepressants, but not antibiotics), a pattern that may be consistent with their amplified responses to other substances.

Optimally, the practitioner should offer options and openly discuss the fact that any drug may cause idiosyncratic reactions, watch carefully for such symptoms, and begin with a low dose where feasible. Many patients say they have learned from experience that certain drugs cause problems for them, for example, dental anesthetics with epinephrine, and alternatives can easily be adopted, for example, using the same drug without epinephrine.

Environmental Evaluation

In a small percentage of cases, air or other environmental sampling may contribute to an understanding of patients' health problems. According to MCS patients, professional industrial hygienists or other indoor air specialists who are familiar with, and sensitive to, their concerns can sometimes provide helpful guidance on the basis of an initial walk-through survey. Preferably they should be nonsmokers with a good sense of smell who are able to identify potential sources of low-level VOCs or moisture for mold growth. The levels of indoor air contaminants of potential concern for chemically sensitive persons are orders of magnitude below those prescribed by law for occupational environments. Unless they are exceeded, which is rare in these circumstances, legally adequate occupational exposure levels have no relevance for susceptible individuals and should not be invoked.

Psychological Therapies

Given the competing hypotheses that MCS is psychogenic versus physiological/organic in origin, it is incumbent upon physicians to explain the current rift in the medical community over these alternative explanations to patients and their families. An unbiased approach to treatment would involve offering both avoidance and appropriate psychological interventions. Psychological support is a reasonable and recognized therapeutic adjunct, whether the illness is psychogenic or organic in origin (e.g., cancer, heart disease). This support can be provided by psychologists, psychiatrists, social workers, or primary care doctors. Many MCS patients experience enormous disruption of their work, family, and social lives. Divorces occur, for example, from spouses who smoke and refuse to quit. Families pull away. Suicides have been reported. Psychological support has been viewed as "very helpful" by some MCS patients although the majority have said it did not alter their underlying sensitivities (Miller 1995).

Of course, patients who are actively psychotic or suicidal, or who in any way constitute a threat to themselves or others, need to be identified, treated, and protected accordingly. A grassroots survey of 243 MCS patients found that of those who had tried taking particular antidepressants, roughly 10 to 20 percent found them a "major" or "enormous" help; however, 50 to 65 percent of those who said they had tried the same drugs reported harmful effects, and 10 to 30 percent reported that the drugs did not help, or that their effect was unclear (Johnson 1996). It is uncertain what percentage of patients who participated in this survey found benefit from at least one of the various antidepressants they tried, as it is widely recognized by psychiatrists that different people respond differently to different antidepressants; nor do we know what percentage continued the medication long enough for improvement to occur, or what percentage stopped the drugs because of intolerable side effects. Controlled studies are needed to clarify which patients might benefit from these and other drugs. However, MCS patients, many of whom say they have reacted adversely to several drugs, become increasingly reluctant to try new ones. Recruiting patients into therapeutic drug trials thus may prove difficult. "Placebo effects" and spontaneous remission could also account for the small percentage of patients who improve with various drugs; therefore, future drug studies should be double-blinded and placebo-controlled.

A number of investigators have claimed success in using various psychological or psychiatric interventions, but none with more than anecdotal data to support their claims (Amundsen et al. 1996; Bolla-Wilson et al. 1988; Guglielmi et al. 1994; Schottenfeld and Cullen 1985; Spyker 1995). The efficacy of psychological interventions, beyond psychological support, has not been adequately demonstrated. Several authors suggest that their patients improved following psychotherapeutic interventions, but essential follow-up is not provided; for example, "[f]or unknown reasons, [Mr. A.] did not continue with therapy" (Bolla-Wilson et al. 1988). In another example, Spyker (1995) states, "Results with three patients [using behavioral desensitization] were encouraging . . . ," yet no follow-up information is provided on two of the three patients described. The one patient who appears to have improved was a 40-year-old electronics assembly technician who was also evaluated for Xanax habituation/withdrawal and ethanol abuse—hardly a classical MCS patient.

Amundsen et al. (1996), at Mayo Clinic, published a paper with the promising title of "Odor Aversion or Multiple Chemical Sensitivities: Recommendation for a Name Change and Description of Successful Behavioral Medicine Treatment." The entire description of the two cases reported by these authors appears in one sentence in the abstract of their paper: "Two subjects [out of 34 patients] with typical odor-triggered symptoms have been treated, using behavioral medicine techniques, with marked improvement in both cases." Unfortunately, no further description of the treatment or outcome in these cases is offered, and the reader is left to wonder what happened to the other 32 cases.

A paper entitled "Occupation-Induced Posttraumatic Stress Disorders" (Schottenfeld and Cullen 1985) describes a 34-year-old man who had acute respiratory distress following an occupational exposure in which he was initially blinded by a solvent. He was treated with desipramine and psychotherapy and "was able gradually to expose himself to potentially dangerous situations and to desensitize himself . . . ." No long-term follow-up is provided, nor is it clear from the paper whether the patient was eventually able to return to work.

In an article entitled "Behavioral Treatment of Phobic Avoidance in

Multiple Chemical Sensitivity," Guglielmi et al. (1994) present three MCS cases—two women exposed to a "bug bomb" at work and a 40-year-old woman exposed to chlorine gas [evidently the same 40-year old electronics assembly technician evaluated for Xanax habituation/withdrawal and ethanol abuse described by Spyker (see above)]. Following treatment, only the latter patient responded to these authors' follow-up contacts. The two exposed to the bug bomb were lost to follow-up. In this paper, Guglielmi et al. lament clinical ecology's failure to use elementary forms of experimental control, stating, "Subjective impression, anecdotal evidence, and testimonials from 'satisfied customers' cannot be accepted as substitutes for sound scientific methodology." They accuse clinical ecologists of wishing to be exempted from the requirements of the scientific method. Yet, Guglielmi et al.'s own work suffers from precisely the same subjective impressions, anecdotal evidence, and testimonials: They discuss the fact that two of their three patients failed to respond to their letters and "are assumed to have relapsed." Yet, in the very next paragraph, the authors proclaim, "Our experience with these three patients suggests that MCS can be effectively treated with an intensive and relatively brief course of biofeedback-assisted in vivo exposure and cognitive reframing."

Staudenmeyer et al. (1993b) similarly attest, without providing follow-up data or an adequate scientific study design, that among their MCS patients who agreed to undergo psychotherapy, 75 percent had a successful outcome (for further analysis of this paper, see Chapter 8, under "Mechanisms").

A major logic error in each of these papers when they proclaim treatment efficacy is: *post hoc, ergo propter hoc* [after it, therefore, because of it]— that is, attributing improvement or resolution of illness to their treatment just because improvement happened to occur after that treatment. The use of blinding, placebos, treatment control groups, and application of the scientific method should apply no less to psychiatrists and psychologists than to clinical ecologists. In summary, data supporting the efficacy of psychotherapeutic interventions in treating the chemical, food, and other intolerances reported by MCS patients are meager.

Help from Governmental Agencies and Other Organizations

In addition to seeking guidance form physicians and other health care professionals, many affected individuals consult various organizations and publications designed to help laypersons. Listed below are the names of government agencies and other organizations that patients and physicians may be able to contact for further information on MCS. This information is provided in an effort to be helpful. The listing of an organization does not constitute endorsement of its views.

Governmental Agencies

For Problems Related to Occupational and Environmental Exposures

Environmental Protection Agency
Indoor Air Quality Information Clearinghouse
P.O. Box 37133
Washington, DC 20013-7133
Phone: (800) 438-4318; (202) 484-1307
Fax: (202) 484-1510
E-mail: IAQINFO@aol.com

National Institute of Environmental Health Sciences
2605 Meridian Parkway, Suite 115
Durham, NC 27713
Phone: (800) 643-4794
Fax: 919-361-9408

National Institute for Occupational Safety and Health
4676 Columbia Parkway
Cincinnati, OH 45226
Phone: (800) 356-4674; (513) 841-4382
Fax: (513) 841-4488

Occupational Safety and Health Administration
Office of Health Compliance
U.S. Department of Labor
200 Constitution Ave., NW
Washington, DC 20210
Phone: (202) 219-8036
Fax: (202) 219-9187

For Problems Related to Consumer Products

Consumer Product Safety Commission
Washington, DC 20207
Phone: (800) 638-2772
(for carpet information, press 1, and then 129)

For Problems Related to Pesticides

Environmental Protection Agency
Mail Code 7506C
Office of Pesticide Programs
Public Information Center
401 M St., SW
Washington, DC 20460
Phone: (202) 260-2080

For Publications on the Safe Use of Pesticides

National Center for Environmental Publications and Information
P.O. Box 42419
Cinncinati, OH 45242-2419
Phone: (513) 489-8190
Fax: (513) 489-8695

National Pesticide Telecommunications Network
Agricultural Chemistry Extension
Oregon State University
333 Weniger
Corvallis, OR 97331-6502
Phone: (800) 858-7378

For Problems Related to Sanitizers, Disinfectants, and Sterilants

National Pesticide Telecommunications Network
Agricultural Chemistry Extension
Oregon State University
333 Weniger
Corvallis, OR 97331-6502
Phone: (800) 447-6349

For Problems Related to Community-Based Contamination

Agency for Toxic Substances and Disease Registry
Office of Policy and External Affairs
1600 Clifton Road
Mail Stop E28
Atlanta, GA 30333
Phone: (404) 639-0501
Fax: (404) 639-0715

Environmental Protection Agency Superfund and Right-to-Know Hotline
Phone: (800) 424-9346

For Problems Related to the Need for Workplace Accommodation

Office of the Americans with Disability Act
Office of Legal Counsel
Equal Employment Opportunity Commission
1801 L St., NW
Washington, DC 20507
Phone: (202) 663-4503
Fax: (202) 663-4639

For Problems Related to the Need for Accommodation in Housing

Office of Fair Housing and Equal Opportunity
Department of Housing and Urban Development
451 7th St., SW
Washington, DC 20410
Phone: (202) 619-8041
Fax: (202) 708-1425

For Problems Related to Disability
Social Security Administration
Phone: (800) 772-1213
Or call your state's disability determination service.

Office of the Americans with Disabilities Act
Civil Rights Division, U.S. Department of Justice
P.O. Box 66118
Washington, DC 20035-6118
Phone: (800) 514-0301; (202) 514-0301

Other Organizations

National Coalition against the Misuse of Pesticides
701 E St., SE
Washington, DC 20003
Phone: (202)543-5450
Fax: (202)543-4791
E-mail: NCAMP@igc.atc.org
Contact: Jay Feldman

National Organization for Social Security Claimants' Representatives
Phone: (800) 431-2804

Useful References

The following have been mentioned frequently by patients and others as useful references relevant to MCS. As with the above listing of organizations, the listing of the following references does not constitute endorsement.

Building Air Quality: A Guide for Building Owners and Facility Managers, EPA/NIOSH, December 1991. Available from the U.S. Government Printing Office, (202) 512-1803 (phone); (202) 512-2168 (fax).

The Clean Air Guide, Canadian Mortgage and Housing Corporation, Ottawa, Ontario, Canada. 1993. May be ordered from the CMHC at (613) 748-2003 for $3.99.

This Clean House (videotape), Canadian Mortgage and Housing Corporation, Ottawa, Ontario, Canada, 1995. May be ordered from the CMHC at (613) 748-2003 for $12.99.

The Healthy School Handbook, Norma Miller, ed., Washington, DC: National Education Association, 1995. To order, call (800) 229-4200.

Indoor Air Pollution—A Guide for Health Professionals by the American Lung Association, American Medical Association, U.S. Consumer Product Safety Commission, and the U.S. EPA (phone: 800-438-4318 or 202-484-1307; fax: 202-484-1510).

The Inside Story: A Guide to Indoor Air Quality, EPA 402-K-93-007, EPA/CPSC, September 1993. Available from the Indoor Air Quality Information Clearinghouse, Environmental Protection Agency (phone for Indoor Air Division: 800-638-2772 or 202-233-9030).

Multiple Chemical Sensitivities at Work (MCS workbook and videotape to help workers and train trainers recognize, cope with, and seek medical and legal assistance for problems related to multiple chemical sensitivity). Available from The Labor Institute, 853 Broadway, Room 2014, New York, NY 10003, (phone: 212-674-3322.

Multiple Chemical Sensitivity: A Scientific Overview, edited by Frank Mitchell, U.S. Department of Health and Human Services, Agency for Toxic Substances and Disease Registry, 1995. Published by Princeton Scientific Publishing Co., Inc. A compendium of the papers and proceedings from the first three conferences on MCS sponsored by U.S. government agencies.

Pest Control for Home and Garden: The Safest and Most Effective Methods for You and the Environment, by Michael Hansen and the editors of Consumer Reports Books, Yonkers, NY: Consumers Union, 1993 (phone: 914-378-2000).

Pest Control in the School Environment: Adopting Integrated Pest Management, EPA 735-F-93-012, August 1993. Published by the Office of Pesticide Programs, Environmental Protection Agency (phone: 202-260-2080 or 513-489-8190; fax: 513-489-8695 or 800-858-7378).

Epilogue

There have been many conferences, numerous publications, and increased dialogue on chemical sensitivity in the past five years, but from the perspective of those affected, not enough progress has occurred. Mechanisms for the condition remain unknown, treatments remain empirical, there still is no environmental medical unit for research, and doctors, insurance companies, employers, and patients' own friends and family members continue to impugn their credibility.

Understanding MCS is pivotal to establishing sound environmental policy. If there is a subset of the population that is especially sensitive to low-level chemical exposures, a strategy for protecting this subset must be found. If it were to be determined that certain chemical exposures could lead to MCS, then perhaps these exposures could be better controlled or avoided. Possibly by preventing chemical accidents, forbidding occupancy of buildings prior to finish-out or completion, avoiding the use of cholinesterase-inhibiting pesticides indoors, and other measures, society could protect more vulnerable individuals from becoming sensitized in the first place. Certainly, it would make little sense to regulate chemicals at the parts per billion level or lower if what was required was to keep people from becoming sensitized in the first place. Indeed, by understanding the true nature of MCS and who is at risk, we may prevent unnecessary and costly regulation of environmental exposures in the years to come.

We have dedicated considerable effort to describing the state of the science regarding low-level chemical sensitivity. Although the precise nature of this condition (or conditions) and the underlying mechanisms remain somewhat uncertain, chemical sensitivity seems sufficiently well-recognized to require attention from government, industry, and the medical profession. Their actions are needed for the establishment of regulations to minimize exposures to those chemicals that are suspected of initiating or seriously exacerbating illness, for the notification of sensitive or potentially sensitive populations of past or possible future exposures, for making accommodations in housing and employment, and for compensation for damage to health. The strength of the evidence (the "burden of proof" in legal terms) sufficient to trigger a particular regulatory, legal, or political response will vary according to the area of action, that is, control of exposures, notification, accommodation, or compensation.

Public policy needs to be focused on two distinct groups: (1) those individuals who could become sensitized as a result of an initiating exposure, and (2) those individuals who have already become sensitized and now report sensitivities to chemicals at extremely low levels. Regulations and policies need to be developed to prevent sensitization of individuals in the first place. Some sensitizing events may occur in domestic indoor or white-collar work environments (e.g., exposure to certain pesticides or chemicals used in remodeling). Other sensitizing events may occur in industrial workplaces (e.g., exposure to classical sensitizers such as toluene diisocyanate or to solvents), in contaminated communities, or as a result of exposure to consumer products, pharmaceuticals, or possibly anaesthetic agents. To prevent sensitization we would need to identify possible sensitizers and establish regulatory standards within the appropriate regulatory regime. If, in fact, chemical sensitivity proceeds through a neurotoxic mechanism, attention should be focused on the neurotoxicity of chemicals and the development of appropriate standards. To the extent that immunotoxic mechanisms are involved, attention should be directed toward immunotoxicity. The indoor air environment presents a particularly difficult regulatory challenge because it is not the sole domain of any regulatory agency, even though the Environmental Protection Agency has established an Office of Indoor Environments in its Office of Air and Radiation. Consumer products, building materials, and construction practices, as well as pesticide applications, are but a few of the areas that would need to be considered for regulation.

In addition to establishing regulations minimizing exposures to sensitizing and triggering chemicals, advance notification of possible sensitizing or triggering exposures should be considered. In Massachusetts, for example, advance notice of pesticide applications in both public buildings and apartment buildings is required.

For individuals who are already affected, public policy must focus on ways to accommodate them by minimizing chemical exposures in work environments, schools, and housing. Indeed the recently passed Americans with Disabilities Act requires "reasonable accommodation" for those individuals who are in fact or who are considered disabled. This means that although "proof" that a particular person has chemical sensitivity is absent, discrimination by employers, landlords, and others may be illegal if the person is *regarded* as disabled by the discriminator. This feature of the law may circumvent the need for persons who are the target of such discrimination to prove that they have chemical sensitivity, or that it, in fact, exists as a medical condition.

For a particular case, the accommodation may need to be either temporary or permanent in nature, depending on the person. Accommodation can take the form of providing a workplace with adequate ventilation, removal of offending substances, locating an employee in a temporary office, allowing the person to work at home, the cessation of pesticide application at certain times, changes in cleaning materials, and other approaches. Regulatory and corporate policies are needed to accommodate these individuals in an immediate, humane, and understanding manner.

At first blush, addressing the needs of people who are exquisitely sensitive to chemicals—at levels possibly several orders of magnitude lower than conventional toxic limits—might seem to present an impractical and insurmountable challenge. However, if an aggressive set of initiatives to prevent future sensitization were instituted, then we might be able to eliminate having a succeeding generation of sensitive individuals, and the need to accommodate (or compensate) such individuals would become much less burdensome in practice.

For persons whose health is already damaged, compensation can come from the workers' compensation, tort (court-awarded damages), Social Security, or private insurance disability systems. Under workers' compensation, the worker must prove that the injury was job-related by a preponderance of the evidence, that is, that his or her condition was more likely than not caused by (or exacerbated by) a workplace exposure. The same burden is required in a tort suit for damages in the courts against a product manufacturer, a pesticide applicator, or an owner of a building. The recent *Daubert* decision in the U.S. Supreme Court is having mixed impacts regarding the admissibility of scientific evidence related to chemical sensitivity in the courts in damage suits. The evidentiary burden in Social Security disability awards is less than it is in workers' compensation or the tort system, and the burden for private insurance coverage varies with the insurance company and policy.

In sum, chemical sensitivity is debilitating condition and a serious public health concern, but one that can be addressed by aggressive, coordinated public and private sector efforts. Making serious attempts to improve our understanding of the nature of this difficult and perplexing condition is the first step.

Appendix A: Health Effects Associated with Chemicals or Foods

Many, but by no means all, of the articles summarized here were written by clinical ecologists. Indeed, studies by *nonecologists* link rhinitis to laser printers and diesel exhaust; headaches, panic attacks, and kidney problems to solvents; heart arrhythmias to aerosol propellants; balance and memory difficulties to formaldehyde exposure; asthma to carbonless copy paper, perfume, and tobacco smoke; and connective tissue diseases to rocket fuel, vinyl chloride, and hair dye. By presenting this material, we are not affirming an environmental cause for these diseases but hoping to alert the reader to that possibility and the need for evaluating such patients in an environmental unit when more traditional approaches have failed. These references illustrate the range and diversity of health effects that some have associated with chemicals or foods. In some cases, mechanisms may be those of classical allergic sensitivity or toxicity, but for the most part the mechanisms have not yet been elucidated.

Alcoholism. Termed the "ultimate food addiction" by Randolph in the late 1940s (Randolph 1956, 1976c, 1980, pp. 109–116). According to him, people who drink heavily may be sensitive to the food from which the alcoholic beverage is derived. For example, bourbon drinkers may be sensitive to corn and may need to avoid all sources of corn in order to control their cravings (Randolph 1987, pp. 40–47). Individuals may first become aware of a food intolerance when they react adversely to a particular alcoholic beverage, for example, vodka (potatoes), wine (grapes), or bourbon (corn). Alcohol is quickly absorbed so that the individual is more aware of symptoms being associated with a particular alcoholic beverage than with the corresponding food. For example, a person who develops a headache only 3 minutes after drinking bourbon might have a headache 10 to 12 minutes after eating corn sugar or 20 to 25 minutes after cornstarch, and only a scratchy throat 2 to 3 hours after corn oil. The time for symptoms to become manifest thus depends upon the rate of absorption. According to Randolph (1987, p. 184), the majority of alcoholics in this country are addicted to corn; if they manage to abstain from alcohol, they often substitute corn sugar in the form of candy, ice cream, or some other corn-containing food.

Obesity.   According to Philpott (1976), it "characteristically involves addiction to several foods." Insatiable hunger or food cravings may emerge as a withdrawal symptom from certain foods or chemical exposures, making adherence to a weight-reduction diet exceedingly difficult (Randolph 1956). Diets structured around calorie restriction most often fail because foods that the dieter is addicted to and thus craves have not been eliminated. "Suffice it to say, briefly, that obesity and alcoholism are basically similar illnesses, one dealing with addicting foods in their edible form and the other in their potable form" (Randolph 1980, p. 100).

Tobacco Use.   Philpott and associates (1980) reported that 75 percent of schizophrenics he saw in his practice exhibited psychiatric symptoms when smoking cigarettes. Following a 2- to 3-week period of abstinence (deadaptation), 10 percent became psychotic upon reexposure to tobacco smoke. Many physicians recall with displeasure how smoky psychiatric ward lounges were when they rotated through psychiatry as students and recall being taught to allow a patient who seemed to be decompensating the opportunity to smoke a cigarette.

Tobacco belongs to the nightshade food family along with potato, tomato, eggplant, and green pepper. Smokers, who are addicted to tobacco, may have sensitivities to these foods as well; paradoxically, sensitivity may result in either a strong dislike for these foods or a craving for them, that is, addiction (Randolph 1987, p. 253).

Cardiac and Vascular Disease

Rea (1975, 1977, 1981, 1987c), a thoracic surgeon prior to his involvement with clinical ecology, reviews cardiovascular disease from the clinical ecologists' perspective.

Arrhythmias. See the Rea references for cardiac and vascular disease and also Boxer (1976) and Seyal (1986b, 1986d).

Vasculitis. See Cardiac and Vascular Disease.

Thrombophlebitis. See Cardiac and Vascular Disease.

Hypertension. Increased blood pressure, heart rate, and arrhythmias are attributed to the wearing of synthetic clothing versus cotton clothing by Seyal (1986b, 1986c).

Angina and Myocardial Infarctions. Numerous studies outside clinical ecology support the idea that environmental agents may trigger cardiac symptoms. Kalsner and Richards (1984) reported in *Science* that histamine levels are increased in coronary arteries of cardiac patients, suggesting that "an 'allergic' response as occurs in an antigen-antibody type reaction could induce a powerful contraction or spasm of a coronary vessel segment and precipitate a cardiac crisis such as angina or rhythm disruption." Nitroglycerin is widely recognized to be able to provoke angina as well as mitigate it (see Chapter 2). Fluorocarbons in aerosol propellants may precipitate arrhythmias (Speizer et al. 1975; Taylor and Harris 1970). See also the Rea references for cardiac and vascular disease and Seyal (1987/88).

Edema and Fluid-Retention Syndromes. See Cardiac and Vascular Disease.

Eye, Ear, Nose, and Throat Disorders

Ear, nose, and throat (ENT) symptoms secondary to environmental triggers may be a common early warning sign of environmental hyperreactivity, with heightened awareness and intolerance of odors being one of the most common symptoms. Roughly 20 percent of ENT physicians practice allergy themselves and perhaps about one third of these physicians are interested in chemical sensitivity. Some refer their patients with multisystem complaints to clinical ecologists. See Rea (1979a, 1979b).

Eye Disorders. For conjunctivitis, eczema of the eyelids, blurring of vision, tearing, light sensitivity (photophobia), and other eye problems, see Rapp and Bamberg (1986); Rapp (1978b); Sandberg (1987b); also indices of Randolph and Moss (1980) for individual cases.

Laryngeal Edema. Using videoendoscopy of the larynx, LaMarte and co-workers (1988) documented laryngeal edema in a patient exposed to the alkylphenol novolac resin used in making carbonless carbon paper. Concomitantly, plasma histamine levels rose sixfold from prechallenge levels. Similarly Selner and Staudenmayer (1985b) showed spasm of the pharyngeal constrictor muscles in a woman exposed to copy machine emissions when challenged in a blinded fashion with sham challenges as controls.

Meniere's Disease. See Eye, Ear, Nose, and Throat Disorders.

Otitis Media. Bernstein (1988) reported 2 to 3 times more serum IgE directed against milk, eggs, and wheat among a group of 10 otitis-prone children (six or more episodes in first 2 years of life) compared to 18 controls (less than four episodes in the first 2 years). See also Rea (1979a, 1979b), Pelikan (1987), Shambaugh (1983), and Boris et al. (1985b).

Rhinitis, Frequent Colds, Chronic Nasal Obstruction. Skoner and associates (1990) describe "laser printer rhinitis" in a 51-year-old man who experienced nasal congestion, skin irritation, headache, and chest and stomach discomfort when a new computer and laser printer were installed at his work station. Blinded challenge confirmed that laser printed paper, which emits various volatile organic compounds, including combustion products from styrene-butadiene toners, caused a three- to four-fold increase in nasal airway resistance as measured by computerized posterior rhinomanometry. See also Rea (1979a, 1979b) and Pelikan (1987).

Salivary Gland Disorders. See Eye, Ear, Nose, and Throat Disorders.

Sinusitis. See Eye, Ear, Nose, and Throat Disorders.

Vertigo, Hearing Loss, Tinnitus, Pressure in Ear. See Rea (1979a, 1979b), Odkvist et al. (1985), and Powers (1976).

Endocrine Disorders

See generally Saifer and Becker (1987).

Thyroid Dysfunction. Gaitan and associates (1985) hypothesize that organic and microbial water pollutants may be responsible for an increased incidence of goiter and autoimmune thyroiditis in certain regions. Polychlorinated biphenyls (PCBs) and polybrominated biphenyls (PBBs) may interfere with thyroid hormone secretion (Bastomsky 1985). Of workers at a plant manufacturing PBBs, 11 percent were hypothyroid and showed increased titers of thyroid antimicrosomal antibodies, perhaps from a PBB-induced autoimmune response (Bahn et al. 1980).

Premenstrual Syndrome. See Rea (1988b), Mabray et al. (1982), and Mabray (1982/1983).

Fatigue

Randolph's (1945, 1947, 1980, pp. 138–146) writings on this subject extending back to the 1940s suggest that fatigue syndrome, an illness currently the subject of great discussion among physicians, might be investigated on this basis (environmental unit, fasting). Indeed, Randolph reported seeing many atypical lymphocytes in the peripheral blood smears of chronic allergic patients, resembling mononucleosis, which has been linked by some to chronic fatigue syndrome (Randolph 1945). Fatigue is reportedly one of the most common manifestations of food and chemical sensitivity and resolves with avoidance of incriminated foods and chemicals. Drowsiness following a meal is said to be a common sign of food sensitivity. Features of chronic fatigue syndrome (Jessop 1990), such as its age of onset; female predominance; polysymptomatic complaints, especially neurological ones; normal examinations and routine lab findings; carbohydrate cravings; recurrent yeast infections; and immunologic abnormalities, such as frequent low titers of autoantibodies and altered cellular immunity, are strikingly similar to the features of multiple chemical sensitivity.

Gastrointestinal Disorders

Certain digestive tract disorders are clearly linked to foods; for example, gluten sensitive enteropathy is associated with wheat consumption.

However, traditional gastroenterologists doubt or remain uncertain about the role of foods in many conditions.

The following papers concern food as triggers of gastrointestinal disorders; however, chemical exposures are also reported (see Randolph's books) to result in increased food intolerance, bloating, heartburn, and other gastrointestinal manifestations.

Oral Manifestations Including Aphthous Ulcers. See Challacombe and Walker-Smith (1987), Ford (1987), and Hindle and Franklin (1986).

Celiac Disease (Gluten-Sensitive Enteropathy). See Mike and Asquith (1987).

Enterocolitis in Infants. See Van Sickle et al. (1985). Lymphocytes from infants with milk or soy intolerance (demonstrated by oral food challenge) had an augmented response to mitogen stimulation when cultured with soy protein or milk protein.

Gastroenteritis. See Trounce and Tanner (1985). Gross and co-workers (1989) describe a physician who was hospitalized four times and underwent biopsies of his esophagus, stomach, small bowel (three times), liver, bone marrow, gums, and skin in an effort to find the cause of his severe abdominal cramps. An elimination diet helped to identify yellow dye #6 as the offending agent. It was confirmed with double-blind challenge.

Inflammatory Bowel Disease. See Shorter (1987).

Chronic Ulcerative Colitis. Food sensitivity was proposed as the principal cause of ulcerative colitis by Rowe in 1942. See Rowe (1949) and McEwen (1987).

Irritable Bowel Syndrome. Jones and Hunter (1987) found that specific foods induced symptoms of irritable bowel syndrome in 14 of 21 patients; double-blind food challenges with six patients confirmed food intolerance. See also Jones (1982).

Gynecological Disorders

See Mabray et al. (1982) and Mabray (1982/1983, 1983).

Dysmenorrhea (Painful Menses). See Gynecological Disorders.

Premenstrual Syndrome (PMS).   See the Mabray references and also Rea (1988b).

Infertility.   See Gynecological Disorders.

Fibrocystic Breast Disease (Breast Tenderness).   Russell (1989) demonstrated that breast pain can be mitigated by eliminating caffeine from the diet. A study published in the *Journal of Allergy and Clinical Immunology* found that the total methyl xanthine content in the diet (including tea, coffee, chocolate, colas, and theophylline, which is used to treat asthma) was predictive of fibrocystic breast disease severity (Hindi-Alexander et al. 1985). Other papers confirm this relationship (Boyle et al. 1984) or dispute it (Levinson and Dunn 1986; Lubin et al. 1985). Jacobson and Liebman (1986) discussed the limitation of case-controlled studies of fibrocystic breast disease; such studies can easily miss an association if not all cases have the same disease or if all individuals are not equally sensitive to methyl xanthines. Interestingly, Jacobson and Liebman felt further case-controlled studies would be futile and suggested instead double-blind controlled challenges to resolve the debate once and for all.

More recently, Russell (1989) counseled 138 women with documented fibrocystic breast disease to reduce their caffeine intake. Of 113 women (81.9 percent) who decreased their caffeine consumption significantly, 69 (61 percent) experienced lessening of breast pain versus 21 percent of women who did not decrease their caffeine intake. See also the Mabray references.

Hematological Abnormalities

Anemia.   See Heiner et al. (1962). IgE is clearly involved in Heiner's syndrome, in which milk consumption results in anemia, poor weight gain, gastrointestinal symptoms, severe recurrent lung disease, and upper respiratory tract symptoms; other concomitant mechanisms play a role in this syndrome.

Thrombocytopenia.   See Caffrey and others (1981).

Neurobehavioral and Psychiatric Manifestations

Randolph (1980) describes the stimulatory and withdrawal effects of environmental incitants (see Chapter 2) and their psychiatric correlates, ranging from hyperactivity (+ +), autism, anxiety, mania, panic attacks, and seizures (+ + + +) at the furthest extreme, to withdrawal levels, including "brain fag," that is, impaired thinking ability (− − −), and severe depression (− − −, − − − −). King (1981) performed double-blind sublingual testing on 30 patients who complained of at least one psychological symptom by using conventional allergy food extracts as well as tobacco smoke extract and was able to provoke cognitive-emotional symptoms more frequently using these incitants than placebo (p = 0.001). See also Pearson and Rix (1987), Bell and King (1982), and Bell (1982, 1987a, 1987b).

Affective Disorders (Depression, Mania).   See Bell (1987a) and Randolph (1980, pp. 147–155).

Anxiety and Somatoform Disorder.   Bell (1987a) cautions: "Psychological diagnoses such as somatization disorder are always diagnoses of exclusion of organic factors. If a double-blind study is inadequately designed, researchers might miss a true biological effect and mistakenly conclude a psychogenic basis for the presenting complaints. In addition, the finding of psychological factors does not rule out biological components to a phenomenon." (See also Chapter 4.)

Sexual Dysfunction.   See Randolph (1976f).

Eating Disorders.   See Bell (1987a).

Hyperactivity.   Using a blinded study design, Kaplan and associates (1989) provided a replacement diet free of food additives and stimulants and low in simple sugars to 24 hyperactive preschool boys. In addition, any foods implicated by parents were avoided, and an attempt was made to reduce exposure to common environmental inhalants. Forty-two percent of the children showed approximately 50 percent improvement in behavior on the diet as well as having less halitosis, night awakenings, and difficulty getting to sleep. The authors stress that had they eliminated and tested only a *single* type of substance, such as sugar or dyes, as in others' studies, 10 percent or fewer of the children would have shown improvement.

See also Egger (1987b) and Rapp and Bamberg (1986).

Schizophrenia.   See "Tobacco Use" in this section. Philpott (1976, p. 16) wrote that the schizophrenic patient usually is sensitive to a wide assortment of substances. Foods most likely to provoke reactions included wheat (64 percent), corn (51 percent), and milk (50 percent). Tobacco and coffee also produced symptoms frequently. Avoidance of smoking improved psychosis in some schizophrenic patients, and rechallenge exacerbated their psychotic symptoms. King (1985) reviewed studies of wheat gluten as a factor in schizophrenia and reported that the studies with more adequate statistical power were positive and suggest that wheat gluten may provoke schizophrenic symptoms. See also Bell (1987a).

Panic Disorder.   Panic disorder has been associated with organic solvent exposure (Dager 1987) and caffeine consumption (Boulenger 1984). In one study, 20 patients with panic disorder were exposed to 5.5 percent carbon dioxide–enriched air, a mixture that provokes attacks in most patients with panic disorder (Sanderson 1989). All patients were told that illumination of a light in front of them would signal that they could dial downward the amount of CO_2 they were receiving. In fact, the dials were nonfunctional. For ten of the patients, the light was illuminated the entire time CO_2 was administered; for the other ten, it was not illuminated at all. Patients who believed they had control of the CO_2 experienced fewer and less intense panic disorder symptoms, suggesting that psychological factors (the illusion of control) can influence the biological response to a physical stressor.

Neurological Disorders

The occupational health literature is replete with reports of neurological impairment attributed to various chemicals, many of which were not previously recognized for these effects. Whether certain of these neurological sequelae represent a facet of the multiple chemical sensitivity syndrome or are subtle toxic effects not previously recognized remains to be determined. For example, formaldehyde, which for decades was considered an irritant, has now been linked with protracted impairment of memory, equilibrium, and dexterity in histology technicians (Kilburn 1987). Knowing if those technicians most affected by formaldehyde also experience central nervous system or multisystem effects from other chemical exposures such as perfume and diesel exhaust would be useful.

Headaches of almost any description (tension, migraine, "sinus") are considered by ecologists to be common manifestations of food and chemical intolerance (Randolph 1979). Randolph cautions that "headache diets" most often do not relieve the patient's symptoms either because they fail to exclude certain key foods or because important chemical exposures are not avoided. No single diet works for *all* patients; foods must be tested for each individual. Frequently patients note that a particular food relieves their headache yet are unaware that the same food may also be the cause (Randolph 1980, pp. 123–128).

Migraine.   See Monro (1987), Egger (1987a), Egger and co-workers (1983, 1989), and Mansfield and associates (1985). See also "Seizure Disorders."

Seizure Disorders.   Egger (1989) found improvement in 40 of 45 children with epilepsy *and* migraine placed on an oligo-antigenic elimination diet; complete control of seizures was achieved with 25 patients; double-blind, placebo-controlled challenges conducted in eight patients provoked seizures. None of 18 patients with epilepsy alone improved. Alternation between seizures (+ + + +) and headache (− −) (see Table 2-2) in some individuals is recognized by neurologists, but the mechanism is not known. Randolph's (1980) description of the levels of addiction provides a possible context for understanding this phenomenon. See also Bell (1987a).

A Swedish study found that 7.7 percent of 104 subjects with focal epilepsy had significant exposure to organic solvents in their jobs, for example, painters (Littorin 1988).

Sleep Disorders.   Sleep apnea, hypersomnia, narcolepsy, and restless legs syndrome are discussed in Bell (1987a).

Pulmonary Disorders

Asthma.   Bronchospasm in certain workers exposed to toluene di-isocyanate and certain other industrial chemicals is undisputed among medical practitioners; however, such responses to tobacco smoke or perfume are often questioned or dismissed as irritant reactions. Shim and Williams (1986), pulmonary specialists, challenged four asthmatics with cologne for 10 minutes; their pulmonary function tests (FEV_1) dropped 18 to 58 percent below baseline. Of 60 asthmatics they surveyed, 57 complained of respiratory symptoms with exposure to common "odors": insecticide (85 percent), household cleaners (78 percent), perfume or cologne (72 percent), cigarette smoke (75 percent), fresh paint (73 percent), auto exhaust fumes (60 percent), and cooking smells (37 percent).

Gerdes and Selner (1980) studied a 35-year-old steroid-dependent asthmatic who complained of worsening bronchospasm after eating corn. Double-blind challenges with D_5NS (intravenous normal saline with 5 percent dextrose, a corn-derived sugar) and plain normal saline showed a reproducible decrement in pulmonary function after dextrose only.

Stankus and associates (1988) studied 21 asthmatics (19 atopic) who complained of cough, shortness of breath, and chest tightness with exposure to cigarette smoke. Seven of 21 experienced significant, reproducible reductions in their ability to perform pulmonary function tests (more than 20 percent decrease in FEV_1) when exposed to cigarette smoke for 2 hours. The gradual declines in FEV_1 that occurred were unlike the usual early or late responses induced by classic allergen inhalation testing, and there was no association with serum IgE antibodies or skin tests to tobacco leaf extract. Accordingly, the authors comment that the mechanism of bronchospasm from cigarette smoke is unclear. Of interest is their finding that the other 14 asthmatics who claimed they were sensitive to cigarette smoke did *not* experience bronchospasm with challenge testing. These individuals might have shown positive challenges if the testing had been done while they were in the deadapted state, that is, after an appropriate interval (such as 4 to 7 days) away from cigarette smoke. Some of these individuals may have avoided smoke for weeks or longer and thus have lost their sensitivity.

Rodriguez de la Vega et al. related asthma in Cuba to exposures to kerosene fuel (Rodriguez de la Vega 1990). Of 286 asthmatic women followed for five years, only 15.5 percent of those who improved clinically had contact with kerosene, while 43.9 percent of those who failed to improve used kerosene as a cooking fuel. In 16 of the women, asthma began soon after they began using kerosene.

A variety of food additives, including sulfites (Bush 1986) and monosodium glutamate (Allen 1987), have been shown in blinded, placebo-controlled challenges to provoke asthma in certain individuals. See also Wraith (1987) and Hoj and co-workers (1981).

Pneumonitis.　See Heiner and associates (1962) and also "Anemia."

Renal and Urological Disorders

See Sandberg (1987) and Dickey (1976).

Enuresis (Bedwetting).　See Gerrard and Zaleski (1976).

Glomerulopathy.　Finn et al. (1980) report increased occupational exposure to hydrocarbons among patients whose renal failure resulted from glomerular nephritis. See also McCrory and associates (1986).

Nephrotic Syndrome.   Sandberg et al. (1977) discuss six cases of severe idiopathic nephrotic syndrome that were related to milk ingestion. See also Sandberg (1987a) and Dickey (1976).

Rheumatological Disorders

See Wojtulewski (1987).

Lupus Erythematosus.   Reidenberg and co-workers (1983) report the case of a laboratory worker exposed to hydrazine who developed a lupuslike syndrome with arthralgias, fatigue, malar (cheek) rash, photosensitivity, antinuclear antibody, and antibodies to native DNA. Symptoms cleared away from work and returned when an in-hospital challenge test with hydrazine was performed. Her lymphocytes, but not those of three normal controls, showed inhibition of mitogen-stimulated IgG synthesis following five daily exposures to hydrazine. Two major drugs, procainamide and hydralazine, which contain hydrazine moieties in their chemical structures, are widely recognized as causing lupuslike diseases. Hydrazine also occurs in a wide variety of natural and synthetic substances (over 30 million pounds of hydrazine are used by industry in the United States each year) such as mushrooms, tobacco smoke, plastics, rubber products, herbicides, pesticides, photographic supplies, textiles, dyes, and drugs. Tartrazine (FD&C yellow #5), which is found in thousands of foods and drugs, can be metabolized to hydrazine compounds and has been linked to one case of a lupuslike syndrome.

Scleroderma.   Scleroderma-like syndromes (scleroderma is a connective tissue disorder that can affect the skin, lung, esophagus, and other tissues) have been linked to a variety of environmental exposures including vinyl chloride, silica dust, organic solvents, epoxy resins, and ingestion of toxic cooking oil in Spain (Black 1988). Specific features of the illnesses resulting from each type of exposure vary somewhat, but overlap is significant. Other exposures that have been related to scleroderma-like illnesses include various drugs, breast augmentation (paraffin and silicone), and use of hair dyes (Fremi-Titulaer 1989). Hair dyes contain aromatic amines that are absorbed through the scalp and metabolized by acetylation in the liver. Individuals who are slow acetylators, that is, those whose enzymes do not break down these amines as readily, may be at greater risk for the disease.

Myalgia and Arthralgia.   According to Randolph (1976d), myalgias, arthralgias, and pain associated with both osteoarthritis and rheumatoid arthritis improve when incriminated food and chemical incitants are avoided.

Rheumatoid Arthritis.   In 1976 the American Arthritis Foundation concluded, "No food has anything to do with causing arthritis, and no food is effective in treating or curing it" (Skoldstam 1989). However, some rheumatologists have identified a few patients who seem to benefit from special diets. (See also Randolph 1976d.)

Kroker and associates (1984) and Marshall and associates (1984) describe a multicenter study conducted by clinical ecologists in which 43 patients with rheumatoid arthritis entered an environmental unit and underwent fasting followed by food challenge. Seven parameters of arthritis activity were measured, and all significantly improved during the fast ($p = 0.001$). Following challenge with provoking foods in 27 patients, joint pain and circumference increased, while grip strength decreased ($p = 0.001$).

Panush (1986a, 1986b), a rheumatologist, showed that rheumatoid arthritis improved significantly in 2 of 11 patients on a restricted diet (foods were not individually tested, nor were patients in an environmental unit). In the two patients who improved, symptoms recurred when they deviated from their diet, and double-blind food challenges demonstrated that specific foods exacerbated their symptoms. One wonders if more patients might have improved had foods been tested on an individual basis in a chemically controlled environment, as was done in the clinical ecology study. The clinical ecologists followed a more strict elimination diet (fasting) than did Panush. In addition, chemical exposures were controlled because ecologists' patients were in an environmental unit.

In an interesting animal study, Coombs and Oldham (1981) placed rabbits on cow's milk instead of water for 12 weeks and induced knee joint synovitis, in some cases quite severe. See also Randolph (1976d).

Other Arthritides.   Randolph (1980, p. 130) is of the opinion that Reiter's syndrome, ankylosing spondylitis, psoriatic arthritis, and other types of arthritis may also have environmental bases.

Skin Diseases

Atopic Dermatitis (Eczema).   That IgE-mediated food sensitivities have a role in some cases of atopic dermatitis is gaining wider acceptance by allergists and dermatologists. Studies have shown that at least one third of patients presenting to allergists or dermatologists with this condition have underlying food allergies (Burks et al. 1988). In addition to provoking skin symptoms, 30 percent of positive food challenges also resulted in gastrointestinal symptoms (nausea, vomiting, abdominal pain, or diarrhea) and 52 percent in respiratory symptoms (wheezing, nasal congestion, or sneezing). In select (referred) patients with eczema, Sampson (1985) found that foods provoked symptoms in about 56 percent of those who underwent double-blind, placebo-controlled food challenges. See also Pike and Atherton (1987).

Dermatitis Herpetiformis.   See Leonard and Fry (1987). That dermatitis herpetiformis is associated with gluten-sensitive enteropathy (celiac disease) and that gluten (for example, from wheat) plays a causal role in both this rash and the enteropathy are widely accepted.

Urticaria.   See Winkelmann (1987). Allergists recognize that a wide variety of foods and additives, including caffeine (Pola et al. 1988), can be potential triggers for urticaria and exercise-induced anaphylaxis. Contact urticaria and airway obstruction in response to carbonless copy paper have been reported (Marks et al. 1984). Delayed pressure urticaria has also been observed to clear during fasting and to recur with food challenge (Davis et al. 1986).

Appendix B: Laboratory Diagnostic Tests Used in MCS Studies

(Adapted from Kreutzer and Neutra 1996)

Immunologic Testing

IMMUNOGLOBINS (IgG, IgA, IgM, IgE)

- Results generally normal except for study by McGovern, who found low IgA in patients with food allergies (ACOP 1989).
- Positive antibody titers to formaldehyde-HSA in six chemically sensitive, formaldehyde-exposed cases (IgE in 2/6, IgG in 5/6, and IgM in 3/4 tested) (Thrasher et al. 1988).
- Autoantibody titers did not differentiate physician-diagnosed MCS cases (41) from controls (34) (Simon et al. 1993).
- Scattered out-of-range values for IgG, IgA, IgM, IgE among 11 MCS cases meeting Cullen case definition (Fiedler et al. 1992).

COMPLEMENT COMPONENTS (C3, C4, total hemolytic complement, circulating immune complexes)

- Various reports of low C3 (<80 mg/dL), high C4 (>40mg/dL), low and high hemolytic complement, and circulating immune complexes detected in some patients (ACOP 1989).

- C3, C4, CH50 testing of 11 cases revealed no significant abnormalities (Fiedler et al. 1992).

LYMPHOCYTES (T-cells, B-cells, CD4 helper cells, CD8 suppressor cells, CD14, CD25, CD26, CD45, CD45R, CD56, HLA markers, interleukin)

- Various reports of high, normal, and low T-cell counts; low and normal B-cell counts; normal CD4s; normal and low CD8 counts (ACOP 1989).
- Increased T4/T8 ratio (Johnson and Rea 1989) and decreased T4/T8 ratio (Levin and Byers 1987).
- Interleukin-1 lower among cases (41) than controls (34), but difference appeared attributable to laboratory methods. CD26 significantly higher among controls; otherwise testing did not differentiate between patients and controls (Simon et al. 1993).
- Increased frequency of CD26 in chemically sensitive cases (Madison et al. 1991).
- Circulating lymphocyte, B-cell, T-cell, and T-cell subset counts not significantly abnormal in a group of 50 patients diagnosed with MCS by clinical ecologists (Terr 1986).
- Elevated Ta1 cells (antigen memory cells) in six of six chemically sensitive individuals with a history of chronic formaldehyde exposure (Thrasher et al. 1988).

INFLAMMATORY MEDIATORS (serotonin, histamine, prostaglandin F2 alpha, epinephrine, norepinephrine, dopamine)

- Greater changes in serotonin, histamine, prostaglandin, epinephrine, norepinephrine, and dopamine levels among six patients challenged with foods, phenol, or petroleum distillate compared to controls (McGovern 1983).

OTHER BLOOD-BORNE MEDIATORS—TOXICANTS, ANTIBODIES, PROTEINS (immunoelectrophoresis, isoantibody titers, immunofluorescent monoclonal antibodies, protein electrophoresis, antimyelin antibodies, blood levels of pesticides, organic compounds, heavy metals)

SKIN PRICK TESTING (Type 1 hypersensitivity)

- Three of ten subjects meeting Cullen criteria for MCS reacted to at least one antigen (Fiedler et al. 1992).

DELAYED-TYPE HYPERSENSITIVITY TESTING

- Anergy to four recall antigens in two of eleven subjects with MCS (Fiedler et al. 1992).
- No significant findings (Fiedler et al. 1992).

Respiratory Tests

PULMONARY FUNCTION TESTS

- 62 percent of 78 MCS patients had abnormal findings, typically a decrease of FEF 25–75% to below 70 percent of predicted value, indicating small airway disease (Heuser et al. 1992).
- Seven of eight chlorine dioxide-exposed workers with MCS who were tested showed evidence of airway obstruction (Meggs et al. 1996a).
- Seven of seven patients tested, who met the Cullen case definition for MCS, had normal pulmonary function tests (Meggs and Cleveland 1993).

METHACHOLINE CHALLENGE TEST (assesses airway hyperactivity in response to increasing concentrations of methacholine)

- Negative in two out of two patients who were tested and who met the Cullen case definition for MCS (Meggs and Cleveland 1993).
- Positive in seven of eight chlorine dioxide–exposed workers with MCS-like symptoms who were tested (Meggs et al. 1996a).

ANTERIOR RHINOMANOMETRY

- Cases had significantly higher nasal resistance on both inhalation and exhalation than controls both before and after odor threshold testing for phenyl ethyl alcohol (rose oil) and methyl ethyl ketone (Doty et al. 1988).

FIBEROPTIC RHINOLARYNGOSCOPY

- General pattern of edema, accompanied by abnormal amount of mucus, found in each of ten patients meeting Cullen case definition, including one patient without nasal symptoms. Marked cobblestoning found in the mucosa of the proximal pharynx in six patients; focal areas of blanched mucosa surrounding a large prominent vessel found in eight patients (Meggs and Cleveland 1993).

NASAL BIOPSY

- Greater inflammation and increased number of nerve fibers in nasal biopsies in 13 chlorine dioxide-exposed workers with MCS symptoms than in 3 normal controls; staining for Substance P and VIP nonspecific and inconclusive (Meggs et al. 1996a).

CHEST X RAYS

- Abnormalities noted in 16 percent of 32 patients tested (Heuser et al. 1992).

SINUS X RAYS

- Abnormalities found in one-half of the patients tested (Heuser et al. 1992).

Cardiovascular Tests

HEART RATE CHANGE UNDER STRESS

- 24 persons with self-reported food and chemical intolerances showed greater heart rate acceleration during a timed mental arithmetic task (serial 7 subtraction from 1,000) than during isometric exercise. Controls (15) showed the opposite pattern (Bell et al. 1993c).

Neurological Tests

ELECTROENCEPHALOGRAPHY (EEG)

- The distribution of subjects across eight EEG spectral categories not significantly different for physician-referred MCS cases (58) compared to outpatient psychologic controls (89), but both MCS cases and psychologic controls showed significantly different distributions from nonpsychologic controls. Similar results found for EEG beta activity (Staudenmayer and Selner 1990).
- EEGs (including spontaneous sleep, hyperventilation, and photic stimulation) showed mild abnormalities in 45 percent of MCS patients. Computerized analysis (BEAM) of EEG activity per-

formed on seven patients showed abnormal findings for three of them (Heuser et al. 1992).

SCALP ELECTROMYOGRAPHY (EMG)

- Significantly higher levels of EMG scalp activity observed in a greater number of "universal reactors" (58) compared to psychologic (89) and nonpsychologic controls (55) (Staudenmayer and Selner 1990).

PERIPHERAL TEMPERATURE

- No significant difference between "universal reactors" and outpatient psychologic and nonpsychologic controls (Staudenmayer and Selner 1990).

SKIN RESISTANCE LEVEL (SRL)

- No significant difference between "universal reactors" and outpatient psychologic and nonpsychologic controls (Staudenmayer and Selner 1990).

MAGNETIC RESONANCE IMAGING (MRI)

- 28 percent of 54 MCS patients showed abnormalities, including a definite impression of atrophy (13 percent) or demyelinating disease (7 percent) (Heuser et al. 1992).

VISUAL EVOKED RESPONSE (VER)

- 8 of 32 MCS patients tested had abnormal findings (Heuser et al. 1992).

BRAIN STEM AUDITORY EVOKED RESPONSES (BAER)

- 6 of 18 MCS patients tested had abnormal findings (Heuser et al. 1992).

SINGLE PHOTON EMISSION COMPUTED TOMOGRAPHY (SPECT)

- Findings for three of four MCS patients abnormal (Heuser et al. 1992).

- Six of six Gulf War veterans with MCS who were scanned had abnormal scans (Simon et al. 1994).
- Decreased cerebral blood flow (right side worse than left) with scattered areas of hypoperfusion especially in the dorsal frontal and parietal lobes in 41 MCS patients (15 exposed to pesticides, 29 to solvents) (Heuser et al. 1994).

NERVE CONDUCTION STUDIES

- Over half of 13 MCS patients tested showed abnormal responses (Heuser et al. 1992).

PERCEPTION THRESHOLDS

- Three of seven MCS patients tested with a neurometer had abnormal results (Heuser et al. 1992).

LANTHONY D-15 COLOR VISION TEST

- Of 19 workers who became ill with MCS following exposure to pesticides, 11 had blue-yellow color vision loss and 3 had complex color vision loss (Cone and Sult 1992).

Psychological and Behavioral Tests Used in MCS Studies

Neuropsychological Tests

DIGIT SPAN (test of concentration requiring subjects to repeat digits forward and backward after examiner presentation)

- No significant findings (Fiedler et al. 1992).
- No significant findings (Fiedler et al. 1996b).

DIGIT SYMBOL (test of concentration requiring coding of abstract symbols according to a key)

- No significant findings (Fiedler et al. 1992).
- No significant findings (Fiedler et al. 1996b).

CVLT (CALIFORNIA VERBAL LEARNING TEST) (assesses immediate and delayed verbal memory)

- MCS cases had relatively more difficulty with learning new information compared to recalling information or concentration tasks (Fiedler et al. 1992).

- No significant findings (Fiedler et al. 1996b).

CONTINUOUS VISUAL MEMORY TEST (CVMT) (assesses ability to discriminate between target and nontarget designs)

- MCS subjects (23) whose illness began following a defined exposure less able to discriminate target from nontarget designs (increased False Alarm rate); however, chemically sensitive subjects without a clear date of onset performed normally on CVMT (Fiedler et al. 1996b).

VISUAL REPRODUCTION (assesses immediate and delayed visual memory)

- No significant findings (Fiedler et al. 1992).

GROOVED PEGBOARD (assesses visuomotor coordination)

- No significant findings (Fiedler et al. 1996b).

STROOP COLOR WORD TEST

- No significant findings (Fiedler et al. 1996b).

SIMPLE REACTION TIME

- No significant findings (Fiedler et al. 1996b).

CONTINUOUS PERFORMANCE–MEDIAN REACTION TIME

- No significant findings (Fiedler et al. 1996b).

HAND-EYE COORDINATION (computerized task)

- No significant findings (Fiedler et al. 1996b).

SERIAL SUBTRACTION

- Higher prevalence of subjective difficulty concentrating and trend toward poorer objective performance in patients with food and chemical sensitivities than in normal controls (Bell et al. 1993c).

DIVIDED ATTENTION TASK (DAT) (computer-administered measure of vigilance)

- 15 older adults with high chemical intolerance scores (versus 15 controls with low chemical intolerance scores) showed slower reaction times in registering both centrally and peripherally placed stimuli. No difference in target tracking errors observed (Bell et al. 1996b).
- 8 middle-aged adults with chemical intolerance but no life-style changes performed progressively worse with repeated testing on the DAT than did 8 chemically tolerant adults with life-style changes or 10 normals (Bell et al. 1997a).

Psychological Tests

SCID-III-R (STRUCTURED CLINICAL INTERVIEW FOR THE DSM-III-R) (assesses current and previous psychiatric symptomatology and diagnoses)

- None of 11 subjects tested fulfilled criteria for premorbid psychiatric disorders, although premorbidly, subjects had an average of seven unexplained somatic symptoms and three somatic symptoms for which a medical diagnosis could be given. Four were diagnosed as currently depressed (Fiedler et al. 1992).
- 74 percent of MCS patients (23) with a history of a defined initiating exposure, and 61 percent of CFS patients (18) did not meet criteria for any current psychiatric disorder; 69 percent of chemically sensitive patients without a clear date of onset did meet criteria. Most common symptoms were depression and some symptoms associated with somatization disorder (Fiedler et al. 1996b).

MMPI (MINNESOTA MULTIPHASIC PERSONALITY INVENTORY)

- Before/after testing on 42 subjects showed improved scores on depression and "energy" scales following treatment in clinical ecology facility (Bertschler 1985).
- 7 of 11 subjects had test score profiles associated with somatoform disorder (Fiedler et al. 1992).
- 44 percent of MCS patients (23) with a history of a defined initiating exposure and 42 percent of MCS patients (13) without a clear date of onset, compared with 53 percent of chronic fatigue syndrome patients (18) and 0 percent of matched controls (18), had

significant elevations on scales associated with somatoform disorders (Fiedler et al. 1996b).

CAQ (CLINICAL ANALYSIS QUESTIONNAIRE) (personality inventory)

- Before/after testing on 42 subjects showed improved scores following treatment in a clinical ecology facility (Bertschler 1985).

WAIS (WESCHLER ADULT INTELLIGENCE SCALE) (standard intelligence test)

- Before/after testing on 42 subjects showed improved scores after treatment in clinical ecology facility (Bertschler 1985).

DIS (DIAGNOSTIC INTERVIEW SCHEDULE) (measures current/past major mental disorders using DSM-III criteria)

- 15 of 23 recruited, clinical ecologist–diagnosed MCS patients (65 percent) had current or past mood, anxiety, or somatoform disorder compared to 13 (28 percent) of controls (Black et al. 1990).
- Only major depression and panic disorder occurred frequently enough for analysis among 37 aerospace workers, but neither significantly associated with development of MCS. No subjects endorsed symptoms of post-traumatic stress disorder. However, the reported number of prior medically unexplained symptoms was the strongest predictor of MCS (Simon et al. 1990).

ILLNESS BEHAVIOR QUESTIONNAIRE (assesses somatic concerns/hypochondriacal symptoms)

- Used, but results not reported (Black et al. 1990).

SCL 90-R (SYMPTOMS CHECKLIST 90) (assesses presence and severity of somatic and psychological symptoms)

- Significant difference between subjects (13) and controls (23) on somatization scale (Simon et al. 1990).
- Increased scores on all subscales of SCL-90-R in college students with cacosmia (Bell et al. 1996a).

WHITELY INDEX (abridged version of Pilowsky's Illness Behavior Questionnaire, a yes/no scale that assesses hypochondriacal beliefs and behaviors)

- Difference between subjects (13) and controls (23) approached significance. Subject scores typical of hypochondriasis (Simon et al. 1990).

BARSKY SOMATIC SYMPTOM AMPLIFICATION SCALE (self-reported measure of tendency to amplify bodily sensations and to report physical symptoms)

- Increased score in subjects (13) versus controls (23) (Simon et al. 1990).
- Increased score in subjects (28) versus controls (20) (Bell et al. 1995a).

MCLEAN LIMBIC SYMPTOM CHECKLIST (self-reported measure of limbic system symptomatology related to temporal lobe epilepsy)

- Increased limbic system checklist score in college students with cacosmia (self-reported illness from chemical odors) (Bell et al. 1995b).

WEINBERGER ADJUSTMENT INVENTORY (a 37-item scale that generates subscale scores on depression, anxiety, and defensiveness)

- No significant difference on depression and defensiveness between older adult group with high versus low chemical and/or food sensitivity (Bell et al. 1993a).
- Weak correlation between high cacosmia (self-reported illness from chemical odors) and depression and anxiety among college students (Bell et al. 1993b).

PEARLIN-SCHOOLER MASTERY SCALE (a seven-item scale measuring sense of control over one's own fate)

- No significant difference between older adult group with high versus low chemical and/or food sensitivity (Bell et al. 1993a).

MARLOW-CROWNE SOCIAL DESIRABILITY SCALE (true/false questions on repression and defensiveness as a personality trait)

- No significant difference between groups of college students classified according to degree of self-reported cacosmia (self-reported illness from chemical odors) (Bell et al. 1995a).

CHEEK-BUSS SHYNESS SCALE

- Increased shyness found among cacosomics in a young adult population (Bell et al. 1994a) and in an older population (Bell et al. 1994b).
- Shyness accounted for a very limited portion (5.8 percent) of variance in cacosmia (self-reported illness from chemical odors) among college students (Bell et al. 1993b).
- Persons from the community with cacosmia (self-reported illness from chemical odors) but without health problems were more shy than newsletter-recruited MCS patients (Bell et al. 1995a).

Social Functioning Tests

PAIS-SR (PSYCHOSOCIAL ADJUSTMENT TO ILLNESS SURVEY—SELF-REPORT) (assesses health care orientation, vocational, social, and domestic environment, sexual relationships, extended family relationships, and psychological distress)

- MCS cases exhibited significantly greater concern about health care and psychosocial disruption than diabetic control group (Fiedler et al. 1992).

AS-SR (SOCIAL ADJUSTMENT SCALE—SELF-REPORT) (assesses social functioning compared to psychiatric subjects)

- Cases had psychosocial disruption scores equivalent to scores reported by depressed or alcoholic subjects (Fiedler et al. 1992).

Appendix C: Environmental Exposure and Sensitivity Inventory

The Environmental Exposure and Sensitivity Inventory (EESI) is a clinical instrument originally developed for research on chemically sensitive populations in whom MCS began after a well-defined exposure, for example, to a pesticide or air in a sick building (Miller and Mitzel 1995). This instrument has subsequently been used to evaluate Gulf War veterans (Miller 1994b; 1996b; 1996c) and implant patients and others (Gammage et al. 1996).

The EESI provides physicians with a rapid, broad-based overview of their patients' chemical, food, and drug sensitivities and helps them gauge how severe patients feel their symptoms are, both before and since an exposure event. If there is no history of an initiating event, the physician may opt to have the patient fill out only the "before" section for each question. Physicians also may find it useful to have patients record their particular symptoms in the margins next to questions 18–27 and 33–57. This will expedite history-taking. The EESI can be readministered at intervals in order to help assess any changes in symptom severity over time or following a particular intervention.

Embedded in the questionnaire* are five scales:

1. The EESI Symptom Scale (Q18-Q27). Respondents rate on a 0 to 10 scale the severity of their symptoms in ten categories: Musculoskeletal; air-

---

*UTHSCSA © 1996

371

way/mucous membrane; heart-related; gastrointestinal; cognitive; affective; neuromuscular; head-related; skin; and genitourinary. Maximum possible score = 100.

2. A Chemical Inhalant Sensitivity Scale (Q33-Q42). Respondents rate (0 to 10) the severity of their symptoms in response to ten common chemical inhalants: Diesel/gas engine exhaust; tobacco smoke; insecticide; gasoline vapors; paint or paint thinner; cleaning products; fragrances; fresh tar; nailpolish/remover or hairspray; and odors (outgassing) from new furnishings. Maximum possible score = 100.

3. An "Other" Sensitivity Scale (Q44-Q53). Respondents rate (0 to 10) the severity of their symptoms when exposed to foods, drugs, alcohol, caffeine, skin contactants, and classical allergens, as well as the severity of any withdrawal symptoms associated with caffeine or a small amount of alcohol. Maximum possible score = 100.

4. A Masking Scale (Q58-Q67). Respondents indicate whether they routinely are exposed to tobacco smoke, caffeine, fragrances, or other substances that may reduce their awareness of possible relationships between their symptoms and particular exposures. Maximum possible score =10.

5. A Life Impact Scale (Q68-Q77). Respondents rate (0 to 10) the degree to which their sensitivities have impacted ten aspects of daily living: Diet; ability to work or go to school; home furnishings; clothing; travel/driving; choice of personal care products; social activities; hobbies/recreational; family relationships; and ability to clean/maintain home. Maximum possible score = 100.

NAME: _____

DATE: _____

ENVIRONMENTAL EXPOSURE AND SENSITIVITY INVENTORY (EESI)
(PRE/POST EXPOSURE EVENT)

Q1   Birthdate: ____   ____   ____
                              m m   d d   y y

Q2   Sex (circle one)   1. Male
                              2. Female

Q3   Ethnicity (circle one) 1. Hispanic
                              2. Non-Hispanic white
                              3. Black
                              4. Other

Q4   Education (circle highest   1 2 3 4 5 6 7 8   9 10 11 12
      educational level achieved): Grade School   High School

              13 14 15 16   17 18 19 20 or more
              College   Post-graduate

About Your Exposure

Q 5   You have indicated that you became ill after a particular exposure or time-limited series of exposures.

When did your exposure(s) occur or begin?   ____ ____
mm   yy

Q 6   Did you experience any symptoms   1. Yes 3 **Describe below**
at the time you were exposed or within   2. No
the first 24 hours after your exposure(s)?   3. Don't know

_____

_____

_____

Q 7   Describe where and how the exposure(s) occurred. Include the names of any chemicals or products to which you were exposed. Be as specific as you can.

_____

_____

_____

_____

_____

_____

_____

Q 8   Are you still exposed to any of these   1. Yes
chemicals or products?   2. No
                                               3. Don't know

About Your Health/Ability to Work

Q9 Rate each of the following on a 0–10 scale (circle one):

Q9a Your health status **before** the
exposure(s):

0 1 2 3 4 5 6 7 8 9 10
very poor   excellent

Your health status **since** the
exposure(s):

0 1 2 3 4 5 6 7 8 9 10
very poor   excellent

Q9b Your average energy level
before the exposure(s):

0 1 2 3 4 5 6 7 8 9 10
very poor   excellent

Your average energy level
since the exposure(s):

0 1 2 3 4 5 6 7 8 9 10
very poor   excellent

Q9c Your average level of body pain
before the exposure(s):

0 1 2 3 4 5 6 7 8 9 10
no pain   severe pain

Your average level of body pain
since the exposure(s):

0 1 2 3 4 5 6 7 8 9 10
no pain   severe pain

Q9d Your quality of life **before** the
exposure(s):

0 1 2 3 4 5 6 7 8 9 10
very poor   excellent

Your quality of life **since** the
exposure(s):

0 1 2 3 4 5 6 7 8 9 10
very poor   excellent

Q10   Just prior to the exposure(s),   1. Yes
were you employed? (circle one)   2. No

Q11   Were you physically able to work   1. Yes
at least 40 hours a week **before**   2. No
the exposure(s)?   3. Don't Know

Q12   Are you currently employed?   1. Yes
(circle one)   2. No

Q13   Are you physically able to work   1. Yes
at least 40 hours per week now?   2. No
                                     3. Yes, but I can no longer do work
                                        that is physically demanding.
                                   4. Don't know

Q14   Has your illness affected   1. Yes
your ability to work? (circle one)   2. No → **skip to Q16**
                                   3. Don't know → **skip to Q16**

Q15   How has your illness affected   1. I cannot work at all.
your ability to work? (circle   2. I cannot work as many hours
all that apply)   as before.
                                   3. I changed jobs or occupations
                                       because of my health.
                                   4. I quit or was asked to leave a job
                                       because my health was affecting
                                       my ability to work.
                                 5. My schedule or duties at work
                                       have been changed to
                                     accommodate my health
                                     problems.

Q16   **Since the exposure(s),** how many
doctors have you seen for
conditions you feel may be
related to your exposure(s)?   _____ (number of doctors
                                                                    seen)

Q17   **Since the exposure(s),** how many
times have you gone to an emergency
room or been hospitalized for
conditions you feel may be related
to your exposure(s)?   _____ (total number of
                                                                    emergency room visits
                                                                    and hospitalizations)

Symptoms

The following questions ask about symptoms you may have experienced commonly before or since the exposure(s). Rate the severity of your symptoms on a 0 to 10 scale: 0 = not at all a problem; 5 = moderate symptoms; 10 = disabling symptoms.

0 = not at all a problem
5 = moderate symptoms
10 = disabling symptoms

Q18   Problems with your muscles or joints, such as pain, aching, cramping, stiffness, or weakness?

Before the exposure(s): (circle one)   0 1 2 3 4 5 6 7 8 9 10

Since the exposure(s): (circle one)   0 1 2 3 4 5 6 7 8 9 10

Q19   Problems with burning or irritation of your eyes or problems with your airway or breathing, such as feeling short of breath, coughing, or having a lot of mucus, post-nasal drainage, or respiratory infections?

Before the exposure(s): (circle one)   0 1 2 3 4 5 6 7 8 9 10

Since the exposure(s): (circle one)   0 1 2 3 4 5 6 7 8 9 10

Q20   Problems with your heart or chest, such as a fast or irregular heart rate, skipped beats, your heart pounding, or chest discomfort?

Before the exposure(s): (circle one)   0 1 2 3 4 5 6 7 8 9 10

Since the exposure(s): (circle one)   0 1 2 3 4 5 6 7 8 9 10

Q21   Problems with your stomach or digestive tract, such as abdominal pain or cramping, abdominal swelling or bloating, nausea, diarrhea, or constipation?

Before the exposure(s): (circle one)   0 1 2 3 4 5 6 7 8 9 10

Since the exposure(s): (circle one)   0 1 2 3 4 5 6 7 8 9 10

Q22	Problems with your ability to think, such as difficulty concentrating or remembering things, feeling spacey, or having trouble making decisions?	
	Before the exposure(s): (circle one)	0 1 2 3 4 5 6 7 8 9 10
	Since the exposure(s): (circle one)	0 1 2 3 4 5 6 7 8 9 10
Q23	Problems with your mood, such as feeling tense or nervous, irritable, depressed, having spells of crying or rage, or loss of motivation to do things that used to interest you?	
	Before the exposure(s): (circle one)	0 1 2 3 4 5 6 7 8 9 10
	Since the exposure(s): (circle one)	0 1 2 3 4 5 6 7 8 9 10
Q24	Problems with your balance or coordination, with numbness or tingling in your extremities, or with focusing your eyes?	
	Before the exposure(s): (circle one)	0 1 2 3 4 5 6 7 8 9 10
	Since the exposure(s): (circle one)	0 1 2 3 4 5 6 7 8 9 10
Q25	Problems with your head, such as headaches, or a feeling of pressure or fullness in your face or head?	
	Before the exposure(s): (circle one)	0 1 2 3 4 5 6 7 8 9 10
	Since the exposure(s): (circle one)	0 1 2 3 4 5 6 7 8 9 10
Q26	Problems with your skin, such as a rash, hives, or dry skin?	
	Before the exposure(s): (circle one)	0 1 2 3 4 5 6 7 8 9 10
	Since the exposure(s). (circle one)	0 1 2 3 4 5 6 7 8 9 10
Q27	Problems with your urinary tract or genitals, such as pelvic pain or frequent or urgent urination? (for women: or discomfort or other problems with your menstrual periods?)	
	Before the exposure(s): (circle one)	0 1 2 3 4 5 6 7 8 9 10
	Since the exposure(s): (circle one)	0 1 2 3 4 5 6 7 8 9 10
Q28	Do you consider yourself sensitive to everyday chemicals like those in household cleaning supplies, paints, perfumes, soaps, garden sprays, or things like that? (circle one)	1. Yes 2. No → **Go to Q30** 3. Don't know → **Go to Q30**

Q29 When you are exposed to chemicals, what kinds of symptoms do you **commonly** experience? (circle all that apply)

1. Headaches or head/sinus pressure
2. Dizziness or lightheadedness
3. Breathing problems
4. Burning of your eyes, nose, and/or airway
5. Upset stomach, nausea, or vomiting
6. Fatigue or general weakness
7. Concentration difficulties or confusion
8. Problems with your vision such as focusing your eyes
9. Muscle or joint pain
10. Heart pounding or beating irregularly
11. Feeling irritable, jittery, or nervous
12. Feeling depressed
13. Problems with coordination
14. Speech problems
15. Skin rash or hives
16. Sudden need to urinate or defecate
17. Other _____
18. Other _____

Q30 **Prior to the exposure(s),** did you consider yourself especially sensitive to everyday chemicals like those in household cleaning supplies, paints, perfumes, soaps, garden sprays, or things like that? (circle one)

1. Yes
2. No → **skip to Q33**
3. Don't know → **skip to Q33**

Q31 How old were you when you first noticed your sensitivity to everyday chemicals? (circle one)

1. Age____(approximate)
2. Entire life → **skip to Q33**
3. Don't know → **skip to Q33**

Q32 Was there something that happened when you were that age that you feel caused this sensitivity? (circle one)

1. Yes — What? _____
2. No
3. Don't recall any specific event

Exposures

The following items ask about various odors or chemical exposures. Please indicate whether or not these substances would make you feel sick. By sick we mean you would get a headache, have difficulty thinking, feel weak, have trouble breathing, get an upset stomach, feel dizzy, or something like that. For any substance that makes you feel sick, on a 0 to 10 scale rate the severity of your symptoms with that substance both before and since the exposure(s) you feel caused your chemical sensitivities.

0 = **not at all a problem**
5 = **moderate symptoms**
10 = **disabling symptoms**

Q33	**diesel or gas engine exhaust**	
	Before the exposure(s): (circle one)	0 1 2 3 4 5 6 7 8 9 10
	Since the exposure(s): (circle one)	0 1 2 3 4 5 6 7 8 9 10
Q34	**tobacco smoke**	
	Before the exposure(s): (circle one)	0 1 2 3 4 5 6 7 8 9 10
	Since the exposure(s): (circle one)	0 1 2 3 4 5 6 7 8 9 10
Q35	**insecticide**	
	Before the exposure(s): (circle one)	0 1 2 3 4 5 6 7 8 9 10
	Since the exposure(s): (circle one)	0 1 2 3 4 5 6 7 8 9 10
Q36	**gasoline,** for example, at a service station while filling the gas tank	
	Before the exposure(s): (circle one)	0 1 2 3 4 5 6 7 8 9 10
	Since the exposure(s): (circle one)	0 1 2 3 4 5 6 7 8 9 10
Q37	**paint or paint thinner**	
	Before the exposure(s): (circle one)	0 1 2 3 4 5 6 7 8 9 10
	Since the exposure(s): (circle one)	0 1 2 3 4 5 6 7 8 9 10
Q38	**cleaning products,** such as disinfectants, bleach, bathroom cleansers, or floor cleaners	
	Before the exposure(s): (circle one)	0 1 2 3 4 5 6 7 8 9 10
	Since the exposure(s): (circle one)	0 1 2 3 4 5 6 7 8 9 10

Q39   certain perfumes, air fresheners, or other
fragrances

Before the exposure(s): (circle one)   0 1 2 3 4 5 6 7 8 9 10

Since the exposure(s): (circle one)   0 1 2 3 4 5 6 7 8 9 10

Q40   **fresh tar or asphalt**

Before the exposure(s): (circle one)   0 1 2 3 4 5 6 7 8 9 10

Since the exposure(s): (circle one)   0 1 2 3 4 5 6 7 8 9 10

Q41   **nailpolish, nailpolish remover, or
hairspray**

Before the exposure(s): (circle one)   0 1 2 3 4 5 6 7 8 9 10

Since the exposure(s): (circle one)   0 1 2 3 4 5 6 7 8 9 10

Q42   **new furnishings,** such as new carpeting,
a new soft plastic shower curtain, or the
interior of a new car

Before the exposure(s): (circle one)   0 1 2 3 4 5 6 7 8 9 10

Since the exposure(s): (circle one)   0 1 2 3 4 5 6 7 8 9 10

Q43   forgetting where you are or becoming
confused while **driving**

Before the exposure(s): (circle one)   0 1 2 3 4 5 6 7 8 9 10

Since the exposure(s): (circle one)   0 1 2 3 4 5 6 7 8 9 10

Q44   **chlorinated tap water**

Before the exposure(s): (circle one)   0 1 2 3 4 5 6 7 8 9 10

Since the exposure(s): (circle one)   0 1 2 3 4 5 6 7 8 9 10

Q45   **particular foods,** such as candy, pizza,
milk, fatty foods, meats, barbecue,
onions, garlic, spicy foods, or food
additives such as MSG

Before the exposure(s): (circle one)   0 1 2 3 4 5 6 7 8 9 10

Since the exposure(s): (circle one)   0 1 2 3 4 5 6 7 8 9 10

Q46   unusual **cravings** or eating any foods as
though you were addicted to them; or
feeling ill if you miss a meal

Before the exposure(s): (circle one)   0 1 2 3 4 5 6 7 8 9 10

Since the exposure(s): (circle one)   0 1 2 3 4 5 6 7 8 9 10

Q47 **feeling ill after meals**

Before the exposure(s): (circle one)   0 1 2 3 4 5 6 7 8 9 10

Since the exposure(s): (circle one)   0 1 2 3 4 5 6 7 8 9 10

Q48 **caffeine**, such as coffee, tea, cola drinks, caffeine-containing sodas (such as Big Red, Dr. Pepper, Mountain Dew, Snapple) or chocolate

0 1 2 3 4 5 6 7 8 9 10

Before the exposure(s): (circle one)

0 1 2 3 4 5 6 7 8 9 10

Since the exposure(s): (circle one)

Q49 feeling ill if you drink or eat **less** than your usual amount of coffee, tea, caffeinated soda, or chocolate, or miss it altogether

Before the exposure(s): (circle one)

0 1 2 3 4 5 6 7 8 9 10

Since the exposure(s): (circle one)

Q50 **alcoholic beverages in small amounts such as one beer or a glass of wine**

Before the exposure(s): (circle one)   0 1 2 3 4 5 6 7 8 9 10

Since the exposure(s): (circle one)   0 1 2 3 4 5 6 7 8 9 10

Q51 fabrics, metal jewelry, creams, cosmetics, or other items that touch your **skin**

Before the exposure(s): (circle one)   0 1 2 3 4 5 6 7 8 9 10

Since the exposure(s): (circle one)   0 1 2 3 4 5 6 7 8 9 10

Q52 being unable to tolerate or having **adverse or allergic reactions to any drugs or medications** (such as antibiotics, anesthetics, pain relievers, x-ray contrast dye, vaccines, or birth control pills), or to an implant, prosthesis, dental material, contraceptive chemical or device, or other medical, surgical, or dental material or procedure

Before the exposure(s): (circle one)   0 1 2 3 4 5 6 7 8 9 10

Since the exposure(s): (circle one)   0 1 2 3 4 5 6 7 8 9 10

Q53 problems with **any classical allergic reactions (asthma, nasal symptoms, hives, anaphylaxis, or eczema)** when exposed to allergens such as: tree, grass or weed pollen, dust, mold, animal dander, insect stings, particular foods

Before the exposure(s): (circle one)   0 1 2 3 4 5 6 7 8 9 10

Since the exposure(s): (circle one)   0 1 2 3 4 5 6 7 8 9 10

Q54   Sensitivity to **bright light**

Before the exposure(s): (circle one)   0 1 2 3 4 5 6 7 8 9 10

Since the exposure(s): (circle one)   0 1 2 3 4 5 6 7 8 9 10

Q55   Sensitivity **to noise**

Before the exposure(s): (circle one)   0 1 2 3 4 5 6 7 8 9 10

Since the exposure(s): (circle one)   0 1 2 3 4 5 6 7 8 9 10

TOBACCO USERS ONLY:

Q56   feeling ill when you smoke or dip more tobacco than usual or if you use a stronger brand of tobacco

Before the exposure(s):   0 1 2 3 4 5 6 7 8 9 10

Since the exposure(s):   0 1 2 3 4 5 6 7 8 9 10

Q57   feeling ill if you miss a cigarette or try to quit smoking (or dipping)

Before the exposure(s):   0 1 2 3 4 5 6 7 8 9 10

Since the exposure(s):   0 1 2 3 4 5 6 7 8 9 10

Masking

On the average, how much of the following do you use in a typical day or week?

Q58 **Tobacco:** (circle all that apply) cigarettes,
cigar, pipe, or dips of tobacco   ____ / day or ____ / wk

Q59 **Alcohol:** (circle all that apply) alcoholic
drinks, cans of beer, glasses of wine, or
shots of liquor (total)   ____ / day or ____ / wk

Q60 **Caffeine:** (circle all that apply) total
number of cups or cans of caffeine-
containing beverages (including regular
coffee, regular tea, or sodas containing
caffeine such as cola drinks, Dr. Pepper,
Big Red, Mountain Dew, or flavored teas
such as Snapple: ____ / day or ____ / wk

Q61 Do you routinely (once a week or more) use   1. Yes
perfume, aftershave, hairspray, or other   2. No
scented personal care products?   3. Don't know

Q62 Has either your home or your workplace   1. Yes
been sprayed for insects or fumigated in   2. No
the past year?   3. Don't know

Q63 In your current job or hobby, are you   1. Yes → **Describe below**
routinely (once a week or more) exposed to   2. No
any chemicals, smoke, or fumes?   3. Don't know

Other than yourself, does anyone routinely   1. Yes
smoke inside your home?   2. No
                                               3. Don't know

Q65 Is either a gas or propane stove used for   1. Yes
cooking in your home?   2. No
                                               3. Don't know

Q66 Is a scented fabric softener (liquid or dryer 1. Yes
sheets) routinely used in laundering your   2. No
clothes or bedding?   3. Don't know

Q67 Do you routinely (once a week or more   1. Yes
often) take any of the following: steroid   2. No
pills, such as prednisone; pain medications   3. Don't know
requiring a prescription; medications for
depression, anxiety, or mood disorders;
medications for sleep; or recreational or
street drugs?

Impact of Sensitivities

If you are sensitive to certain chemicals or foods, on a 0-10 scale rate the degree to which **your sensitivities** have affected various aspects of your life:

How much have your sensitivities affected:

0 = not at all
5 = moderately
10 = severely

Q68	Your diet	0 1 2 3 4 5 6 7 8 9 10
Q69	Your ability to work or go to school	0 1 2 3 4 5 6 7 8 9 10
Q70	How you furnish your home	0 1 2 3 4 5 6 7 8 9 10
Q71	Your choice of clothing	0 1 2 3 4 5 6 7 8 9 10
Q72	Your ability to travel to other cities or drive a car	0 1 2 3 4 5 6 7 8 9 10
Q73	Your choice of personal care products, such as soaps, deodorants, or makeup	0 1 2 3 4 5 6 7 8 9 10
Q74	Your ability to be around others and enjoy social activities, for example, going to meetings, church, restaurants, etc.	0 1 2 3 4 5 6 7 8 9 10
Q75	Your choice of hobbies or recreation	0 1 2 3 4 5 6 7 8 9 10
Q76	Your relationship with your spouse or family	0 1 2 3 4 5 6 7 8 9 10
Q77	Your ability to clean your home, iron, mow the lawn, or perform other routine chores	0 1 2 3 4 5 6 7 8 9 10

Please write any comments or additions below:

_____

_____

_____

_____

_____

_____

_____

_____

_____

_____

Bibliography

Abrams, H., et al., "Letter to the World Health Organization, International Labor Office, United Nations Environment Program, and International Program on Chemical Safety," *Archives of Environmental Health* (1996) 51(4):338–342.

ACOP: see American College of Physicans.

Adamec, R., "Does Kindling Model Anything Clinically Relevant?" *Biological Psychiatry* (1990) 27:249–279.

Adamec, R., and Stark-Adamec, C., "Partial Kindling and Emotional Bias in the Cat: Lasting Aftereffects of Partial Kindling of the Ventral Hippocampus: Physiological Changes," *Behavioral and Neural Biology* (1983) 38:205–222.

Agency for Toxic Substances and Disease Registry (ATSDR), *Annual Report for FY 1988,* May (1989).

AIHA: see American Industrial Hygiene Association.

Akiyama, K., et al., "Hapten-modified Basophils: A Model of Human Immediate Hypersensitivity That Can Be Elicited by IgG Antibody," *Journal of Immunology* (1984) 133(6):3286–3290.

Alarie, Y., "Sensory Irritation by Airborne Chemicals," *CRC Critical Reviews in Toxicology* (1973) 2:299–363.

——— , Personal communication, 1996.

Allen, D., et al., "Monosodium L-glutamate-induced Asthma," *Journal of Allergy and Clinical Immunology* (1987) 80(4):530–537.

American Academy of Allergy and Immunology, "Position Statements—Allergen Standardization," *Journal of Allergy and Clinical Immunology* (1980) 66(6):431.

———, "Position Statements—Controversial Techniques," *Journal of Allergy and Clinical Immunology* (1981) 67(5):333–338.

———, "Position Statements—Clinical Ecology," *Journal of Allergy and Clinical Immunology* (1986) 72(8):269–271.

American College of Physicans (ACOP), "Clinical Ecology," *Annals of Internal Medicine* (1989) 111(2):168–178.

American Council on Science and Health, *Multiple Chemical Sensitivity* (1994), New York, 28 pp.

American Industrial Hygiene Association (AIHA), *Odor Thresholds for Chemicals with Established Occupational Health Standards* (1989), AIHA, Fairfax, VA.

American Psychiatric Association (APA), Board of Trustees, "Statement on Memories of Sexual Abuse," *International Journal of Clinical and Experimental Hypnosis* (1994) XLII (4):261–264.

American Society for Testing and Materials (ASTM), *Standard Test Method for Estimating Sensory Irritancy of Airborne Chemicals* (designation E981-84) (1984), Philadelphia, PA.

Amundsen, M., Hanson, N., Bruce, B., Lantz, T., Schwartz, M., and Lukach, B., "Odor Aversion or Multiple Chemical Sensitivities: Recommendation for a Name Change and Description of Successful Behavioral Medicine Treatment," *Regulatory Toxicology and Pharmacology* (1996) 24:S116–S118.

Anderson, R., "Sick Buildings, Sick Occupants: Measuring Indoor Air Quality by Bioassay," *Indoor Air Review* (1992) 1(11):1–4.

———, "Toxic Emissions from Carpets," *Journal of Nutritional and Environmental Medicine* (1995) 5:375–386.

———, Personal communication, 1996.

Angyal, A., *Neurosis and Treatment,* Revised Edition (1981), Hanfman, E., et al. (eds.) Unpublished manuscript.

Anonymous, "Conclusions and Recommendations of a Workshop on Multiple Chemical Sensitivities (MCS)" *Regulatory Toxicology and Pharmacology* (1996) 24:S188–S189.

Antelman, S., "Time-Dependent Sensitization in Animals: A Possible Model of Multiple Chemical Sensitivity," *Toxicology and Industrial Health* (1994) 10(4/5): 335–342.

AOEC: see Association of Occupational and Environmental Clinics.

APA: see American Psychiatric Association.

Appeal of Denise Kehoe (NH Dept. of Labor Compensation Appeals Board, No. 92-723) 648 A.2d 472 (N.H. 1994).

Ashford, N., "New Scientific Evidence and Public Health Imperatives," *Environmental Impact Assessment Review* (1987) 7:203–206.

――――, "Science and Values in the Regulatory Process," *Statistical Science* (1988) 3:377–383.

――――, Letter to the Editor, *American Journal of Industrial Medicine* (1995) 28:611–612.

Ashford, N., et al., "Human Monitoring: Scientific, Legal and Ethical Considerations," *Harvard Environmental Law Review* (1984) 8(2):263–364.

Ashford, N., and Miller, C., *Chemical Sensitivity: A Report to the New Jersey State Department of Health,* December 1989.

――――, *Chemical Exposures: Low Levels and High Stakes,* First Edition (1991), Van Nostrand Reinhold, New York, 232 pp.

――――, "Case Definitions for Multiple Chemical Sensitivity." In: *Multiple Chemical Sensitivities:* Addendum to Biologic Markers in Immunotoxicology, National Research Council, Commission on Life Science, Board on Environmental Studies and Toxicology (1992) National Academy Press, Washington D.C., 41–45.

Ashford, N., Spadafor, C., Hattis, D., and Caldart, C., *Monitoring the Worker for Exposure and Disease: Scientific, Legal and Ethical Considerations in the Use of Biomarkers* (1990), Johns Hopkins University Press, Baltimore, MD, 224 pp.

Ashford, N., Heinzow, B., Lütjen, K., Marouli, C., Mølhave, L., Mönch, B., Papadopoulos, S., Rest, K., Rosdahl, D., Siskos, P., and Velonakis, E., *Chemical Sensitivity in Selected European Countries: An Exploratory Study,* A report to the European Commission (1995).

Association of Occupational and Environmental Clinics (AOEC), "Advancing the Understanding of Multiple Chemical Sensitivity," *Toxicology and Industrial Health* (1992) 8(4):1–257.

ASTM: see American Society for Testing and Materials.

ATSDR: See Agency for Toxic Substances and Disease Registry.

Baggett, B., "The Great War and the Shaping of the 20th Century," Public Broadcasting Service television documentary on World War I, cited in *U.S. News and World Report,* November 11, 1996, p.52.

Bahn, A., et al., "Hypothyroidism in Workers Exposed to Polybrominated Biphenyls," *New England Journal of Medicine* (1980) 302(1):31–33.

Bahura v. S.E.W. Investors L.P., DC Superior Ct, No. 90-Ca-10594 (12/7/95).

Bailey, K., and Vanderslice, R., "Volatilization of Drinking Water Contaminants While Showering" (1987). Unpublished manuscript.

Baker, D., "Social and Organizational Factors in Office Building-Associated Illness." In: Cone, J., and Hodgson, M. (eds.) *Problem Buildings: Building-Associated Illness and the Sick Building Syndrome* (1989), Hanley & Belfus, Philadelphia, 607–624.

Baldwin, C., Bell, I., O'Rourke, M., Nadella, M., and Lebowitz, M., "Personal and Parental Cardiopulmonary and Diabetes Medical Risk Ratios in a Community Sample with and without Sensitivity to Chemical Stressors" (1995), presentation at the 16th Annual Meeting of the Society of Behavior Medicine, San Diego, CA, March 22–25.

Bardana, E., Jr., and Montanaro, A., "'Chemically Sensitive' Patients: Avoiding the Pitfalls," *Journal of Respiratory Diseases* (1989) 10(1):32–45.

Barrett, K., and Metcalfe, D., "The Mucosal Mast Cell and Its Role in Gastrointestinal Allergic Diseases," *Clinical Reviews in Allergy* (1984) 2:39–53.

Barrow, C., et al., "Sensory Irritation and Incapacitation Evoked by Thermal Decomposition Products of Polymers and Comparisons with Known Sensory Irritants," *Archives of Environmental Health* (1978) 33(2):79–88.

Barsky, A., and Borus, J., "Somatization and Medicalization in the Era of Managed Care," *JAMA* (1995) 274:1931–1934.

Bascom, R., "Chemical Hypersensitivity Syndrome Study: Options for Action, a Literature Review, and a Needs Assessment," prepared for the State of Maryland Department of Health, February 7, 1989.

————, "Multiple Chemical Sensitivity: A Respiratory Disorder?" *Toxicology and Industrial Health* (1991) 8(4):221–228.

Bascom, R., and Willes, S., "Nasal Inhalation Challenge Studies: An Approach to the Study of Health Effects of Indoor Air Pollutants," Indoor Air '90, Fifth International Conference on Indoor Air Quality and Climate, Toronto, Ontario (1990) 1:295–300.

Bastomsky, C., "Polyhalogenated Aromatic Hydrocarbons and Thyroid Function," *Clinical Ecology* (1985) 3(3):162–163.

Begley, S., "The Trials of Silicone," *Newsweek*, December 16, 1996, 56–58.

Bekesi, J., et al., "Lymphocyte Function of Michigan Dairy Farmers Exposed to Polybrominated Biphenyls," *Science* (1978) 199:1207–1209.

Bell, I., *Clinical Ecology* (1982), Common Knowledge Press, Bolinas, California.

————, "The Biopersonality of Allergies and Environmental Illness." Paper presented at the Eighth Annual International Symposium on Man and His Environment in Health and Disease, Dallas, TX, February 21, 1990.

————, "Effects of Food Allergy on the Central Nervous System." In: Brostoff, J., and Challacombe, S. (eds.) *Food Allergy and Intolerance* (1987a), Bailliere Tindall, Philadelphia, 709–722.

————, "Environmental Illness and Health: The Controversy and Challenge of Clinical Ecology for Mind-Body Health," *Advances* (1987b) 4(3):45–55.

————, "White Paper: Neuropsychiatric Aspects of Sensitivity to Low Level Chemicals: A Neural Sensitization Model." In: Mitchell, F. (ed.) "Proceedings of the Agency for Toxic Substances and Disease Registry Conference on Low-Level

Exposure to Chemicals and Neurobiologic Sensitivity," *Toxicology and Industrial Health* (1994) 10:277–312.

————, "Clinically Relevant EEG Studies and Psychophysiological Findings: Possible Neural Mechanisms for Multiple Chemical Sensitivity," *Toxicology* (1996) 111(1–3):101–117.

Bell, I., and King, D., "Psychological & Physiological Research Relevant to Clinical Ecology: Overview of Current Literature," *Clinical Ecology* (1982) 1(1):15–25.

Bell, I., Miller, C., and Schwartz, G., "An Olfactory-limbic Model of Multiple Chemical Sensitivity Syndrome: Possible Relationship to Kindling and Affective Spectrum Disorders," *Biological Psychiatry* (1992) 32:218–242.

Bell, I., Schwartz, G., Peterson, J., Amend, D., and Stini, W., "Possible Time-Dependent Sensitization to Xenobiotics: Self-Reported Illness from Chemical Odors, Foods, and Opiate Drugs in an Older Adult Population," *Archives of Environmental Health* (1993a) 48(5):315–327.

Bell, I., Schwartz, G., Peterson, J., and Amend, D., "Self-Reported Illness from Chemical Odors in Young Adults without Clinical Syndromes or Occupational Exposures," *Archives of Environmental Health* (1993b) 48(1):6–13.

Bell, I., Markley, E., King, D., Asher, S., Marby, D., Kayne, H., Greenwald, M., Ogar, D., and Margen, S. "Polysymptomatic Syndromes and Autonomic Reactivity to Nonfood Stressors in Individuals with Reported Adverse Food Reactions," *Journal of the American College of Nutrition* (1993c) 12(3):227–238.

Bell, I., Schwartz, G., Amend, D., Peterson, J., Kazniak, A., and Miller, C., "Psychological Characteristics and Subjective Intolerance for Xenobiotic Agents of Normal Young Adults with Trait Shyness and Defensiveness. A Parkinsonian-like Personality Type?" *Journal of Nervous and Mental Disease* (1994a) 182(7): 367– 374.

Bell, I., Schwartz, G., Amend, D., Peterson, J., and Stini, W., "Sensitization to Early Life Stress and Response to Chemical Odors in Older Adults," *Biological Psychiatry*, (1994b) 35(11):857–863.

Bell, I., Peterson, J., and Schwartz, G., "Medical Histories and Psychological Profiles of Middle-aged Women with and without Self-Reported Illness from Environmental Chemicals," *Journal of Clinical Psychiatry* (1995a) 56(4):151–160.

Bell, I., Hardin, E., Baldwin, C., and Schwartz, G., "Increased Limbic Symptomatology and Sensitizability of Young Adults with Chemical and Noise Sensitivities," *Environmental Research* (1995b) 70:84–97.

Bell, I., Miller, C., Schwartz, G., Peterson, J., and Amend, D., "Neuropsychiatric and Somatic Characteristics of Young Adults with and without Self-Reported Chemical Odor Intolerance and Chemical Sensitivity," *Archives of Environmental Health* (1996a) 51(1):9–21.

Bell, I., Wyatt, J., Bootzin, R., and Schwartz, G., "Slowed Reaction Time Performance on a Divided Attention Task in Elderly with Environmental Chemical Odor Intolerance," *International Journal of Neuroscience* (1996b) 84:127– 134.

Bell, I., Schwartz, G., Baldwin, C., and Hardin, E., "Neural Sensitization and Physi-ological Markers in Multiple Chemical Sensitivity," *Regulatory Toxicology and Phar-macology* (1996c) 24:S39–S47.

Bell, I., Bootzin, R., Davis, T., Hau, V., Ritenbaugh, C., Johnson, K., and Schwartz, G., "Time Dependent Sensitization in Community Elderly with Self-Reported Environmental Chemical Odor Intolerance," *Biological Psychiatry* (1996d) 40:134–143.

Bell, I., Bootzin, R., Schwartz, G., Szarek, M., DiCenso, D., and Baldwin C., "Differ-ing Patterns of Cognitive Dysfunction and Heart Rate Reactivity in Chemically-Intolerant Individuals with and without Lifestyle Changes: Is Prefontal Cortex Involved?" submitted for publication (1997a).

Bell, I., Schwartz, G., Bootzin, R., and Wyatt, J., "Time-Dependent Sensitization of Heart Rate and Blood Pressure over Multiple Laboratory Sessions in Communi-ty Elderly Individuals with Chemical Odor Intolerance," *Archives of Environmental Health* (1997b) 52:6–17.

Bell, I., Schwartz, G., Bootzin, R., Hardin, E., Baldwin, C., and Kline, J., "Differen-tial Resting QEEG Alpha Patterns in Women with Environmental Chemical Intol-erance, Depressives, and Normals," *Biological Psychiatry* (1997c) (in press).

Bell, I., Walsh, M., Goss, A., Gersmeyer, Schwartz, G., and Kanof, P., "Cognitive Dys-function and Disability in Geriatric Veterans with Self-Reported Intolerance to Environmental Chemicals," *Journal of Chronic Fatigue Syndrome* (1997d) (in press).

Bellinger, D., et al., "Longitudinal Analyses of Prenatal and Postnatal Lead Expo-sure and Early Cognitive Development," *New England Journal of Medicine* (1987) 316:1037–1043.

Benedi v. McNeil-P.P.C. Inc. 66 F.3d 1378 (CA 4 1995).

Berglund, B., Berglund, U., Johansson, I., and Lindvall, T. "Mobile Laboratory for Sensory Air Quality Studies in Non-industrial Environments," in *Indoor Air: Pro-ceedings of the Third International Conference on Indoor Air Quality and Climate* (1984) 3:467–472. In: Berglund, B., Lindvall, T., and Sundell, J. (eds.), *Hyperreactivity Reactions to Sick Buildings* (1984), Swedish Council for Building Research, Stock-holm.

Bernstein, D., "Multiple Chemical Sensitivity: State of the Art Symposium: The Role of Chemical Allergens," *Regulatory Toxicology and Pharmacology* (1996) 24:S28–S31.

Bernstein, J., "New Perspectives on Immunologic Reactivity in Otitis Media with Effusion," *Annals of Otology, Rhinology and Laryngology* (1988) 97(3) Part 2, Supple-ment 132, 19–23.

Bertschler, J., et al., "Psychological Components of Environmental Illness: Factor Analysis of Changes During Treatment," *Clinical Ecology* (1985) 3(2):85–94.

Besedovsky, H., et al., "Hypothalamic Changes During the Immune Response," *European Journal of Immunology* (1977) 7:323–325.

Binkley, K.E., and Kutcher, S. "Panic Response to Sodium Lactate Infusion in Patients with Multiple Chemical Sensitivity Syndrome," *Journal of Allergy and Clinical Immunology* (1997) 99(4):570–574.

Black, C., and Welsh, K., "Occupationally and Environmentally Induced Scleroderma-Like Illness: Etiology, Pathogenesis, Diagnosis, and Treatment," *Internal Medicine* (1988) 9(6):135–154.

Black, D., Rathe, A., and Goldstein, R., "Environmental Illness. A Controlled Study of 26 Subjects with '20th Century Disease'" *JAMA* (1990) 264:3166–3170.

Blondell, J., and Dobozy, V.A., EPA Memorandum: Review of Chlorpyrifos Poisoning Data, January 14, 1997, U.S. Environmental Protection Agency.

Bock, S., et al., "Double-Blind, Placebo-Controlled Food Challenge (DBPCFC) as an Office Procedure: A Manual," *Journal of Allergy and Clinical Immunology* (1988) 82(6):986–997.

Bokina, A., et al., "Investigation of the Mechanism of Action of Atmospheric Pollutants on the Central Nervous System and Comparative Evaluation of Methods of Study," *Environmental Health Perspective* (1976) 13:37–42.

Bolla-Wilson, K., et al., "Conditioning of Physical Symptoms after Neurotoxic Exposure," *Journal of Occupational Medicine* (1988) 30(9):684–686.

Boris, M., et al., "Antigen Induced Asthma Attenuated by Neutralization Therapy," *Clinical Ecology* (1985a) 3(2):59–62.

———— , "Association of Otitis Media with Exposure to Gas Fuels," *Clinical Ecology* (1985b) 3(4):195–198.

———— , "Injection of Low-Dose Antigen Attenuates the Response to Subsequent Bronchoprovocative Challenge," *Otolaryngology—Head and Neck Surgery* (1988) 98(6):539–545.

Boulenger, J., et al., "Increased Sensitivity to Caffeine in Patients with Panic Disorder," *Archives of General Psychiatry* (1984) 41:1067–1071.

Boxer, R., "Cardiac Arrhythmias Due to Foods." In: Dickey, L. (ed.) *Clinical Ecology* (1976), Charles C Thomas, Springfield, IL, 193–200.

Boyle, C., et al., "Caffeine Consumption and Fibrocystic Breast Disease: A Case-Controlled Epidemiologic Study," *Journal of the National Cancer Institute* (1984) 72(5):1015–1019.

Bradley v. Brown, 852 F. Supp. 690 (N.D. Ind.), aff'd., 42 F.3d 434 (7th Cir. 1994).

Brautbar, N., "Silicone Breast Implants and Autoimmunity: Causation or Myth?" (editorial), *Archives of Environmental Health* (1994) 49(3):151–153.

Brautbar, N., and Campbell, A., "Silicone Implants and Immune Dysfunction: Scientific Evidence for Causation," *International Journal of Occupational Medicine and Toxicology* (1995)4(1):3–13.

Brautbar, N., Vojdani, A., and Campbell, A., "Multiple Chemical Sensitivities—Fact or Myth?" *Toxicology and Industrial Health* (1992) 8(6):v–xiii.

Bravo, A., Personal communication with Susan Kaplan, November 26, 1996.

Brawer, A., "Chronology of Systemic Disease Development in 300 Symptomatic Recipients of Silicone Gel-Filled Breast Implants," *Journal of Clean Technology, Environmental Toxicology, and Occupational Medicine* (1996) 5(3):223–233.

Brazil, E., "Ex-Casino Dealers Win Toxic Exposure Case," *San Francisco Examiner,* A-8, August 4, 1993.

Briere, J., et al., "Summer in the City: Urban Weather Conditions and Psychiatric Emergency-Room Visits," *Journal of Abnormal Psychology* (1983) 92(1):77–80.

Brodsky, C. M., "Allergic to Everything: A Medical Subculture," *Psychosomatics* (1983) 24:731–732, 734–736, 740–742.

Brodsky, C., "Multiple Chemical Sensitivities and Other 'Environmental Illness': A Psychiatrist's View." In: Cullen, M. (ed.) *Workers with Multiple Chemical Sensitivities, Occupational Medicine: State of the Art Reviews* (1987), Hanley & Belfus, Philadelphia, 2(4):695–704.

Brooks, S., et al., "Reactive Airways Dysfunction Syndrome (RADS). Persistent Asthma Syndrome after High Level Irritant Exposures," *Chest* (1985) 88(3):376–384.

Broughton, A., and Thrasher, J., "Antibodies and Altered Cell Mediated Immunity in Formaldehyde Exposed Humans," *Comments in Toxicology* (1988) 2(3):155–174. Gordon and Breach Science Publishers, Great Britain.

Broughton, A., et al., "Immunological Evaluation of Four Arc Welders Exposed to Fumes from Ignited Polyurethane (Isocyanate) Foam: Antibodies and Immune Profiles," *American Journal of Industrial Medicine* (1988) 13:463–472.

——— , "Biological Monitoring of Indoor Air Pollution: A Novel Approach," *Indoor Air '90*, Fifth International Conference on Indoor Air Quality and Climate, Toronto, Ontario (1990) 2:145–150.

Brown, H., et al., "The Role of Skin Absorption as a Route of Exposure for Volatile Organic Compounds (VOCs) in Drinking Water," *American Journal of Public Health* (1984) 74(5):479–484.

Brown v. Shalala, 15 F.3d 97 (8th Cir. 1994).

Buchwald, D., and Garrity, D., "Comparison of Patients with Chronic Fatigue Syndrome, Fibromyalgia, and Multiple Chemical Sensitivities," *Archives of Internal Medicine* (1994) 154:2049–2053.

Burchfiel, J., and Duffy, F., "Organophosphate Neurotoxicity: Chronic Effects of Sarin on the Electroencephalogram of Money and Man," *Neurobehavioral Toxicology and Teratology* (1982) 4(6):767–778.

Burks, A., et al., "Atopic Dermatitis: Clinical Relevance of Food Hypersensitivity Reactions," *Journal of Pediatrics* (1988) 113(3):447–451.

Bush, R., et al., "A Critical Evaluation of Clinical Trials in Reactions to Sulfites," *Journal of Allergy and Clinical Immunology* (1986) 78(1):191–202.

Butcher, B., et al., "Development and Loss of Toluene Diisocyanate Reactivity: Immunologic, Pharmacologic, and Provocative Challenge Studies," *Journal of Allergy and Clinical Immunology* (1982) 70(4):231–235.

Byers, V., et al., "Association between Clinical Symptoms and Lymphocyte Abnormalities in a Population with Chronic Domestic Exposure to Industrial Solvent-Contaminated Domestic Water Supply and a High Incidence of Leukaemia," *Cancer Immunology Immunotherapy* (1988) 27:77–81.

Caffrey, E., et al., "Thrombocytopenia Caused by Cow's Milk," *Lancet* (1981)2 (8241):316.

Calabrese, E., *Pollutants and High Risk Groups: The Biological Basis of Increased Human Susceptibility to Environmental and Occupational Pollutants* (1978), John Wiley & Sons, New York.

California Legislature, Senate Subcommittee on the Rights of the Disabled, "Access for People with Environmental Illness/Multiple Chemical Sensitivity and Other Related Conditions," September 30, 1996.

Callender, T., Morrow, L., Subramanian, K., Duhon, D., and Ristow, M. "Three-Dimensional Brain Metabolic Imaging in Patients with Toxic Encephalopathy," *Environmental Research* (1993) 60:295–319.

Campbell, A., Brautbar, N., and Vojdani, A., "Suppressed Natural Killer Cell Activity in Patients with Silicone Breast Implants: Reversal Upon Explantation," *Toxicology and Industrial Health* (1994) 10(3):149–154.

Canadian Mortgage and Housing Corporation (CMHC), *The Clean Air Guide,* Ottawa, Ontario, Canada, 1993.

———, *This Clean House* (videotape), Ottawa, Ontario, Canada, 1995.

Caplin, I., "Report of the Committee on Provocative Food Testing," *Annals of Allergy* (1973) 31:375–381.

Carmichael, P., et al., "Sudden Death in Explosives Workers," *Archives of Environmental Health* (1963) 7:50–65.

Carpet and Rug Institute (CRI), "Carpet and Indoor Air Quality in Commercial Installations—A Guide for Specifiers and Designers," Dalton, GA, 1994.

———, "Industry in Depth—Carpet: The Environmentally Responsible Floor Covering of Choice," Dalton, GA, Spring 1995.

———, "Technical Bulletin—Carpet and Indoor Air Quality," Dalton, GA, September 10, 1996.

Carter, K., "Ignaz Semmelweiss, Carl Mayrhofer, and the Rise of the Germ Theory," *Medical History* (1985) 29:33–53.

Casarett and Doull's Toxicology: The Basic Science of Poisons, 5th ed. Claassen, C. (ed.) (1996), McGraw-Hill, New York, 27.

Cavallo v. Star Enterprise, 892 F. Supp. 756 (E.D. Va. 1995).

CDC: see Centers for Disease Control and Prevention.

Centers for Disease Control and Prevention, (CDC) "Unexplained Illness among Persian Gulf War Veterans in an Air National Guard Unit: Preliminary Report," *Mortality and Morbidity Weekly Report* (1995) 44(23):443–447.

Challacombe, S., "Oral Manifestations of Food Allergy and Intolerance." In: Brostoff, J., and Challacombe, S. (eds.) *Food Allergy and Intolerance* (1987), Bailliere Tindall, Philadelphia, 511–520.

Chaudhuri, A., Majeed, T., Dinan, T. and Behan, P., "Chronic Fatigue Syndrome: A Disorder of Central Cholinergic Transmission," *Journal of Chronic Fatigue Syndrome* (1997) 3(1):4–16.

Chester, A., and Levine, P., "Concurrent Sick Building Syndrome and Chronic Fatigue Syndrome: Epidemic Neuromyasthenia Revisited," *Clinical Infectious Diseases* (1994) 18 (Suppl. 1):543–548.

City of Greenville v. W. R. Grace & Co. 827 F.2d 975 (CA 4 1987).

Claussen, E., Letter to Earon S. Davis, 7 January (1988).

CMHC: see Canadian Mortgage and Housing Corporation.

Code of Federal Regulations, 28 CFR Ch. 1, Pt. 35, App. A (7-1-96 Edition).

Cohn, J., "Multiple Chemical Sensitivity or Multi-organ Dysesthesia," *Journal of Allergy and Clinical Immunology* (1994) (114) 93:953–954.

Cole, D., Tarasuk, V., Frank, J., and Eyles, J. "Research Responses to Outbreaks of Concern about Local Environments," *Archives of Environmental Health* (1996) 51(5): 352–358.

Cone, J., and Hodgson, M. (eds.) "Problem Buildings: Building Associated Illness and the Sick Building Syndrome", *Occupational Medicine: State of the Art Reviews* (1989) 4(4):575–802.

Cone, J., and Sult, T., "Acquired Intolerance to Solvents Following Pesticide/Solvent Exposure in a Building: A New Group of Workers at Risk for Multiple Chemical Sensitivities?" *Toxicology and Industrial Health* (1992) 8(4):29– 39.

Cone, J., et al., "Patients with Multiple Chemical Sensitivities: Clinical Diagnostic Subsets among an Occupational Health Clinic Population." In: Cullen, M. (ed.) *Workers with Multiple Chemical Sensitivities, Occupational Medicine: State of the Art Reviews* (1987), Hanley & Belfus, Philadelphia, 2(4):721–738.

Connolly, D., "Toxics: Too Risky to Insure? Yes, No, Maybe So," *Toxics Law Reporter* (1988) 3(23):715–723.

Coombs, R., and Oldham, G., "Early Rheumatoid-Like Joint Lesions in Rabbits Drinking Cow's Milk," *International Archives of Allergy and Applied Immunology* (1981) 64:287–292.

Corcoran, M., "Characteristics and Mechanisms of Kindling." In: Kalivas, P., and Barnes, C. (eds.), *Sensitization in the Nervous System* (1988), Telford Press, Caldwell, NJ, 84.

Cornfeld, R., and Schlossman, S., "Immunologic Laboratory Tests: A Critique of the Alcolac Decision," *Toxics Law Reporter* (1989) 4(14):381–390.

Corrigan, F., MacDonald, S., Brown, A., Armstrong, K., and Armstrong, E., "Neurasthenic Fatigue, Chemical Sensitivity and GABAa Receptor Toxins," *Medical Hypotheses* (1994) 43(4):195–200.

Corwin, A., "A Chemist Looks at Health and Disease," *Proceedings of the Society for Clinical Ecology*, 12th Advanced Seminar, Key Biscayne, FL (1978).

———, "The Allergy Fallacy: Clinical Consequences," *Clinical Ecology* (1985) 3(4):177–182.

Cosentino, C., and Collins, M., "Sexual Abuse of Children: Prevalence, Effects, and Treatment." In: Sechzer, J., et al. (eds.), *Women and Mental Health, Annals of the New York Academy of Sciences* (1996) 789:45–65.

Courpass, M., *Indoor Air Pollution: Cause for Concern?* (1988) Congressional Research Service, Washington, D.C., 88–745ENR.

Crandall, M., and Sieber, W., "The National Institute for Occupational Safety and Health Indoor Environmental Evaluation Experience. Part One: Building Environmental Evaluations," *Applied Occupational and Environmental Hygiene* (1996) 11(6):533–539.

Crayton, J., "Adverse Reactions to Foods: Relevance to Psychiatric Disorders," *Journal of Allergy and Clinical Immunology* (1986) 78(1):243–250.

CRI: see Carpet and Rug Institute.

Cross, C., et al., "Oxygen Radicals and Human Disease," *Annals of Internal Medicine* (1987) 107:526–545.

Cullen, M., "The Worker with Multiple Chemical Sensitivities: An Overview." In: Cullen, M. (ed.), "Workers with Multiple Chemical Sensitivities," *Occupational Medicine: State of the Art Reviews* (1987a), Hanley & Belfus, Philadelphia, 2(4):655–662.

———, "Multiple Chemical Sensitivities: Summary and Directions for Future Investigators." In: Cullen, M. (ed.) "Workers with Multiple Chemical Sensitivities," *Occupational Medicine: State of the Art Reviews* (1987b), Hanley & Belfus, Philadelphia, 2(4):801–804.

Cullen, M., and Redlich, C., "Significance of Individual Sensitivity to Chemicals. Elucidation of Host Susceptibility by Use of Biomarkers in Environmental Health Research," *Clinical Chemistry* (1995) 41(12):1809–1813.

Cullen, M., Pace, P., and Redlich, C., "The Experience of the Yale Occupational and Environmental Medicine Clinics with Multiple Chemical Sensitivities, 1986–1991," *Toxicology and Industrial Health* (1992) 8:15–19.

Custer, W., "Multiple Chemical Sensitivity Syndrome: The Wavering Influence of the Courts on Public Policy," *Regulatory Toxicology and Pharmacology* (1996) 24:S183.

Dager, S., et al., "Panic Disorder Precipitated by Exposure to Organic Solvents in the Work Place," *American Journal of Psychiatry* (1987) 144(8):1056–1058.

Dantzer, R., and Kelley, K., "Stress and Immunity: An Integrated View of Relationships Between the Brain and Immune System," *Life Sciences* (1989) 44: 1995–2008.

Datta, M., "You Cannot Exclude the Explanation You Have Not Considered," *Lancet* (1993) 342:345–347.

Daubert v. Merrill Dow Pharmaceuticals, Inc., 509 U.S. 579, 113 S. Ct. 2786, 125 L.Ed.2d 469 (1993).

Daum, S., "Nitroglycerin and Alkyl Nitrates." In: Rom, W. (ed.) *Environmental and Occupational Medicine* (1983), Little, Brown and Co., Boston, 639–648.

Davidoff, A., and Fogarty, L., "Psychogenic Origins of Multiple Chemical Sensitivities Syndrome: A Critical Review of the Research Literature," *Archives of Environmental Health* (1994) 49(5):316–325.

Davidoff, A., and Keye, P., "Symptoms and Health Status in Individuals with Multiple Chemical Sensitivities Syndrome from Four Reported Sensitizing Exposures and a General Population Comparison Group," *Archives of Environmental Health* (1996) 51(3):201–213.

Davies, H., Richter, R., Keifer, M., Broomfield, C., Sowalla, J., and Furlong, C., "The Effect of the Human Serum Paraoxonase Polymorphism Is Reversed with Diazoxon, Soman and Sarin," *Nature Genetics* (1996) 14:334–336.

Davis, E., Personal communication, March 28, 1989.

Davis, K., et al., "Possible Role of Diet in Delayed Pressure Urticaria," *Journal of Allergy and Clinical Immunology* (1986) 77(4):566–569.

Dayal, H., Gupta, S., Trieff, N., Maierson, D., and Reich, D., "Symptom Clusters in a Community with Chronic Exposure to Chemicals in Two Superfund Sites," *Archives of Environmental Health* (1995) 50(2):108–111.

Descotes, J., *Immunotoxicology of Drugs and Chemicals* (1986), Elsevier, New York.

Dickey, L., "Ecology and Urology." In: Dickey, L. (ed.) *Clinical Ecology* (1976), Charles C Thomas, Springfield, IL, 702–707.

Dilsaver, S., "Cholinergic Mechanisms in Depression," *Brain Research* (1986) 11:385–316.

Djuric, V., Overstreet, D., Bienenstock, J., and Perdue, M., "Immediate Hypersensitivity in the Flinders Rat: Further Evidence for a Possible Link Between Susceptibility to Allergies and Depression," *Brain, Behavior and Immunity* (1995) 9:196–206.

Djuric, V., Overstreet, D., Crosswaithe, D., Dunn, E., and Steiner, M., "Continuous Access to Sucrose Induces Depressive-like Behavior in the Rat" (1996), Presentation at the 26th Annual meeting of the Society for Neuroscience, Washington, D.C., November 16–21.

Doane, B., "Clinical Psychiatry and the Physiodynamics of the Limbic System." In: Doane, B., and Livingston, K. (eds.), *The Limbic System: Functional Organization and Clinical Disorders* (1986), Raven Press, New York, 285–315.

Donnay, A., Bibliography of all articles, editorials, book chapters, letters and reports on or related to Multiple Chemical Sensitivity published through February 1997, MCS Referral and Resources, Baltimore, Maryland (1997).

Donnay, A., and Ziem, G., "Protocol for Diagnosing Disorders of Porphyrin Metabolism in Chemically Sensitive Patients," *MCS Referral & Resources,* March 1995.

Doss, M., "Porphyrinurias and Occupational Disease." In: Silbergeld, E., and Fowler, B. (eds.), *Mechanisms of Chemical-Induced Porphyrinopathies, Annals of the New York Academy of Sciences,* (1987) 514:204–218.

Doty, R., et al., "Olfactory Sensitivity, Nasal Resistance, and Autonomic Function in Patients with Multiple Chemical Sensitivities," *Archives of Otolaryngology—Head and Neck Surgery* (1988) 114:1422–1427.

Doty, R., Deems, D., Frye, R., Pelberg, R., and Shapiro, A., "Olfactory Sensitivity, Nasal Resistance, and Autonomic Function in Patients with Multiple Chemical Sensitivities," *Archives of Otolaryngology—Head and Neck Surgery* (1988) 114:1422–1427.

Downey, D., "Fatigue Syndromes Revisited: The Possible Role of Porphyrins," *Medical Hypotheses* (1994) 42:285–290.

DSM III, *Diagnostic and Statistical Manual of Mental Disorders,* Edition 3, American Psychiatric Association (1980), Washington, DC.

Duffy, F., Burchfiel, J., Bartels, P., Gaon, M., and Sim, V., "Long-Term Effects of an Organophosphate upon the Human Electroencephalogram," *Toxicology and Applied Pharmacology* (1979) 47:161–176.

Dunstan, R., Donohoe, M., Taylor, W., Roberts, T., Murdoch, R., Watkins, J., and McGregor, N., "A Preliminary Investigation of Chlorinated Hydrocarbons and Chronic Fatigue Syndrome," *Medical Journal of Australia* (1995) 163:294–297.

Düsseldorf, Hearing of 16 September 1988 in Report of the Ministerium für Arbeit, Gesundheit und Soziales des Landes Nordheim-Westfalen: Luftverunreinigungen in Innenraümen, Düsseldorf, Germany, January 1990.

Dyer, R., and Sexton, K., "What Can Research Contribute to Regulatory Decisions about the Health Risks of Multiple Chemical Sensitivity?" *Regulatory Toxicology and Pharmacology* (1996) 24:S139–S151.

Ecology House, Inc., press release, November 28, 1995.

Edgar, R., et al., "Air Pollution Analysis Used in Operating an Environmental Control Unit," *Annals of Allergy* (1979) 42(3):166–173.

Egger, J., "Food Allergy and the Central Nervous System in Childhood." In: Brostoff, J., and Challacombe, S. (eds.) *Food Allergy and Intolerance* (1987a), Bailliere Tindall, Philadelphia, 666–673.

———, "The Hyperkinetic Syndrome." In: Brostoff, J., and Challacombe, S., (eds.) *Food Allergy and Intolerance* (1987b), Bailliere Tindall, Philadelphia, 674–687.

Egger, J., et al., "Oligoantigenic Diet Treatment of Children with Epilepsy and Migraine," *Journal of Pediatrics* (1989) 114(1):51–58.

———, "Is Migraine Food Allergy? A Double-blind Controlled Trial of Oligoantigenic Diet Treatment," *Lancet* (1983) 2:865–869.

Ellefson, R., and Ford, R., "The Porphyrias: Characteristics and Laboratory Tests," *Regulatory Toxicology and Pharmacology* (1996) 24:S119–S125.

Elliott, E., "The Future of Toxic Torts: Of Chemophobia, Risk as a Compensable Injury and Hybrid Compensation Systems," *Houston Law Review* (1988) 25:781–786.

Environmental Protection Agency, Office of Air and Radiation, *Report to Congress on Indoor Air Quality,* August, 1989.

———, "EPA Moves to Reduce Chemical Emissions from Carpets," *Environmental News,* April 18, 1990.

Environmental Protection Agency (EPA), *Report to Congress on Indoor Air Quality,* Volume II, *Assessment and Control of Indoor Air Pollution* (1989).

———, Workshop on Animal Models of Nervous System Susceptibility to Indoor Air Contaminants, Summary Report, October 19–21, Chapel Hill, N.C.: National Health and Environmental Effects Research Laboratory.

———, *Pest Control in the School Environment: Adopting Integrated Pest Management* (1993), EPA Office of Pesticide Programs, August.

EPA: see Environmental Protection Agency.

Equal Employment Opportunity Commission and U.S. Department of Justice, *Americans with Disabilities Act Handbook* (1991): 1–31.

Estes, E., Jr., Coble, Y., et al., "Clinical Ecology," *JAMA* (1992) 268:3465–3467.

Evans, G., et al., "Psychological Reactions to Air Pollution," *Environmental Research* (1988) 45:1–15.

Fabig, K., "Suppression of Regional Cerebral Blood Flow (rCBF) in Human Beings after Exposure to PCDD/PCDF by Inhalation" (1988), poster session at Dioxin '88—the 8th International Symposium on Chlorinated Dioxins and Related Compounds, Umeå, Sweden, August 21–26.

Fabig, K. R., Physician/medical practitioner, Hamburg, Germany, Personal communication with Birger Heinzow, European Study Team (1994) (see Ashford et al. 1995).

Falk, J., Department of Otorhinolaryngology, Stockholm, Sweden, Personal communication with Didi Rosdahl, European Study Team (1994) (see Ashford et al. 1995).

Fauci, A., "The Revolution in the Approach to Allergic and Immunologic Diseases," *Annals of Allergy* (1985) 55:632–633.

Ferebee v. Chevron Chemical Co. 736 F.2d 1535 (CA DC 1984).

Fiedler, N., Maccia, C., and Kipen, H., "Evaluation of Chemically Sensitive Patients," *Journal of Occupational Medicine* (1992) 34:529–538.

Fiedler, N., Kipen, H., Deluca, J., Kelly-McNeil, K., and Natelson, B., "Neuropsychology and Psychology of MCS," *Toxicology and Industrial Health* (1994) 10 (4/5):545–554.

Fiedler, N., Kipen, H., Natelson, B., and Ottenweller, J., "Chemical Sensitivities and the Gulf War: Department of Veterans Affairs Research Center in Basic and Clinical Science Studies of Environmental Hazards," *Regulatory Toxicology and Pharmacology* (1996a) 24:S129–S138.

Fiedler, N., Kipen, H., DeLuca, J., Kelly-McNeil, K., Natelson, B., "A Controlled Comparison of Multiple Chemical Sensitivities and Chronic Fatigue Syndrome," *Psychosomatic Medicine* (1996b) 58(1):38–49.

Finn, R., et al., "Hydrocarbon Exposure and Glomerulonephritis," *Clinical Nephrology* (1980) 14(4):173–175.

Finnegan, M., et al., "The Sick Building Syndrome: Prevalence Studies," *British Medical Journal* (1984) 289:1573–1575.

Fiore, M., et al., "Chronic Exposure to Aldicarb-Contaminated Groundwater and Human Immune Function," *Environmental Research* (1986) 41(2):633–645.

Fletcher, J., "More States Cracking Down on Use of Home Pesticides," *Wall Street Journal*, October 18, 1996.

Ford, R., and Walker-Smith, J., "Paediatric Gastrointestinal Food-Allergic Disease." In: Brostoff, J., and Challacombe, S. (eds.) *Food Allergy and Intolerance* (1987), Bailliere Tindall, Philadelphia, 570–582.

Foster, S., and Chrostowski, P., *Inhalation Exposures to Volatile Organic Contaminants in the Shower* (1987), presented at the 80th annual meeting of APCA (June 21–26).

Fox, R., Medical Director, Nova Scotia Environmental Health Center, Personal communication, 1996.

Fregly, M., "Comments on Cross-Adaptation," *Environmental Research* (1969) 2:435–441.

Freni-Titulaer, L., et al., "Connective Tissue Disease in Southeastern Georgia: A Case-Control Study of Etiologic Factors," *American Journal of Epidemiology* (1989) 130(2):404–409.

Freud, S., *The Problem of Anxiety*, Turner, H., trans. (1936), Psychoanalytic Quarterly Press, New York.

Friedman, R., *Sensitive Populations and Environmental Standards* (1981), The Conservative Foundation, Washington, D.C.

Fukuda, K., Straus, S., Hickie, H., Sharpe, M., Dobbins, J., Domaroff, A., "The Chronic Fatigue Syndrome: A Comprehensive Approach to Its Definition and Study," *Annals of Internal Medicine* (1994) 121(12):953–959.

Gaitan, E., et al., "Simple Goiter and Auto-immune Thyroiditis: Environmental and Genetic Factors," *Clinical Ecology* (1985) 3(3):158–161.

Galland, L., "Biochemical Abnormalities in Patients with Multiple Chemical Sensitivities." In: Cullen, M. (ed.) *Workers with Multiple Chemical Sensitivities, Occupational Medicine: State of the Art Reviews* (1987) Hanley & Belfus, Philadelphia, 2(4):755–777.

Gammage, R., and Kaye, S. (eds.) *Indoor Air and Human Health* (1985), Lewis Publishers, Chelsea, MI.

Gammage, R., Miller, C., and Jankovic, J., "Exacerbation of Chemical Sensitivity Associated with New Construction: A Case Study," *Proceedings of Indoor Air '96*, Nagoya, Japan (1996) 2:291–296.

Gerdes, K., and Selner, J., "Bronchospasm Following IV Dextrose," abstract in *Annals of Allergy* (1980) 44:48.

Gerrard, J., and Zaleski, A., "Functional Bladder Capacities in Children with Enuresis and Recurrent Urinary Infections." In: Dickey, L. (ed.) *Clinical Ecology* (1976), Charles C Thomas, Springfield, IL, 224–232.

Gershon, S., and Shaw, F., "Psychiatric Sequelae of Chronic Exposure to Organophosphorus Insecticides," *Lancet* (1961) 1:1371–1374.

Gilbert, M., "The Phenomenology of Limbic Kindling," *Toxicology and Industrial Health* (1994) 10(4/5):343–358.

——— , "Repeated Exposure to Lindane Leads to Behavioral Sensitization and Facilitates Electrical Kindling," *Neurotoxicology and Teratology* (1995) 17(2):131–141.

Gilbert, M., and Mack, C., "Seizure Thresholds in Kindled Animals Are Reduced by the Pesticides Lindane and Endosulfan," *Neurotoxicology and Teratology* (1995) 17(2):143–150.

Gilman, S., and Winans, S., *Manter and Gatz's Essentials of Clinical Neuroanatomy and Neurophysiology* (1982), F. A. Davis Company, Philadelphia.

Gilovich, T. and Savitsky, K., "Like Goes with Like: The Role of Representativeness in Erroneous and Pseudoscientific Beliefs" *The Skeptical Inquirer*, March/April 1996, 34–40.

Girgis, M., "Biochemical Patterns in Limbic System Circuitry: Biochemical-Electrophysiological Interactions Displayed by Chemitrode Techniques." In: Doane B., and Livingston, K., (eds.), *The Limbic System: Functional Organization and Clinical Disorders* (1986), Raven Press, New York, 55–65.

Gliner, J., et al., "Pre-exposure to Low Ozone Concentrations Does Not Diminish the Pulmonary Function Responses to Higher Ozone Concentrations," *American Review of Respiratory Diseases* (1983) 127:51–55.

Good, C., and Dadd, D., *Healthful Houses* (1988), Guaranty Press, Bethesda, Maryland.

Gordon, M., "Reactions to Chemical Fumes in Radiology Department," *Radiography* (1987) 52(607):85–89.

——— , "Dangers in the Darkroom," *The Radiographer* (1989) 36:114–115.

——— , "Danger—Toxic Fumes," *Radiography News*, April 1995.

Göthe, C., Molin, C., and Nilsson, C., "The Environmental Somatization Syndrome," *Psychosomatics* (1995) 36(1):1–11.

Gots, R., "Multiple Chemical Sensitivities—Public Policy" (editorial), *Journal of Toxicology—Clinical Toxicology* (1995) 33(2):111–113.

———, "Multiple Chemical Sensitivities: Distinguishing between Psychogenic and Toxicodynamic," Part 2 of 2 Parts, *Regulatory Toxicology and Pharmacology*, August (1996) 24(1):S8–S15.

Gots, R., Hamosh, T., Flamm, W., and Carr, C., "Multiple Chemical Sensitivities: A Symposium on the State of the Science," *Regulatory Toxicology and Pharmacology* (1993) 18:61–78.

Griffiths, R., and Woodson, P., "Caffeine Physical Dependence: A Review of Human and Laboratory Animals," *Psychopharmacology* (1988) 94:4.'7–451.

Gross, P., et al., "Additive Allergy: Allergic Gastroenteritis Due to Yellow Dye Number Six," *Annals of Internal Medicine* (1989) 111(1):87–88.

Guglielmi, R., et al., "Behavioral Treatment of Phobic Avoidance in Multiple Chemical Sensitivity," *Journal of Behavior Therapy and Experimental Psychiatry* (1994) 25(3):197–209.

Gurka, G., and Rocklin, R., "Immunologic Responses During Allergen-Specific Immunotherapy for Respiratory Allergy," *Annals of Allergy* (1988) 61:239–243.

Hackney, J., et al., "Adaptation to Short-term Respiratory Effects of Ozone in Men Exposed Repeatedly," *Journal of Applied Physiology* (1977a) 43(1):82–85.

———, "Effects of Ozone Exposure in Canadians and Southern Californians, Evidence for Adaptation?" *Archives of Environmental Health* (1977b) 32(3):110–116.

Hafler, D., "In Vivo Activated T Lymphocytes in the Peripheral Blood and Cerebrospinal Fluid of Patients with Multiple Sclerosis," *New England Journal of Medicine* (1985) 312(22):1405–1411.

Hagman, J., President and CEO DowElanco, Letter to Lynn R. Goldman, M.D., Assistant Administrator, EPA, January 16, 1997.

Hahn, M., and Bonkovsky, H.L., "Multiple Chemical Sensitivity Syndrome and Porphyria," *Archives of Internal Medicine* (1997) 157:281–285.

Hall v. Baxter Healthcare Corp., DC Ore, No. 92–182, December 18, 1996

Hallén, H., and Juto, J., "Nasal Mucosa Reaction. A Model for Mucosal Reaction during Challenge," *Rhinology* (1992) 30:129–133.

Harvey's Wagon Wheel, Inc. DBA Harvey's Resort Hotel v. Joan Amann, et al. No. 25155 (Nevada Supreme Court, 1/25/95).

Hattis, D., et al., "Human Variability in Susceptibility to Toxic Chemicals: A Preliminary Analysis of Pharmacokinetic Data from Normal Volunteers," *Risk Analysis* (1987) 7:415–426.

Health Canada, *Multiple Chemical Sensitivities and Their Relevance to Psychiatric Disorders*, Workshop Proceedings, Ottawa, Canada, December 7, 1992.

Heiner, D., et al., "Multiple Precipitins to Cow's Milk in Chronic Respiratory Disease," *American Journal of Disease in Children* (1962) 103:40–60.

Hendrick, D., and Bird, A., "Alveolitis." In: Brostoff, J., and Challacombe, S. (eds.) *Food Allergy and Intolerance* (1987), Bailliere Tindall, Philadelphia, 498–510.

Heuser, G., "Diagnostic Markers in Clinical Immunotoxicology and Neurotoxicology" (editorial), *Journal of Occupational Medicine and Toxicology* (1992) 1(4):v–x.

Heuser, G., Vojdani, A., and Heuser, S., "Diagnostic Markers of Multiple Chemical Sensitivity." In: *Multiple Chemical Sensitivity, Addendum to Biologic Markers in Immunotoxicology,* (1992) National Academy Press, Washington, DC, 117–138.

Heuser, G., Mena, I., and Alamos, F., "Neurospect Findings in Patients Exposed to Neurotoxic Chemicals," *Toxicology and Industrial Health* (1994) 10(4/5):561–572.

Hileman, B., "Multiple Chemical Sensitivity," *Chemical and Engineering News* (1991) 69:26–42.

Hill, A., "The Environment and Disease: Association or Causation?" *Proceedings of the Royal Society of Medicine* (1965) 58:295–300.

Hill, D., et al., "A Study of 100 Infants and Young Children with Cow's Milk Allergy," *Clinical Review of Allergy* (1984) 2:125–142.

Hindi-Alexander, M., et al., "Theophylline and Fibrocystic Breast Disease," *Journal of Allergy and Clinical Immunology* (1985) 75(6):709–715.

Hindle, M., and Franklin, C., "Food Allergy or Intolerance in Severe Recurrent Aphthous Ulceration of the Mouth," *British Medical Journal* (1986) 292:1237–1238.

Hirzy, J., Personal communication, April 5, 1989.

Hirzy, J., and Morison, R., "Analysis of Problems Related to Installation of Carpet at EPA Headquarters in 1987–8" (unpublished).

———— , *Carpet/4-Phenylcyclohexane Toxicity: The EPA Headquarters Case,* paper presented at the Annual Meeting of the Society for Risk Analysis, October 1989b, San Francisco.

Hodgson, A., Wooley, J., and Daisey, J., "Emissions of Volatile Organic Compounds from New Carpets Measured in a Large-Scale Environmental Chamber," *Journal of the Air and Waste Management Association* (1993) 43:316–324.

Hoffman, D., Stockdale, S., Hicks, L., Schwaninger, J., "Neurocognitive Symptoms and Quantitative EEG Results in Women Presenting with Silicone-Induced Autoimmune Disease," *International Journal of Occupational Medicine and Toxicology* (1995) 4(1):91–98.

Hoj, L., et al., "A Double-Blind, Controlled Trial of Elemental Diet in Severe, Perennial Asthma," *Allergy* (1981) 36(4):257–262.

Hooks, M., Jones, G., Neill, B., and Justice, J., "Individual Differences in Amphetamine Sensitization: Dose-Dependent Effects," *Pharmacology, Biochemistry and Behavior* (1992) 41(1):203–210.

Horrobin, D., "The Philosophical Basis for Peer Review and the Suppression of Innovation," *Journal of the American Medical Association* (1990) 263(10):1438–1441.

Horvath, S., et al., "Adaptation to Ozone: Duration of Effect," *American Review of Respiratory Diseases* (1981) 123:496–499.

Housing and Urban Development (HUD), policy statement (letter from Timothy L. Coyle, HUD Assistant Secretary, to U.S. Senator Frank Lautenberg), October 26, 1990.

————, Technical Guidance Memorandum 91-3: "Multiple Chemical Sensitivity Disorder," June 6, 1991.

————, memorandum from George L. Weidenfeller, HUD Deputy General Counsel (Operations) to all Regional Counsel, April 14, 1992.

Howard, J., Executive Administrator, American Academy of Environmental Medicine, letter dated May 15, 1989.

Huber, W., Maletz, J., Fonfara, J., and Daniel, W., "Zur pathogenität des CKW-(chlorierte kohlenwasserstoffe) syndroms am beispiel des pentachlorphenol (PCP)," *Klinische Laboratorium* (1992) 38:456.

HUD: see Housing and Urban Development.

Huebner, W., et al., "Oral and Pharyngeal Cancer and Occupation: A Case-Control Study," *Epidemiology* (1992) 3(4):300–308.

Hummel, T., Roscher, S., Jaumann, M., and Kobal, G., "Intranasal Chemoreception in Patients with Multiple Chemical Sensitivities: A Double-Blind Investigation," *Regulatory Toxicology and Pharmacology* (1996) 24:S79–S86.

Hunter, D., *The Diseases of Occupations* (1978), Hodder and Stoughton, London.

Husman, K., "Symptoms of Car Painters with Long-term Exposure to a Mixture of Organic Solvents," *Scandinavian Journal of Work, Environment and Health* (1980) 6:19–32.

Immerman, F., and Schaum, J., *Final Report of the Nonoccupational Pesticide Exposure Study*, U.S. EPA, Research Triangle Park, January 23, 1990.

Institute of Medicine, *Role of the Primary Care Physician in Occupational and Environmental Medicine* (1988), National Academy Press, Washington, D.C.

Isaacson, R., *The Limbic System* (1982), Plenum Press, New York.

Jacobson, M., and Liebman, B., "Caffeine and Benign Breast Disease," *Journal of the American Medical Association* (1986) 255(11):1438–1439.

Jackson, R., et al., "Increased Circulating Ia-Antigen-Bearing T cells in Type I Diabetes Mellitus, *New England Journal of Medicine* (1982) 306(13):785–788.

Janowsky, D., Overstreet, D., and Nurnberger, J., "Is Cholinergic Sensitivity a Genetic Marker for the Affective Disorders?" *American Journal of Medical Genetics and Neuropsychiatric Genetics* (1994) 54:335–344.

Jasanoff, S., "Science on the Witness Stand," *Issues in Science and Technology* (1989) 6(1):80–87.

Jessop, C., "The Chronic Fatigue Syndrome," presented at the American Academy of Otolaryngic Allergy Post Graduate Symposium in Allergy and Immunology, Phoenix, Arizona, January, 1990.

Jewett, D., et al., "A Double-Blind Study of Symptom Provocation to Determine Food Sensitivity," *New England Journal of Medicine* (1990) 323(7):429–433.

Johnson, A., MCS Information Exchange, Brunswick, Maine, 1996.

Johnson, A., and Rea, W., *Review of 200 Cases in the Environmental Control Unit, Dallas,* presented at the Seventh International Symposium on Man and His Environment in Health and Disease, February 25–26, 1989, Dallas.

Jones, V., "Food Intolerance: A Major Factor in the Pathogenesis of Irritable Bowel Syndrome," *Lancet* (1982) 1115–1117.

Jones, V., and Hunter, J., "Irritable Bowel Syndrome and Crohn's Disease." In: Brostoff, J., and Challacombe, S. (eds.) *Food Allergy and Intolerance* (1987), Bailliere Tindall, Philadelphia, 555–569.

Kaiser, K., et al., "A Comparison of Self Reported Symptoms among Active Duty Seabees: Gulf War Veterans versus Controls" (1995), 123rd Meeting and Exhibition of the American Public Health Association, Session 2198, San Diego, CA.

Kalsner, S., and Richards, R., "Coronary Arteries of Cardiac Patients Are Hyperreactive and Contain Stores of Amines: A Mechanism for Coronary Vasospasm," *Science* (1984) 223:1435–1437.

Kammuller, M., et al., "Chemical-Induced Autoimmune Reactions and Spanish Toxic Oil Syndrome. Focus on Hydantoins and Related Compounds," *Clinical Toxicology* (1988) 26(3–4):157–174.

Kang, H., and Bullman, T., "Mortality among U.S. Veterans of the Persian Gulf War," *New England Journal of Medicine* (1996) 335(2a):1498–1504.

Kaplan, B., et al., "Dietary Replacement in Preschool-Aged Hyperactive Boys," *Pediatrics* (1989) 83(1):7–17.

Kappas, A., "Mechanisms of Chemical-Induced Porphyrinopathies—Introductory Comments." In: Silbergeld, E. and Fowler, B. (eds.), *Mechanisms of Chemical-Induced Porphyrinopathies, Annals of the New York Academy of Sciences,* (1987) 514:1–6.

Kare, M., "Direct Pathways to the Brain," *Science* (1968) 163:952–953.

Kavanaugh, A., "Fibromyalgia or Multi-organ Dysesthesia?" *Arthritis and Rheumatism* (1996) 39(1):180–181.

Kay, L., "Support for the Kindling Hypothesis in Multiple Chemical Sensitivity Syndrome (MCSS) Induction" (1996), presentation at the 26th Annual meeting of the Society for Neuroscience, Washington, D.C., November 16–21.

Kehrl, H., Personal communication, 1996.

Keplinger, H., "Patient Statement: Chemically Sensitive," *Toxicology and Industrial Health* (1994) 10(4/5):313–317.

Kerr, F., and Pozuelo, J., "Suppression of Physical Dependence and Induction of Hypersensitivity to Morphine by Stereotopic Hypothalamic Lesions in Addicted Rats," *Mayo Clinic Proceedings* (1971) 46:653–665.

Kessler, D., FDA Statement, Food and Drug Administration, Rockville, MD, April 16, 1992.

Kilburn, K., "Is the Human Nervous System Most Sensitive to Environmental Toxins?" *Archives of Environmental Health* (1989a) 44(6):343–344.

———, "Formaldehyde Impairs Memory, Equilibrium and Dexterity in Histology Technicians: Effects which Persist for Days After Exposure," *Archives of Environmental Health* (1987) 42(2):117–120.

———, "Symptoms, Syndrome and Semantics: Multiple Chemical Sensitivity and Chronic Fatigue Syndrome" (editorial), *Archives of Environmental Health* (1993a) 48(5):368–369.

———, "How Should We Think about Chemically Reactive Patients?" (editorial), *Archives of Environmental Health* (1993b) 48(1):4–5.

———, "Appraising Health Effects of Local Environments: Deciding if They Are Adverse" *Archives of Environmental Health* (1996) 51(5):351.

Kilburn, K. et al., "Neurobehavioral Dysfunction in Firemen Exposed to Polychlorinated Biphenyls (PCBs): Possible Improvement after Detoxification," *Archives of Environmental Health* (1989b) 44(6):345–350.

King, D., "Can Allergic Exposure Provoke Psychological Symptoms? A Double-Blind Test," *Biological Psychiatry* (1981) 16(1):3–19.

———, "Psychological and Behavioral Effects of Food and Chemical Exposure in Sensitive Individuals," *Nutrition and Health* (1984) 3:137–151.

———, "Statistical Power of the Controlled Research on Wheat Gluten and Schizophrenia," *Biological Psychiatry* (1985) 20:785–787.

———, "The Reliability and Validity of Provocative Food Testing: A Critical Review," *Medical Hypotheses* (1988) 25:7–16.

King, W., et al., "Provocation-Neutralization: A Two-Part Study; Part I. The Intracutaneous Provocative Food Test: A Multi-center Comparison Study," *Otolaryngology—Head and Neck Surgery* (1988a) 99(3):263–271.

———, "Provocation-Neutralization: A Two-Part Study; Part II. Subcutaneous Neutralization Therapy: A Multi-Center Study," *Otolaryngology—Head and Neck Surgery* (1988b) 99(3):272–277.

———, "Efficacy of Alternative Tests for Delayed-Cyclic Food Hypersensitivity," *Otolaryngology—Head and Neck Surgery* (1989) 101(3):385–387.

Kipen, H., Hallman, W., Kelly-McNeil, K., and Fiedler, N., "Measuring Chemical Sensitivity Prevalence: A Questionnaire for Population Studies," *American Journal of Public Health* (1995) 85(4):574–577.

Klerman, G., and Weissman, M., "Increasing Rates of Depression," *Journal of the American Medical Association* (1989) 261(15):2229–2235.

Kline, G., et al., "Ascorbic Acid Therapy for Atopic Dermatitis," presented at the annual meeting of the American Academy of Allergy and Immunology, February 1989, San Antonio, Texas.

Kniker, W., "Deciding the Future for the Practice of Allergy and Immunology," *Annals of Allergy* (1985) 55(2):106–113.

Konopacki, J., and Goldbiewski, H., "Theta-like Activity in Hippocampal Formation Slices: Cholinergic–GABAergic Interaction," *Neuro Report* (1993) 4(7):963–966.

Koren, H., Personal communication, 1996.

Kornock v. Harris, 648 F.2d 525 (9th Cir. 1980).

Korpela, M., and Tahti, H., "The Effect of Selected Organic Solvents on Intact Human Red Cell Membrane Acetylcholinesterase in Vitro," *Toxicology and Applied Pharmacology* (1986a) 85:257–262.

———, "Effect of Organic Solvents on Human Erythrocyte Membrane Acetylcholinesterase Activity in Vitro," *Archives of Toxicology* (1986b) 9(Suppl): 320–323.

Kouril v. Bowen, 912 F.2d 971 (8th Cir. 1990).

Kreiss, K., "The Epidemiology of Building-Related Complaints and Illness." In: Cone, J., and Hodgson, M. (eds.), *Problem Buildings: Building-Associated Illness and the Sick Building Syndrome, Occupational Medicine: State of the Art Reviews* (1989), Hanley & Belfus, Philadelphia 4(4):575–592.

Kreutzer, R., Health Investigations Branch, Department of Health Services, State of California, Personal communication, 1996.

Kreutzer, R., and Neutra, R., *Evaluating Individuals Reporting Sensitivities to Multiple Chemicals*, Agency for Toxic Substances and Disease Registry, National Technical Information Service, Springfield, VA (publication #PB96-187646), June 1996.

Kroker, G., et al., "Acrylic Denture Intolerance in Multiple Food and Chemical Sensitivity," *Clinical Ecology* (1982) 1(1):48–52.

———, "Fasting & Rheumatoid Arthritis: A Multicenter Study," *Clinical Ecology* (1984) 2(3):137–144.

Kuhn, T., *The Structure of Scientific Revolutions*, Third Edition (1996), University of Chicago Press, Chicago, IL.

Kuntz, C., "Darkroom Disease," *RT Image*, August 31, 1992, 4–10, 59.

Kurt, T., "Multiple Chemical Sensitivities—a Syndrome of Pseudotoxicity Manifest as Exposure Perceived Symptoms," *Journal of Toxicology—Clinical Toxicology* (1995) 33(2):101–105.

Kurt, T., Sullivan, T., "Toxic Agoraphobia," *Annals of Internal Medicine* (1990) 112(3):231–232.

LaMarte, F., et al., "Acute Systemic Reactions to Carbonless Carbon Paper Associated with Histamine Release," *Journal of the American Medical Association* (1988) 260(2):242–243.

Lamielle, M., "Center Calls for Consumer Warnings to Avert Carpet-induced Illnesses and Research to Evaluate Carpet's Role in Causing Multiple Chemical Sensitivities (MCS)," National Center for Environmental Health Strategies, press release, September 30, 1992.

Landrigan, P. "Disclosure of Interest: A Time for Clarity," *American Journal of Industrial Medicine* (1995) 26:281–282.

Lappé, M., "Causal Inference in Syndromes Associated with Silicone Breast Implants: Psychogenic and Environmental Factors," *International Journal of Occupational Medicine and Toxicology* (1995) 4(1):165–175.

Laseter, J., et al., "Chlorinated Hydrocarbon Pesticides in Environmentally Sensitive Patients, *Clinical Ecology* (1992) 2(1):3–12.

La Via, M., and La Via, D., "Phenol Derivatives Are Immunosuppressive in Mice," *Drug and Chemical Toxicology* (1979) 2(1):167–177.

Lawson v. Sullivan, 1990 U.S. Dist. LEXIS 18758 (N.D. Ill. 1990) (magistrate's decision), adopted, 1991 U.S. Dist. LEXIS 1560 (N.D. Ill. 1991).

Lax, M., and Henneberger, P., "Patients with Multiple Chemical Sensitivities in an Occupational Health Clinic: Presentation and Follow-up," *Archives of Environmental Health* (1995) 50(6):425–431.

Lebowitz, M., Personal communication, 5 June 1990.

Lehman, C., "A Double-Blind Study of Sublingual Provocative Food Testing: A Study of Its Efficacy," *Annals of Allergy* (1980) 45:144.

Leonard J., and Fry, L., "Dermatitis Herpetiformis." In: Brostoff, J., and Challacombe, S. (eds.) *Food Allergy and Intolerance* (1987) Bailliere Tindall, Philadelphia, 618–631.

LeRoy, J., Davis, T., and Jason, L., "Treatment Efficacy: A Survey of 305 MCS Patients," *The CFIDS Chronicle,* Winter 1996, 52–53.

Levin, A., Personal communication, July 14, 1989.

Levin, A., and Byers, V., "Environmental Illness: A Disorder of Immune Regulation." In: Cullen, M. (ed.) *Workers with Multiple Chemical Sensitivities* (1987), Hanley & Belfus, Philadelphia, 2(4):669–682.

Levine, S., and Reinhardt, J., "Biochemical Pathology Initiated by Free Radicals, Oxidant Chemicals, and Therapeutic Drugs in the Etiology of Chemical Hypersensitivity Disease," *Orthomolecular Psychiatry* (1983) 12:166–183.

Levinson, W., and Dunn, P., "Nonassociation of Caffeine and Fibrocystic Breast Disease," *Archives of Internal Medicine* (1986) 146(9):1773–1775.

Leznoff, A., "Multiple Chemical Sensitivity: Myth or Reality?" *Practical Allergy and Immunology* (1993) 8(2):48–52.

Leznoff, A., "Provocative Challenges in Patients with Multiple Chemical Sensitivity," *Journal of Allergy and Clinical Immunology* (1997) 99(4): 438–442.

Lincoln Realty Management Co. v. Pennsylvania Human Relations Commission, 598 A.2d 594 (Pa. Commw. 1991).

Lipowski, Z., "Psychiatry: Mindless or Brainless, Both or Neither?" *Canadian Journal of Psychiatry* (1988) 34:249–254.

Little, C., "Mediators in Food Allergy." In: Brostoff, J., and Challacombe, S. (eds.) *Food Allergy and Intolerance* (1987), Bailliere Tindall, Philadelphia, 771–780.

Littorin, M., et al., "Focal Epilepsy and Exposure to Organic Solvents: A Case-Referent Study," *Journal of Occupational Medicine* (1988) 30(10):805–808.

Lockey, R., "Fatalities from Immunotherapy and Skin Testing," *Journal of Allergy and Clinical Immunology* (1987) 79(4):660–677.

Louis, T., et al., "Findings for Public Health from Meta-Analysis," *Annual Review of Public Health* (1985) 6:1–20.

Lubin, F., et al., "A Case-Control Study of Caffeine and Methylxanthines in Benign Breast Disease," *Journal of the American Medical Association* (1985) 253(16): 2388–2392.

Mabray, C., "Obstetrics and Gynecology and Clinical Ecology, Part I," *Clinical Ecology* (1982/1983) 1(2):103–114.

——— , "Obstetrics and Gynecology and Clinical Ecology, Part II," *Clinical Ecology* (1983) 1(3, 4):155–163.

Mabray, C., et al., "Treatment of Common Gynecologic-Endocrinologic Symptoms by Allergy Management Procedures," *Obstetrics and Gynecology* (1982) 59(5):560–564.

MacLean, P., "The Brain in Relation to Empathy and Medical Education," *Journal of Nervous and Mental Disease* (1967) 144(5):374–382.

——— , "A Triune Concept of the Brain and Behavior," In: Boag, T., and Campbell, D., *The Hincks Memorial Lectures* (1973), University of Toronto Press, Toronto, Ontario, 6–66.

——— , "Culminating Developments in the Evolution of the Limbic System: The Thalamocingulate Division." In: Doane, B., and Livingston, K. (eds.), *The Limbic System: Functional Organization and Clinical Disorders* (1986), Raven Press, New York, 1–28.

MacQueen, G., et al., "Pavlovian Conditioning of Rat Mucosal Mast Cells to Secrete Rat Mast Cell Protease II," *Science* (1989) 243:83–85.

Madison, R., Broughton, A., and Thrasher, J., "Immunologic Biomarkers Associated With Acute Exposure to Exothermic Byproducts of a Ureaformaldehyde Spill," *Environmental Health Perspectives* (1991) 94:219–223.

Mage, D., and Gammage, R., "Evaluation of Changes in Indoor Air Quality Occurring over the Past Several Decades." In: Gammage, R., and Kaye, S. (eds.) *Indoor Air and Human Health* (1985), Lewis Publishers, Chelsea, MI, 5–36.

Malkin, R., Wilcox, T., and Sieber, W., "The National Institute for Occupational Safety and Health Indoor Environmental Evaluation Experience. Part Two:

Symptom Prevalence," *Applied Occupational and Environmental Hygiene* (1996) 11(6):540–545.

Maller, O., et al., "Movement of Glucose and Sodium Chloride from the Oropharyngeal Cavity to the Brain," *Nature* (1967) 213(2):713–714.

Mandell, M., and Scanlon, L., *Dr. Mandell's 5-Day Allergy Relief System* (1979), Thomas Y. Crowell, New York.

Mansfield, L., et al., "Food Allergy and Adult Migraine: Double Blind and Mediator Confirmation of an Allergic Etiology," *Annals of Allergy* (1985) 55(2):126–129.

Marcus, D.M., "An Analytical Review of Silicone Immunology," *Arthritis and Rheumatism* (1996) 39(10): 1619–1626.

Marks, J., et al., "Contract Urticaria and Airway Obstruction from Carbonless Copy Paper," *Journal of the American Medical Association* (1984) 252(8):1038–1040.

Marshall, R., et al., "Food Challenge Effects on Fasted Rheumatoid Arthritis Patients: A Multicenter Study," *Clinical Ecology* (1984) 2(4):181–190.

Maryland State Department of Education, *Indoor Air Quality: Maryland Public Schools* (1987).

Maschewsky, W. *Handbuch Chemikalien-Unverträglichkeit* (MCS), ISBN 3-9803957-4-X medi (1996), Verlagsgesellschaft für Wissenschaft und Medizin mbH, 271.

Massachusetts, Commonwealth of, *Indoor Air Pollution in Massachusetts,* April 1, 1989.

Mayberg, H. "Critique: SPECT Studies of Multiple Chemical Sensitivity," *Toxicology and Industrial Health* (1994) 10(4/5):661–666.

Mayo Medical Laboratories, *Test Catalog* (1995), Rochester, MN.

McConnachie, P., and Zahalsky, A., "Immunological Consequences of Exposure to Pentachlorophenol," *Archives of Environmental Health* (1991) 46(4):249–253.

———, "Immune Alterations in Humans Exposed to Technical Chlordane," *Archives of Environmental Health* (1992) 47(4):295–301.

McCrory, W., et al., "Immune Complex Glomerulopathy in a Child with Food Hypersensitivity," *Kidney International* (1986) 30(4):592–598.

McFadden, S., "Phenotypic Variation in Xenobiotic Metabolism and Adverse Environmental Response: Focus on Sulfur-Dependent Detoxification Pathways," *Toxicology* (1996) 111:43–65.

McGovern, J., et al., "Food and Chemical Sensitivity: Clinical and Immunologic Correlates," *Archives of Otolaryngology* (1983) 109:292–297.

———, "The Role of Naturally Occurring Haptens in Allergy," *Annals of Allergy* (1981–82 Supplement) 47:123.

McEwen, L., "A Double-Blind Controlled Trial of Enzyme Potentiated Hyposensitization for the Treatment of Ulcerative Colitis," *Clinical Ecology* (1987) 5(2): 47–51.

McLellan, R., "Biological Interventions in the Treatment of Patients with Multiple Chemical Sensitivities." In Cullen, M. (ed.) *Workers with Multiple Chemical Sensitivities, Occupational Medicine: State of the Art Reviews* (1987), Hanley & Belfus, Philadelphia, 2(4):755–777.

MCS Referral & Resources, "Who Recognizes Multiple Chemical Sensitivity?" October 1996.

MCS Referral and Resources, "Recognition of Multiple Sensitivity," February 1997.

Mealey's Litigation Reports, Volume 4, May 19, 1995, 17–18. (The settlement agreement and judgment are available as Document 15-950519-016 by calling 800-925-4123.)

Meggs, W., Personal communication (1989a).

———, Letter submitted to the editor, *Journal of Occupational Medicine* (1989b).

———, "RADS and RUDS—The Toxic Induction of Asthma and Rhinitis," *Clinical Toxicology* (1994) 32(5):487–501.

Meggs, W., and Cleveland, C., "Rhinolaryngoscopic Examination of Patients with the Multiple Chemical Sensitivity Syndrome," *Archives of Environmental Health* (1993) 41(1):14–18.

Meggs, W., Elsheik, T., Metzger, W., Albernaz, M., and Bloch, R., "Nasal Pathology and Ultrastructure in Patients with Chronic Airway Inflammation (RADS and RUDS) Following an Irritant Exposure," *Clinical Toxicology* (1996a) 34(4):383–396.

Meggs, W., Dunn, K., Bloch, R., Goodman, P., and Davidoff, L., "Prevalence and Nature of Allergy and Chemical Sensitivity in a General Population," *Archives of Environmental Health* (1996b) 51(4):275–282.

Melanie Marie Zanini v. Orkin Exterminating Company Inc. and Kenneth Johnston, Circuit Court of the 17th Judicial Circuit in and for Broward County, FLA, No. 94011515–07, final judgment December 28, 1995.

Mercier, M., Willis, J., and Pinnagoda, C., Letter to the Editor, *Archives of Environmental and Occupational Health,* July/August (1996) 51(4):340–342.

Metcalfe, D., "Summary and Recommendations," *Journal of Allergy and Clinical Immunology* (1986) 78(1):250–252.

———, quoted in "Pavlov's Rats," *Discover,* May 1989.

Metz, S., "Anti-inflammatory Agents as Inhibitors of Prostaglandin Synthesis in Man," *Medical Clinics of North America* (1981) 65(4):713–53.

Mike, N., and Asquith, P., "Gluten Toxicity in Coeliac Disease and Its Role in Other Gastrointestinal Disorders." In: Brostoff, J., and Challacombe, S. (eds.) *Food Allergy and Intolerance* (1987), Bailliere Tindall, Philadelphia, 521–548.

Miller, C., "Mass Psychogenic Illness or Chemically Induced Hypersusceptibility?" presented at the H.E.W. Symposium on the Diagnosis and Amelioration of Mass Psychogenic Illness, Chicago, May 30–June 1, 1979.

————, "Possible Models for Multiple Chemical Sensitivity: Conceptual Issues and Role of the Limbic System, Advancing the Understanding of Multiple Chemical Sensitivity," Association of Occupational and Environmental Clinics, *Toxicology and Industrial Health* (1992) 8(4):181–202.

————, "White Paper: Chemical Sensitivity: History and Phenomenology," *Toxicology and Industrial Health* (1994a) 10(4/5):253–276.

————, "Multiple Chemical Sensitivity and the Gulf War Veterans" (1994b), presentation at the Persian Gulf Experience and Health NIH Technology Assessment Workshop, Bethesda, MD, April 27–29.

————, "Multiple Chemical Sensitivity Syndrome" (Letter to the Editor), *Journal of Occupational and Environmental Medicine* (1995) 37(12):1323.

————, "Chemical Sensitivity: Symptom, Syndrome or Mechanism for Disease?" *Toxicology* (1996a) 11:69–86.

————, Invited Presentation before the Presidential Advisory Committee on Gulf War Veterans' Illnesses, San Antonio, Texas, February 27 (1996b).

————, Testimony before the U.S. House of Representatives Committee on Government Reform and Oversight, Subcommittee on Human Resources and Intergovernmental Relations, Hearing on Status of Efforts to Identify Persian Gulf War Syndrome, Hearing IV, Washington, DC, September 19 (1996c).

————, "Toxicant-Induced Loss of Tolerance—An Emerging Theory of Disease?" *Environmental Health Perspectives* (1997) 105 (Suppl.2): 445–453.

Miller, C., and Mitzel, H., "Chemical Sensitivity Attributed to Pesticide Exposure versus Remodeling," *Archives of Environmental Health* (1995) 50(2):119–129.

Miller, C., Ashford, N., Doty, R., Lamielle, M., Otto, D., Rahill, A., and Wallace, L., "Empirical Approaches for the Investigation of Toxicant-Induced Loss of Tolerance," *Environmental Health Perspectives* (1997) 105 (Suppl.2): 515–519.

Miller, J., *Food Allergy: Provocative Testing and Injection Therapy* (1972), Charles C Thomas, Springfield, IL.

————, "A Double-Blind Study of Food Extract Injection Therapy: A Preliminary Report," *Annals of Allergy* (1977) 38:185–191.

Mølhave, L., "Indoor Air Pollution Due to Organic Gases and Vapors of Solvents in Building Materials," *Environmental International,* Pergamon Press, Elmhurst, NY (1982) 8:117–127.

————, "Dose-Response Relation of Volatile Organic Compounds in the Sick Building Syndrome," *Clinical Ecology* (1986/1987) IV(2):52–56.

Mølhave, L., et al., "Human Reactions to Low Concentrations of Volatile Organic Compounds," *Environmental International* (1986) 12:167–175.

Mølhave, L., Zunyong, L., Jørgensen, A.N., Pedersen, O.F., and Kjærgaard, S.K., "Sensory and Physiological Effects on Humans of Combined Exposures to Air Temperature and Volatile Organic Compounds," *Indoor Air* (1993) 3:155–169.

Molloy, S., Personal communication, 1996.

412   *Bibliography*

Monk, J., "Farmers Fight Chemical War," *Chemistry and Industry,* February 5, 1996, 108.

Monro, J., "Food-Induced Migraine." In: Brostoff, J., and Challacombe, S. (eds.), *Food Allergy and Intolerance* (1987), Bailliere Tindall, Philadelphia, 632–665.

Monroe, R., "Episodic Behavioral Disorders and Limbic Ictus." In: Doane, K., and Livingston, K. (eds.), *The Limbic System: Functional Organization and Clinical Disorders* (1986), Raven Press, New York, 251–266.

Morey, P., and Shattuck, D., "Role of Ventilation in the Causation of Building-associated Illness," *Problem Buildings: Building-associated Illness and the Sick Building Syndrome, Occupational Medicine: State of the Art Reviews* (1989), Hanley & Belfus, Philadelphia, 4(4):625–642.

Morris, C., and Cabral, J. (eds.), "Reduction of the Human Body Burdens of Hexachlorobenzene and Polychlorinated Biphenyls." In: *Hexachlorobenzene: Proceedings of an International Symposium* (1986), International Agency for Research on Cancer Scientific Publications, Lyon, France, 597–603.

Morrow, L., et al., "PET and Neurobehavioral Evidence of Tetrabromoethane Encephalopathy," *Journal of Neuropsychiatry and Clinical Neurosciences* (1990, forthcoming).

Montville (Conn.) Board of Education, 1 National Disability Law Reporter para. 123, p. 515 (July 6, 1990).

Morton, W.E., "Redefinition of Abnormal Susceptibility to Environmental Chemicals" (1995), presented at the Second International Congress on Hazardous Waste: Impact on Human Ecological Health, Atlanta, Georgia, June 6.

Muller, W., and Black, M., "Sensory Irritation in Mice Exposed to Emissions from Indoor Products," *American Industrial Hygiene Association* (1995) 56:794–803.

Muller, W., and Schaeffer, V., "A Strategy for the Evaluation of Sensory and Pulmonary Irritation Due to Chemical Emissions from Indoor Sources," *Journal of the Air and Waste Management Association* (1996) 46:808–812.

Muranaka, M., et al., "Adjuvant Activity of Diesel-Exhaust Particulates for the Production of IgE Antibody in Mice," *Journal of Allergy and Clinical Immunology* (1986) 77(4):616–623.

Mustafa, M., and Tierney, D., "Biochemical and Metabolic Changes in the Lung with Oxygen, Ozone and Nitrogen Dioxide Toxicity," *American Review of Respiratory Diseases* (1978) 118:1061–1090.

Namba, T., et al., "Poisoning Due to Organophosphate Insecticides," *American Journal of Medicine* (1971) 50:475–492.

Nasr, S., et al., "Concordance of Atopic and Affective Disorders," *Journal of Affective Disorders* (1981) 3:291–296.

National Center for Environmental Health Strategies, "Major Chicago Developer Approves Rental Accommodations for Chemically Sensitive Residents," press release, May 1, 1992.

————, "Comparison of EEOC Statistics on MCS with Other Charges," poster, American Public Health Association meeting, New York, November 19, 1996.

National Council on Disability, "Achieving Independence: The Challenge for the 21st Century: A Decade of Progress in Disability Policy, Setting an Agenda for the Future," July 26, 1996.

National Disability Law Reporter, 1 NDLR paragraph 123, p.515, July 6, 1990.

National Federation of Federal Employees, Local 2050, press release, December 5, 1989.

National Foundation for the Chemically Hypersensitive, *Cheers* (1989) 1:6, Wrightsville Beach, NC.

National Institute of Environmental Health Sciences, "Chemical Sensitivity," *Environmental Health Perspectives* (1997) 105 (Suppl.2.): 405–547.

National Institutes of Health (NIH), Technology Assessment Workshop Panel, "The Persian Gulf Experience and Health," *JAMA* (1994) 272:391–395.

National Research Council (NRC), *Toxicity Testing: Strategies to Determine Needs and Priorities* (1984), National Academy Press, Washington, D.C.

————, *Multiple Chemical Sensitivities: Addendum to Biologic Markers in Immunotoxicology,* National Research Council, Commission on Life Sciences, Board on Environmental Studies and Toxicology (1992a), National Academy Press, Washington, DC, 202 pp.

————, *Biologic Markers in Immunotoxicology,* National Research Council Commission on Life Sciences, Board on Environmental Studies and Toxicology, Committee on Biologic Markers, Subcommittee on Immunotoxicology (1992b), National Academy Press, Washington, DC, 206 pp.

National Research Council, Board of Environmental Studies and Toxicology, Workshop on Health Risks from Exposure to Common Indoor Household Products in Allergic or Chemically Diseased Persons, July 1, 1987.

NDLR: see *National Disability Law Reporter.*

Nelson, N., "Perspectives on Diesel Emissions Health Research." In: Lewtas, J. (ed.) *Toxicological Effect of Emissions from Diesel Engines* (1982), Elsevier Biomedical, New York, 371–375.

Nero, A., "Controlling Indoor Air Pollution," *Scientific American* (1988) 258(5):42–48.

Nethercott, J., Davidoff, L., Curbow, B., and Abbey, H., "Multiple Chemical Sensitivities Syndrome: toward a Working Case Definition," *Archives of Environmental Health* (1993) 48(1):19–26.

Netherton, R.A., and Overstreet, D.H., "Genetic and Sex Differences in the Cholinergic Modulation of Thermoregulation," *Environment, Drugs and Thermoregulation,* 5th International Symposium on Pharmacologic Thermoregulation (1983), Karger, Basel, 74–77.

Neutra, R., "Some Preliminary Thoughts on the Potential Contribution of Epidemiology to the Question of Multiple Chemical Sensitivity," *Public Health Reviews* (1994) 22(3–4):271–278.

News USA, in *The Merchandiser,* Spring Grove, PA, February 1996.

NIEHS: see National Institute of Environmental Health Sciences

NIH: see National Institutes of Health.

NRC: see National Research Council.

O'Banion, D., *Ecological and Nutritional Treatment of Health Disorders* (1981), Charles C Thomas, Springfield, IL.

O'Brien, T., and Decouflé, P., "Cancer Mortality among Northern Georgia Carpet and Textile Workers," *American Journal of Industrial Medicine* (1988) 14:15–24.

Odell, R., *Environmental Awakening* (1980), Ballinger, Cambridge, MA.

Odkvist, L., et al., "Solvent-induced Central Nervous System Disturbances Appearing in Hearing and Vestibulo-Oculomotor Tests," *Clinical Ecology* (1985) 3(3):149–153.

Ohm, M., and Juto, J., "Nasal Hyper-reactivity. A Histamine Provocation Model," *Rhinology* (1993) 31:53–55.

Olson, L., et al., "Aldicarb Immunomodulation in Mice: An Inverse Dose-Response to Parts Per Billion Levels in Drinking Water," *Archives of Environmental Contamination and Toxicology* (1987) 16(4):433–439.

OTA: see U.S. Congress, Office of Technology Assessment.

Otto, D., et al., "Neurobehavioral and Sensory Irritant Effects of Controlled Exposure to a Complex Mixture of Volatile Organic Compounds," *Neurotoxicology and Teratology* (1990) 12(6): 649–652.

Overstreet, D., et al., "Potential Animal Model of Multiple Chemical Sensitivity with Cholinergic Supersensitivity," *Toxicology* (1996) 111:119–134.

Ozonoff, D., Personal communication, March 24, 1989.

Ozonoff, D., et al. "Health Problems Reported by Residents of a Neighborhood Contaminated by a Hazardous Waste Facility," *American Journal of Industrial Medicine* (1987) 11:581–597.

Pan, Y., et al., "Aliphatic Hydrocarbon Solvents in Chemically Sensitive Patients," *Clinical Ecology* (1987–1988) 5(2):126–131.

Panush, R., "Delayed Reactions to Foods. Food Allergy and Rheumatic Disease," *Annals of Allergy* (1986) 56:500–503.

Panush, R., et al., "Food-induced (Allergic) Arthritis: Inflammatory Arthritis Exacerbated by Milk," *Arthritis and Rheumatism* (1986) 29(2):220–226.

Patterson, R., et al., "IgG Antibody Against Formaldehyde Human Serum Proteins: A Comparison with Other IgG Antibodies Against Inhalant Proteins and Reactive Chemicals," *Journal of Allergy and Clinical Immunology* (1989) 84(3):359–366.

Payan, D., "Substance P: A Modulator of Neuroendocrine-Immune Function," *Hospital Practice* (Feb. 15, 1989); 24(2):67–80.

Payan, D., et al., "Neuroimmunology," *Advances in Immunology* (1986) 39:299–3 23.

Pearson, D., and Rix, K., "Psychological Effects of Food Allergy." In: Brostoff, J., and Challacombe, S. (eds.) *Food Allergy and Intolerance* (1987), Bailliere Tindall, Philadelphia, 688–708.

Peereboom, J. (ed.), *Basisboek millieu en gezondheid,* ISBN 90-5352-048-1 (1994), Boom Publishers, Amsterdam, 443 pp.

Pelikan, Z., "Rhinitis and Secretory Otitis Media; A Possible Role of Food Allergy." In: Brostoff, J., and Challacombe, S. (eds.) *Food Allergy and Intolerance* (1987), Bailliere Tindall, Philadelphia, 467–485.

Pennebaker, J., "Psychological Bases of Symptoms Reporting: Perceptual and Emotional Aspects of Chemical Sensitivity" (review), *Toxicology and Industrial Health* (1994) 10(4/5): 497–511.

Perlroth, M., "The Porphyrias." In *Scientific American Medicine* (1988), Volume 9, *Metabolism,* V-1–V-12.

Peterson, W., "*Helicobacter pylori* and Peptic Ulcer Disease," *New England Journal of Medicine* (1991) 324:1043–1048.

Philpott, W., "Allergy and Ecology in Orthomolecular Psychiatry." In: Dickey, L. (ed.) *Clinical Ecology* (1976), Charles C Thomas, Springfield, IL, 729–737.

Philpott, W., et al., "Four-day Rotation of Foods According to Families." In: Dickey, L. (ed.) *Clinical Ecology* (1976), Charles C Thomas, Springfield, IL, 472–486.

———, *Brain Allergies: The Psycho-Nutrient Connection* (1980), Keats Publishing, New Canaan, CT.

Pierson, T. K., et al., "Risk Characterization Framework for Noncancer End-points." Proceedings of a Conference on Methodology for Assessing Health Risks from Complex Mixtures in Indoor Air, Arlington, Virginia, April 16–19, 1990.

Pike, M., and Atherton, D., "Atopic Eczema." In: Brostoff, J., and Challacombe, S. (eds.) *Food Allergy and Intolerance* (1987), Bailliere Tindall, Philadelphia, 583–601.

Pinel, J., and Van Oot, P., "Intensification of the Alcohol Withdrawal Syndrome by Repeated Brain Stimulation," *Nature* (1975) 254:510–511.

Pola, J., et al., "Urticaria Caused by Caffeine," *Annals of Allergy* (1988) 60(3):207–208.

Pope, H., and Hudson, J., "Can Memories of Childhood Sexual Abuse Be Repressed?" *Psychological Medicine* (1995) 25(1):121–126.

Powers, "Metabolic and Allergic Aspects of Inner Ear Dysfunction." In: Dickey, L. (ed.) *Clinical Ecology* (1976), Charles C Thomas, Springfield, IL, 637–644.

Rafuse, J., "Practical Application of Air Quality Research Incorporated in CMHC's Research House," *Canadian Medical Association Journal* (1995) 152(8):1310–1311.

Randolph, T., "Fatigue and Weakness of Allergic Origin (Allergic Toxemia) to Be Differentiated from Nervous Fatigue or Neurasthenia," *Annals of Allergy* (1945) 3:418–430.

————, "Allergy as a Causative Factor of Fatigue, Irritability, and Behavior Problems of Children," *Journal of Pediatrics* (1947) 31(1):560–572.

————, "Allergic Type Reactions to Industrial Solvents and Liquid Fuels," abstract in *Journal of Laboratory and Clinical Medicine* (1954a) 44(6):910–911.

————, "Allergic Type Reactions to Mosquito Abatement Fogs and Mists," abstract in *Journal of Laboratory and Clinical Medicine* (1954b) 44(6):911–912.

————, "Allergic Type Reactions to Motor Exhaust," abstract in *Journal of Laboratory and Clinical Medicine* (1954c) 44(6):912.

————, "Allergic Type Reactions to Indoor Utility Gas and Oil Fumes," abstract in *Journal of Laboratory and Clinical Medicine* (1954d) 44(6):913.

————, "Allergic Type Reactions to Chemical Additives of Foods and Drugs," abstract in *Journal of Laboratory and Clinical Medicine* (1954e) 44(6):913–914.

————, "Allergic Type Reactions to Synthetic Drugs and Cosmetics," abstract in *Journal of Laboratory and Clinical Medicine* (1954f) 44(6):914.

————, "Depressions Caused by Home Exposures to Gas and Combustion Products of Gas, Oil, and Coal," abstract in *Journal of Laboratory and Clinical Medicine* (1955) 46(6):942.

————, "The Descriptive Features of Food Addiction," *Quarterly Journal of Studies on Alcohol* (1956) 17:198–224.

————, "A Third Dimension of the Medical Investigation," *Clinical Physiology* (1960) 2(1):42–47.

————, *Human Ecology and Susceptibility to the Chemical Environment* (1962), Charles C Thomas, Springfield, IL.

————, "Ecologic Orientation in Medicine: Comprehensive Environmental Control in Diagnosis and Therapy," *Annals of Allergy* (1965) 23:7–22.

————, "Adaptation to Specific Environmental Exposures Enhanced by Individual Susceptibility." In: Dickey, L. (ed.) *Clinical Ecology* (1976a), Charles C Thomas, Springfield, IL, 46–66.

————, "Stimulatory and Withdrawal Levels of Manifestations." In: Dickey, L. (ed.) *Clinical Ecology* (1976b), Charles C Thomas, Springfield, IL, 169–170.

————, "Ecologically Oriented Rheumatoid Arthritis." In: Dickey, L. (ed.) *Clinical Ecology* (1976c), Charles C Thomas, Springfield, IL, 201–212.

————, "Ecologically Oriented Myalgia and Related Musculoskeletal Painful Syndromes." In: Dickey, L. (ed.) *Clinical Ecology* (1976d), Charles C Thomas, Springfield, IL, 213–223.

————, "The Role of Specific Alcoholic Beverages." In: Dickey, L. (ed.) *Clinical Ecology* (1976e), Charles C Thomas, Springfield, IL, 321–333.

——— , "The Enzymatic, Acid, Hypoxia, Endocrine Concept of Allergic Inflammation." In: Dickey, L. (ed.) *Clinical Ecology* (1976f), Charles C Thomas, Springfield, IL, 577–596.

——— , *Environmental Medicine: Beginnings and Bibliographies of Clinical Ecology* (1987) Clinical Ecology Publications, Fort Collins, CO.

Randolph, T., and Moss, R., *An Alternative Approach to Allergies* (1980), Lippincott & Crowell, New York.

Rapp, D., "Double-Blind Confirmation and Treatment of Milk Sensitivity," *Medical Journal of Australia* (1978a) 1:571–572.

——— , "Weeping Eyes in Wheat Allergy," *Transactions of the American Society of Ophthalmologic and Otolaryngic Allergy* (1978b) 18:149–150.

——— , "Comments on the Position Paper on Clinical Ecology by the American Academy of Allergy and Immunology" (1985). In: Randolph, T., *Environmental Medicine: Beginnings and Bibliographies of Clinical Ecology* (1987), Clinical Ecology Publications, Fort Collins, CO, 301–307.

Rapp, D., and Bamberg, D., *A Guide for Caring Teachers and Parents; The Impossible Child—In School—At Home* (1986), Practical Allergy Research Foundation, Buffalo, NY.

Rea, W., "Environmentally Triggered Small Vessel Vasculitis," *Annals of Allergy* (1977) 38:245–51.

——— , "Environmentally Triggered Cardiac Disease," *Annals of Allergy* (1978) 40:243–251.

——— , "The Environmental Aspects of Ear, Nose, and Throat Disease: Part I," *Journal of Clinical Ecology, Oto-Rhino-Laryngology & Allergy Digest* (1979a) 41(7):41–56.

——— , "The Environmental Aspects of Ear, Nose, and Throat Disease: Part II," *Journal of Clinical Ecology, Oto-Rhino-Laryngology & Allergy Digest* (1979b) 41(8/9):41–54.

——— , "Review of Cardiovascular Disease in Allergy." In: Frazier, C. (ed.) *Bi-Annual Review of Allergy 1979–1980* (1981), Medical Examination Publishing Co., Garden City, NY, 282–347.

——— , *Outpatient Information Manual* (1984a) (1988 revision), Environmental Health Center, Dallas.

——— , *Clinical Ecology: A Role in Diagnosing Environmental Illness,* presented at the New England Occupational Medicine Association Conference, December 3–4, 1987, Boston.

——— , "Chemical Hypersensitivity and the Allergic Response," *Ear, Nose and Throat Journal* (1988a) 67(1):50–56.

——— , "Inter-relationships between the Environment and Premenstrual Syndrome." In: Brush, M., and Goudsmit, E. (eds.) *Functional Disorders of the Menstrual Cycle,* (1988b), John Wiley & Sons, New York, 135–157.

——— , Letter to American College of Physicians, July 14, 1988c.

————, Personal communication, March 10, 1989.

————, *Chemical Sensitivity,* Volume 1, *Principles and Mechanisms* (1992), Lewis Publishers, Boca Raton, FL, 533 pp. See also Volume 2, *Sources of Total Body Load* (1993), 592 pp.; Volume 3, *Clinical Manifestation of Pollutant Overload* (1995), 929 pp.; and Volume 4, *Tools for Diagnosis and Methods of Treatment* (1996) (ca. 912 pp.)

Rea, W., and Brown, O. "Mechanisms of Environmental Vascular Triggering," *Clinical Ecology* (1985) 3(3):122–128.

————, "Cardiovascular Disease in Response to Chemicals and Foods." In: Brostoff, J., and Challacombe, S. (eds.) *Food Allergy and Intolerance* (1987c); Bailliere Tindall, Philadelphia, 737–754.

Rea, W., et al., "Environmentally Triggered Large-Vessel Vasculitis." In: Johnson, J., and Spencer, J. (eds.) *Allergy and Medical Treatment* (1975), Symposia Specialists, Chicago, 185–198.

————, "Elimination of Oral Food Challenge Reaction by Injection of Food Extracts," *Archives of Otolaryngology* (1984b) 110(4):248–252.

————, "T&B Lymphocyte Parameters Measured in Chemically Sensitive Patients and Controls," *Clinical Ecology* (1986a) 4(1):11–14.

————, "Magnesium Deficiency in Patients with Chemical Sensitivity," *Clinical Ecology* (1986b) 4(1):17–20.

————, "Toxic Volatile Organic Hydrocarbons in Chemically Sensitive Patients," *Clinical Ecology* (1987) 5(2):70–74.

Reidenberg, M., et al., "Lupus Erythematosus-like Disease Due to Hydrazine," *American Journal of Medicine* (1983) 75:363–370.

Rest, K., "Advancing the Understanding of Chemical Sensitivity (MCS): Overview and Recommendations from an AOEC Workshop," *Toxicology and Industrial Health* (1992) 8(4): 1–13.

Reynolds, R., and Natta, C., "Depressed Plasma Pyridoxal Phosphate Concentrations in Adult Asthmatics," *American Journal of Clinical Nutrition* (1985) 41:684–688.

Richter, E., "Multiple Chemical Sensitivity: Respect the Observations . . . Suspect the Interpretations?" *Archives of Environmental Health* (1993) 48(5):366–367.

Riedel, F., et al., "Effects of SO_2 Exposure on Allergic Sensitization in the Guinea Pig," *Journal of Allergy and Clinical Immunology* (1988) 82(4):527–534.

Riihimaki, V., and Savolainen, K., "Human Exposure to m-Xylene. Kinetics and Acute Effects on the Central Nervous System," *Annals of Occupational Hygiene* (1980) 23:411–432.

Rinkel, H., "Food Allergy: The Role of Food Allergy in Internal Medicine," *Annals of Allergy* (1944) 2:115–124.

Rinkel, H., Randolph, T., and Zeller, M., *Food Allergy* (1951), Charles C Thomas, Springfield, IL.

Rinsky, R. A., et al., "Benzene and Leukemia: An Epidemiologic Risk Assessment," *New England Journal of Medicine* (1987) 316:1044–1050.

Roback, G., Randolph, L., Seidman, B., and Pasko, T., *Physician Characteristics and Distribution in the U.S.* (1994), American Medical Association, Chicago, IL.

Robb, N., "The Environment Was Right for Nova Scotia's New Environmental Health Clinic," *Canadian Medical Association Journal* (1995) 152(8):1292–1294.

Robertson, A., et al., "Comparison of Health Problems Related to Work and Environmental Measurements in Two Office Buildings with Different Ventilation Systems," *British Medical Journal* (1985) 291(6492):373–376.

Rodriguez de la Vega, A., et al., "Kerosene-induced Asthma," *Annals of Allergy* (1990) 64(4):362–363.

Roehm, D., "Effects of a Program of Sauna Baths and Megavitamins on Adipose DDE and PCBs and on Clearing of Symptoms of Agent Orange (Dioxin) Toxicity," *Clinical Research* (1983) 32(2):243a.

Rogers, S., "A Practical Approach to the Person with Suspected Indoor Air Quality Poblems," *Indoor Air '90*, Fifth International Conference on Indoor Air Quality and Climate, Toronto, Ontario (1990) 5:345–349.

Root, D., and Lionelli, G., "Excretion of a Lipophilic Toxicant Through the Sebaceous Glands: A Case Report," *Journal of Toxicology Cutaneous and Ocular Toxicology* (1987) 6(1):13–17.

Rosenberg, C., *The Cholera Years* (1962), University of Chicago Press, Chicago, 199–200.

Rosenberg, S., Freedman, M., Schmaling, K., and Rose, C., "Personality Styles of Patients Asserting Environmental Illness, *Journal of Occupational Medicine* (1990) 32:678–681.

Rosenthal, N., "Multiple Chemical Sensitivity: Lessons from Seasonal Affective Disorder," *Toxicology and Industrial Health* (1994) 10(4/5):623–632.

Rosenthal, N., and Cameron, C., "Exaggerated Sensitivity to an Organophosphate Pesticide" (Letter), *American Journal of Psychiatry* (1991) 148(2):270.

Rosenstock, L., "Hospital-based, Academically-Affiliated Occupational Medicine Clinics," *American Journal of Industrial Medicine* (1984) 6:155–158.

Rossi, J., "Sensitization Induced by Kindling and Kindling-Related Phenomena as a Model for Multiple Chemical Sensitivity," *Toxicology* (1996) 111:87–100.

Rotton, J., and Frey, J., "Air Pollution, Weather, and Violent Crimes: Concomitant Time-Series Analysis of Archival Data," *Journal of Personality and Social Psychology* (1985) 49(5):1207–1220.

Rousseau, D., et al., *Your Home, Your Health and Well-Being* (1988). Hartley and Marks, Vancouver, British Columbia.

Rowe, A., "Chronic Ulcerative Colitis—An Allergic Disease," *Annals of Allergy* (1949) 7(6):727–819.

Russell, L., "Caffeine Restriction as Initial Treatment for Breast Pain," *Nurse Practitioner* (1989) 14(2):36–40.

Ruth Elliott, et al. v. San Joaquin County Public Facilities Financing Corp. et al., California Superior Court, San Joaquin County, No. 244601, October 31, 1966.

Ryan, C., et al., "Cacosmia and Neurobehavioral Dysfunction Associated with Occupational Exposure to Mixtures of Organic Solvents," *American Journal of Psychiatry* (1988) 145(11):1442–1445.

Saifer, P., and Becker, N., "Allergy and Autoimmune Endocrinopathy: APICH Syndrome." In: Brostoff, J., and Challacombe, S. (eds.) *Food Allergy and Intolerance* (1987), Bailliere Tindall, Philadelphia, 781–796.

Salvaggio, J., "Allergy and Clinical Immunology—2001," *Journal of Allergy and Clinical Immunology* (1986) 78(2):253–268.

Salvaggio, J., and Aukrust, L., "Mold-induced Asthma," *Journal of Allergy and Clinical Immunology* (1981) 68(5):327–346.

Sampson, H., "Food Hypersensitivity and Atopic Dermatitis: Evaluation of 113 Patients," *Journal of Pediatrics* (1985) 107(5):669–675.

San Diego (Cal.) Unified School District, 1 National Disability Law Reporter para. 61, p. 311 (May 24, 1990).

Sandberg, D., "Food Sensitivity: The Kidney and Bladder." In: Brostoff, J., and Challacombe, S. (eds.) *Food Allergy and Intolerance* (1987a), Bailliere Tindall, Philadelphia, 755–767.

――――, "Food Sensitivity: The Eye." In: Brostoff, J., and Challacombe, S. (eds.) *Food Allergy and Intolerance* (1987b), Bailliere Tindall, Philadelphia, 768–770.

Sandberg, D., et al., "Severe Steroid Responsive Nephrosis Associated with Hypersensitivity," *Lancet* (1977) 1(8008):388–391.

Sanderson, W., et al., "The Influence of an Illusion of Control on Panic Attacks Induced via Inhalation of 5.5 percent Carbon Dioxide-Enriched Air," *Archives of General Psychiatry* (1989) 46:157–162.

SB Latex Council, "Use of SB Latex in Carpets," Washington, D.C., November 1996.

Scadding, G., and Brostoff, J., "Low Dose Sublingual Therapy in Patients with Allergic Rhinitis Due to House Dust Mite," *Clinical Allergy* (1986) 16:483–491.

Scadding, L., et al., "Poor Sulphoxidation Ability in Patients with Food Sensitivity," *British Medical Journal* (1988) 297(6641):105–107.

Schaeffer, V., Consumer Products Safety Commission, Personal communication, 1996.

Schaeffer, V., Bhooshan, B., Shing-Bong, C., Sonenthal, J., and Hodgson, A., "Characterization of Volatile Organic Chemical Emissions from Carpet Cushions," *Journal of the Air and Waste Management Association* (1996) 46:813–820.

Schaper, M., "Development of a Database for Sensory Irritants and Its Use in Establishing Occupational Exposure Limits," *American Industrial Hygiene Association Journal* (1993) 54:488–544.

Schimmelpfennig, W., "Zur problematik der begutachtung umweltbedingter toxischer gesundheitsschäden," *Bundesgesundheitsblatt* (1994) 37:377.

Schnare, D., and Robinson, P., "Reduction of the Human Body Burdens of Hexachlorobenzene and Polychlorinated Biphenyls," International Agency for Research on Cancer, Scientific Publications (1986) 77:597–603.

Schnare, D., et al., "Evaluation of a Detoxification Regimen for Fat Stored Xenobiotics," *Medical Hypotheses* (1982) 9(3):265–282.

———, "Body Burden Reductions of PCBs, PBBs and Chlorinated Pesticides in Human Subjects," *Ambio* (1984) 13(5–6):378–380.

Schottenfeld, R., "Workers with Multiple Chemical Sensitivities: A Psychiatric Approach to Diagnosis and Treatment." In: Cullen, M. (ed.) *Workers with Multiple Chemical Sensitivities* (1987), Hanley & Belfus, Philadelphia, 2(4):739–754.

Schottenfeld, R., and Cullen, M., "Occupation-Induced Posttraumatic Stress Disorders," *American Journal of Psychiatry* (1985) 142(2):198–202.

Schwartz, G., Bell, I., Dikman, Z., et al., "EEG Responses to Low-Level Chemicals in Normals and Cacosmics," *Toxicology and Industrial Health* (1994) 10(4/5):633–643.

Selner, J., "Chemical Sensitivity." In: *Current Therapy in Allergy, Immunology and Rheumatology* (1988), B.C. Decker, Philadelphia, 48–52.

Selner, J., and Staudenmayer, H., "The Relationship of the Environment and Food to Allergic and Psychiatric Illness." In: Young, S., and Rubin, J. (eds.), *Psychobiology of Allergic Disorders* (1985a), Praeger, New York, 102–146.

———, "The Practical Approach to the Evaluation of Suspected Environmental Exposures: Chemical Intolerance," *Annals of Allergy* (1985b) 55(5):665–673.

Selye, H., "The General Adaptation Syndrome and the Diseases of Adaptation," *Journal of Allergy* (1946) 17:231–247, 289–323, 358–398.

———, *The Stress of My Life* (1977), McClelland and Stewart, Toronto.

Seyal, A., et al., "Psychosomatic Cardiovascular Disorders: An Elusive Relationship to the Type of Clothing Worn," *Clinical Ecology* (1986a) 4(1):26–30.

———, "Systolic Blood Pressure, Heart Rate and Premature Ventricular Contractions in a Population Sample: Relationship to Cotton and Synthetic Clothing," *Clinical Ecology* (1986b) 4(2).69–74.

———, "A Relationship of Quality of Garments to Blood Pressure in Young School Children," *Clinical Ecology* (1986c) 4(3):115–119.

———, "Premature Ventricular Contractions: The Relationship of Synthetic vs. Natural Fabrics Worn Next to the Skin," *Clinical Ecology* (1986d) 4(4):149–154.

———, "The Influence of Cotton and Synthetic Fabrics on Nocturnal Angina," *Clinical Ecology* (1987/88) 5(3):121–125.

Shakman, R., "Nutritional Influences on the Toxicity of Environmental Pollutants," *Archives of Environmental Health* (1974) 28:105–113.

Shambaugh, G., "Serous Otitis: Are Tubes the Answer?" *Pediatric Otology* (1983) 5(1):63–65.

Sharma, R., *Immunologic Considerations in Toxicology*, Vols. I and II (1981), CRC Press, Boca Raton, FL.

Sharp, D., "Delayed Health Hazards of Pesticide Exposure," *Annual Review of Public Health* (1986) 7:441–471.

———, "Low-Cost Dwellings for the Environmentally Hypersensitive," *The Canadian Architect,* June 1994, 29–32.

Sherman, J., "Organophosphate Pesticides—Neurological and Respiratory Toxicity," *Toxicology and Industrial Health* (1995) 11(1):33–39.

Shim, C., and Williams, M., "Effect of Odors in Asthma," *American Journal of Medicine* (1986) 80:18–22.

Shipley, M., "Transport of Molecules from Nose to Brain: Transneuronal Antigrade and Retrograde Labeling in the Rat Olfactory System by Wheat Germ Agglutinin—Horseradish Peroxidase Applied to Nasal Epithelium," *Brain Research Bulletin* (1985) 15:129–142.

Shorter, R., "Idiopathic Inflammatory Bowel Disease: A Form of Food Allergy?" In: Brostoff, J., and Challacombe, S. (eds.) *Food Allergy and Intolerance* (1987), Bailliere Tindall, Philadelphia, 549–554.

Sieber, W., Staynor, L., Malkin, R., Peterson, M., Mendel, M., Wallingford, K., Crandall, M., Wilcox, T., and Reed, L., "The National Institute for Occupational Safety and Health Indoor Environmental Evaluation Experience. Part Three: Associations Between Environmental Factors and Self-Reported Health Conditions," *Applied Occupational and Environmental Hygiene* (1996) 11(12):1387–1392.

Silbergeld, E., "Role of Altered Heme Synthesis in Chemical Injury to the Nervous System." In: *Mechanisms of Chemical-Induced Porphyrinopathies, Annals of the New York Academy of Sciences* (1987) 514:297–308.

Silverman, K., Evans, S., Stain, E., and Griffiths, R., "Withdrawal Syndrome after the Double-Blind Cessation of Caffeine Consumption," *New England Journal of Medicine* (1992) 327(16):1109–1114.

Simon, G., "Psychiatric Symptoms in Multiple Chemical Sensitivity" (Review), *Toxicology and Industrial Health* (1994a) 10(4/5):487–496.

———, quoted in Proceedings of the Conference on Low-Level Exposure to Chemicals and Neurobiologic Sensitivity, *Journal of Toxicology and Industrial Health* (1994b) 10:526–527.

Simon, G., Daniell, W., Stockbridge, H., Claypoole, K., and Rosenstock, L., "Immunological, Psychological, and Neuropsychological Factors in Multiple Chemical Sensitivity: A Controlled Study," *Annals of Internal Medicine* (1993) 19(2):97–103.

Simon, G., Katon, W., and Sparks, P.J., "Allergic to Life: Psychological Factors in Environmental Illness," *American Journal of Psychiatry* (1990) 147:901–906.

Simon, T., Hickey, D., Rea, W., Johnson, A., and Ross, G., "Breast Implants and Organic Solvent Exposure Can Be Associated with Abnormal Cerebral SPECT Studies in Clinically Impaired Persons" (abstract), *Radiology,* November (1992)185:234.

Simon, T., Hickey, D., Fincher, C., Johnson, A., Ross, G., and Rea, W., "Single Photon Emission Computed Tomography of the Brain in Patients with Chemical Sensitivities," *Toxicology and Industrial Health* (1994)10(4/5):573–577.

Skoldstam, L., "Effects of Diet on Rheumatoid Arthritis," *Internal Medicine for the Specialist* (1989) 10(5):128–137.

Sly, R.M., "Mortality from Asthma," *Journal of Allergy and Clinical Immunology* (1988) 82(5):705–717.

Snow, J., *Snow on Cholera* (1965; republished from 1936 original), Hoffner Publishing Company, New York, xxxvi.

Sorg, B., "Proposed Animal Model for Multiple Chemical Sensitivity in Studies with Formalin," *Toxicology* (1996a) 111:135–145.

——— , Personal communication (1996b).

Sparks, P., Daniell, W., Black, D., Kipen, H., Altman, L., Simon, G., and Terr, A., "Multiple Chemical Sensitivity Syndrome: A Clinical Perspective I: Case Definition, Theories of Pathogenesis, and Research Needs," *Journal of Occupational Medicine* (1994a) 36(7):718–730.

Sparks, P., Daniell, W., Black, D., Kipen, H., Altman, L., Simon, G., and Terr, A., "Multiple Chemical Sensitivity Syndrome: A Clinical Perspective II: Evaluation. Diagnostic Testing, Treatment, and Social Considerations," *Journal of Occupational Medicine* (1994b) 36(7):731–737.

Speizer, F., et al., "Palpitation Rates Associated with Fluorocarbon in a Hospital Setting," *New England Journal of Medicine* (1975) 292(12):624–626.

Spence, J., President, Environmental Illness Society of Canada, Personal communication with Susan Kaplan, November 1996.

Spengler, J., Testimony Before the Ways and Means Committee, California State Legislature, February 22, 1988.

——— , "Indoor Chemical Pollution," presentation given at the 46th Annual Meeting of the American College of Allergy and Immunology, Orlando, FL, November 15, 1989.

Spengler, J., and Sexton, K., "Indoor Air Pollution: A Public Health Perspective," *Science* (1983b) 221(4605):9–17.

Spengler, J., et al., "Nitrogen Dioxide inside and outside 137 Homes and Implications for Ambient Air Quality Standards and Health Effects Research," *Environmental Science and Technology* (1983a) 17(3):164–168.

Spiegelberg, V., "Psychopathologisch-neurologische Schäden nach Einwirkung Synthetischer Gifte." In: *Wehrdienst und Gesundheit*, Volume III (1961), Wehr und Wissen Verlagsgesellshaft, Darmstadt, Germany.

Spyker, D., "Multiple Chemical Sensitivities—Syndrome and Solution," *Journal of Toxicology—Clinical Toxicology* (1995) 33(2):95–99.

Stankus, R., et al., "Cigarette Smoke-Sensitive Asthma: Challenge Studies," *Journal of Allergy and Clinical Immunology* (1988) 82(3, Pt 1):331–338.

Statistical Abstract of the United States (1994), Bernan Press, Lanham, MD.

Staudenmayer, H., presentation made at the Annual Meeting of the American College of Allergy and Immunology, Orlando, FL, November 11, 1989.

——— , "Clinical Consequences of the EI/MCS 'Diagnosis': Two Paths," *Regulatory Toxicology and Pharmacology* (1996) 24:S96–S110.

——— , "Multiple Chemical Sensitivities or Idiopathic Environmental Intolerances: Psychophysiologic Foundation of Knowledge for a Psychogenic Explanation" (editorial), *Journal of Allergy and Clinical Immunology* (1979) 99(4):434–437.

Staudenmayer, H., and Camazine, M., "Sensing Type Personality, Projection and Universal 'Allergic' Reactivity," *Journal of Psychological Types* (1989).

Staudenmayer, H., and Selner, J., "Post-Traumatic Stress Syndrome (PTSS): Escape in the Environment," *Journal of Clinical Psychology* (1987) 43(1):156–157.

——— , "Neuropsychophysiology during Relaxation in Generalized, Universal 'Allergic' Reactivity to the Environment: A Comparison Study," *Journal of Psychosomatic Research* (1990) 34(3):259–270.

Staudenmayer, H., Selner, J., and Buhr, M., "Double-Blind Provocation Chamber Challenges in 20 Patients Presenting with 'Multiple Chemical Sensitivity,'" *Regulatory Toxicology and Pharmacology* (1993a) 18:44–53.

Staudenmayer, H., Selner M., and Selner J., "Adult Sequelae of Childhood Abuse Presenting as Environmental Illness," *Annals of Allergy* (1993b) 71:538–546.

Stein, M., et al., "The Hypothalamus and the Immune Response." In: Weiner, H., et al. (eds.), *Brain, Behavior and Bodily Disease* (1981), Raven Press, New York, 45–65.

Stephens, R., Spurgeon, A., Calvert, I., et al., "Neuropsychological Effect of Long-Term Exposure to Organophosphates in Sheep Dip," *Lancet* (1995) 345:1135–1139.

Stewart, D., and Raskin, J., "Psychiatric Assessment of Patients with 20th-Century Disease ('Total Allergy Syndrome')," *Canadian Medical Association Journal* (1985) 133:1001–1006.

Stokinger, H., "Ozone Toxicology," *Archives of Environmental Health* (1965) 10:719–731.

Strahilevitz, M., et al., "Air Pollutants and the Admission Rate of Psychiatric Patients," *American Journal of Psychiatry* (1979) 136(2):205–207.

Sullivan, T., Paper presented at the 45th Annual Meeting of the American Academy of Allergy and Immunology, San Antonio, Texas, February (1989).

——— , "Management of Patients Allergic to Antimicrobial Drugs," *Allergy Proceedings* (1991) 12(6):361–361.

Tabershaw, I., and Cooper, C., "Sequelae of Acute Organic Phosphate Poisoning," *Journal of Occupational Medicine* (1966) 8:5–20.

Taylor, E., "Environmental Medicine Practice Threatened," *Peace and Environment News*, Ottawa Peace and Environment Resource Centre, June 1995, 6.

Taylor, G., and Harris, W., "Cardiac Toxicity of Aerosol Propellants," *Journal of the American Medical Association* (1970) 214(1):81–85.

Teicher, M., Glod, C., Surrey, J., and Swett, C., "Early Childhood Abuse and Limbic System Ratings in Adult Psychiatric Outpatients," *Journal of Neuropsychiatry and Clinical Neurosciences* (1993) 5(3):301–306.

Tepper, J., Moser, V., Costa, D., Mason, M., Roache, N., Guo, Z., and Dyer, R., "Toxicological and Chemical Evaluation of Emissions from Carpet Samples," *American Industrial Hygiene Association Journal* (1995) 56:158–170.

Terr, A., "Environmental Illness: A Clinical Review of 50 Cases," *Archives of Internal Medicine* (1986) 146:145–149.

———, "'Multiple Chemical Sensitivities': Immunological Critique of Clinical Ecology Theories and Practice." In: Cullen, M. (ed.) *Workers with Multiple Chemical Sensitivities* (1987), Hanley & Belfus, Philadelphia, 2(4):683–694.

———, "Clinical Ecology in the Workplace," *Journal of Occupational Medicine* (1989a) 31(3):257–261.

———, Remarks presented at the 45th Annual Meeting of the American Academy of Allergy and Immunology, San Antonio, Texas, February (1989b).

———, Personal communication, December 19, 1989 (1989c).

Thomson, G., *Report of the Ad Hoc Committee on Environmental Hypersensitivity Disorders,* Ontario, Canada, 1985.

Thrasher, J., et al., "Evidence for Formaldehyde Antibodies and Altered Cellular Immunity in Subjects Exposed to Formaldehyde in Mobile Homes," *Archives of Environmental Health* (1987) 42(6):347–350.

———, "Building Related Illness and Antibodies to Albumin Conjugates of Formaldehyde, Toluene Diisocyanate, and Trimellitic Anhydride," *American Journal of Industrial Medicine* (1989) 15:187–195.

Thrasher, J., Broughton, A., and Micevich, "Antibodies and Immune Profiles of Individuals Occupationally Exposed to Formaldehyde: Six Case Reports," *American Journal of Industrial Medicine* (1988) 14(4):479–488.

Thrasher, J., Madison, R., and Broughton, A., "Immunologic Abnormalities in Humans Exposed to Chloryrifos: Preliminary Observations," *Archives of Environmental Health* (1993) 48(2):89–93.

Tollerud, D., et al., "The Effects of Cigarette Smoking on T Cell Subsets," *American Review of Respiratory Diseases* (1989) 139(6):1446–1451.

Toxics Law Reporter, "District of Columbia Judge Overturns Verdict for Four Plaintiffs with Chemical Sensitivity" (1995), 10(29):827–828, December 20, 1995.

———, "Dow Corning Seeks Common Causation Trial, Sets $2 Billion Tort Fund in Bankruptcy Plan" (1996b), 11(28): 750–751, December 4, 1996.

———, "Plaintiffs Seek Review of Panel's Ruling Dismissing Claim for Heightened Sensitivity" (1996a), 11(28):769–770, December 11, 1996.

————, "High Court to Review Appellate Standard for Admissibility of Expert Testimony" (1997a) 11(41) 1139, March 19, 1997.

————, "Federal Judge Excludes Causation Evidence As Failing "Daubert" Validity Requirement" (1997b) 11(30):815–816, January 2, 1997.

————, "Michigan Court Dismisses Dow Chemical; Finds No Duty to Breast Implant Plaintiffs" (1997c) 11(43): 1194–95, April 2, 1997.

Tretjak, Z., et al., "Occupational, Environmental, and Public Health in Semic: A Case Study of Polychlorinated Biphenyl (PCB) Pollution." In: Gunnerson, C. (ed.) *Post-Audits of Environmental Programs and Projects* (1989), American Society of Civil Engineers, 57–72.

Trounce, J., and Tanner, M., "Eosinophilic Gastroenteritis," *Archives of Diseases in Childhood* (1985) 60(12):1186–1188.

Tuite, J., Testimony before the U.S. House of Representatives Committee on Government Reform and Oversight, Subcommittee on Human Resources and Intergovernmental Relations, Hearings on Status of Efforts to Identify Persian Gulf War Syndrome, Hearing IV, Washington, DC, September 19 (1996).

United States Congress, Office of Technology Assessment. *Neurotoxicity: Identifying and Controlling Poisons of the Nervous System,* OTA-BA-436, Washington, D.C., April 1990.

United States Department of Labor, *An Interim Report to Congress on Occupational Diseases* (1980), Washington, D.C.

United States District Court for the District of Hawaii, Consent order, Civil No. 92-00641 DAE, August 25, 1993.

Urich, R., et al., "Does Suggestibility Modify Acute Reactions to Passive Cigarette Smoke?" *Environmental Research* (1988) 47:34–47.

U.S. Congress, Office of Technology Assessment (OTA), "Case Studies: Exposure to Lead, Pesticides in Agriculture, and Organic Solvents in the Workplace." In: *Neurotoxicity: Identifying and Controlling Poisons of the Nervous System—New Developments in Neuroscience,* OTA-BA-436, (1990), U.S. Government Printing Office, Washington, DC, April, 281–311.

U.S. Environmental Protection Agency, Peer Review Reports of the EPA Anderson Carpet Studies, 1993 (unpublished).

Van Metre, T., Jr., and Adkinson, N., Jr., "Immunotherapy for Aeroallergen Disease." In: Middleton, E., Jr., et al. (eds.) *Allergy Principles and Practice,* Vol II, 3d ed., (1988), C.V. Mosby Co., St. Louis, 1327–1343.

Van Sickle, G., et al., "Milk- and Soy Protein-Induced Enterocolitis: Evidence for Lymphocyte Sensitization to Specific Food Proteins," *Gastroenterology* (1985) 88(6):1915–1921.

Vasey, F., "Rheumatologists' View of Silicone," *International Journal of Occupational Medicine and Toxicology* (1995) 4(1):203–209.

Vobecky, J., Devroede, G., and Caro, J., "Risk of Large Bowel Cancer in Synthetic Fiber Manufacture," *Cancer* (1984) 54:2537–2542.

Vojdani, A., Campbell, A., and Brautbar, N., "Immune Functional Impairment in Patients with Clinical Abnormalities and Silicone Breast Implants," *Toxicology and Industrial Health* (1992a) 8(6):415–429.

Vojdani, A., Ghoneum, M., and Brautbar, N., "Immune Alteration Associated with Exposure to Toxic Chemicals," *Toxicology and Industrial Health* (1992b) 8(5): 239–253.

Wachter, K., "Disturbed by Meta-Analysis?" *Science* (1988) 241:1407–1408.

Wada, J., and Osawa, T., "Spontaneous Recurrent Seizure State Induced by Daily Electric Amygdaloid Stimulation in Senegalese Baboons (*Papio papio*)," Neurology (1976) 26:273–286.

Wada, J., Sato, M., and Corcoran, M., "Persistent Seizure Susceptibility and Recurrent Spontaneous Seizures in Kindled Cats," *Epilepsia* (1974) 15:465–478.

Waddell, W., "The Science of Toxicology and Its Relevance to MCS," *Regulatory Toxicology and Pharmacology* (1993) 18:13–22.

Wallace, L., "Overview." In: Gammage, R., and Kaye, S. (eds.) *Indoor Air and Human Health* (1985), Lewis Publishers, Chelsea, MI, 331–333.

————, "The TEAM Study: Personal Exposures to Toxic Substances in Air, Drinking Water, and Breath of 400 Residents of New Jersey, North Carolina, and North Dakota," *Environmental Research* (1987) 43:290–307.

Wallace, L., et al., "Organic Chemicals in Indoor Air: A Review of Human Exposure Studies and Indoor Air Quality Studies." In: Gammage, R., and Kaye, S. (eds.) *Indoor Air and Human Health* (1985), Lewis Publishers, Chelsea MI, 361–378.

Walsh, T., and Emerich, D., "The Hippocampus as a Common Target of Neurotoxic Agents," *Toxicology* (1988) 49:137–140.

Warner, M., "Hunting the Yellow Fever Germ: The Principle and Practice of Etiological Proof in Late Nineteenth-Century America," *Bulletin of the History of Medicine* (1985) 59:361–382.

Washington Post, reproduced in the *Boston Globe* December 19, 1996.

Wasner, C., et al., "The Use of Unproven Remedies," *Scientific American* (1980) 23(1):759–760.

Weaver, V., "Medical Management of the Multiple Chemical Sensitivity Patient," *Regulatory Toxicology and Parmacology* (1996) 24:S111–S115.

Webster's Third New International Dictionary of the English Language (Unabridged) (1986), Merriam-Webster, Springfield, MA.

Weiss, B., "Low-Level Chemical Sensitivity: A Perspective from Behavioral Toxicology," *Toxicology and Industrial Health* (1994) 10(4/5):605–617.

Wessely, S., "The Measurement of Fatigue and Chronic Fatigue Syndrome," *Journal of the Royal Society of Medicine* (1992) 85:189–190.

Wilson, C., "Landmark Pesticide Ordinance," *Our Toxic Times,* December 1996a, 23.

——— , "EPA Must Disclose Inerts," *Our Toxic Times,* December 1996b, 10.

——— , "Porphyrinopathies in the MCS Community," *Our Toxic Times,* (1996c) 7(3):1–4.

Winkelmann, R., "Food Sensitivity and Urticaria or Vasculitis." In: Brostoff, J., and Challacombe, S. (eds.) *Food Allergy and Intolerance* (1987), Bailliere Tindall, Philadelphia, 602–617.

Wittmer, J., "Selected Reviews of the Literature. Multiple Chemical Sensitivities," *Journal of Occupational and Environmental Medicine,* November (1996) 38(11):1085.

Wojtulewski, J., "Joints and Connective Tissue." In: Brostoff, J., and Challacombe, S. (eds.) *Food Allergy and Intolerance* (1987), Bailliere Tindall, Philadelphia, 723–736.

Wolff, M., et al., "Equilibrium of Polybrominated Biphenyl (PBB) Residues in Serum and Fat of Michigan Residents," *Bulletin Environmental Contamination and Toxicology* (1979) 21(6):775–781.

——— , "Human Tissue Burdens of Halogenated Aromatic Chemicals in Michigan," *Journal of the American Medical Association* (1982) 247(15):2112–2116.

World Health Organization, *Public Health Impact of Pesticides Used in Agriculture,* (1990), World Health Organization, Geneva, Switzerland, 128 pp.

Wraith, D., "Asthma." In: Brostoff, J., and Challacombe, S. (eds.) *Food Allergy and Intolerance* (1987), Bailliere Tindall, Philadelphia, 486–497.

Wyon, D. P., "Sick Buildings and the Experimental Approach," *Environmental Technology* (1992) 133:313–322.

Yu, D., et al., "Peripheral Blood Ia-Positive T cells," *Journal of Experimental Medicine* (1980) 151:91–100.

Zamm, A., and Gannon, R., *Why Your House May Endanger Your Health* (1980), Simon and Schuster, New York.

Ziem, G., Personal communication, November 8, 1989.

——— , "Multiple Chemical Sensitivity: Treatment and Followup with Avoidance and Control of Chemical Exposures, Advancing the Understanding of Multiple Chemical Sensitivity," Association of Occupational and Environmental Clinics, *Toxicology and Industrial Health* (1992) 8(4):181–202.

Ziem, G., and McTamney, J., "Profile of Patients with Chemical Injury and Sensitivity," *Environmental Health Perspectives* (1997) 105(Suppl. 2):417–436.

Index